Feathered Marvels

Feathered Marvels

*The Natural History
and Extraordinary Lives of Birds*

DOMINIC F. SHERONY
with RANDI MINETOR

McFarland & Company, Inc., Publishers
Jefferson, North Carolina

LIBRARY OF CONGRESS CATALOGUING-IN-PUBLICATION DATA

Names: Sherony, Dominic F., 1942– author. | Minetor, Randi, author.
Title: Feathered marvels : the natural history and extraordinary lives of birds /
Dominic F. Sherony with Randi Minetor.
Description: Jefferson, North Carolina : McFarland & Company, Inc., Publishers, 2024. |
Includes bibliographical references and index.
Identifiers: LCCN 2023045619 | ISBN 9781476691886 (paperback : acid free paper) ∞
ISBN 9781476650531 (ebook)
Subjects: LCSH: Birds—Popular works. | Birds—Evolution—Popular works. |
Birds—Speciation—Popular works. | Birds—Behavior—Popular works. | Birds—Habitat—
Popular works. | Birds—Conservation—Popular works. | BISAC: NATURE / Animals / Birds
Classification: LCC QL676 .S547 2023 | DDC 598—dc23/eng/20231031
LC record available at https://lccn.loc.gov/2023045619

BRITISH LIBRARY CATALOGUING DATA ARE AVAILABLE

ISBN (print) 978-1-4766-9188-6
ISBN (ebook) 978-1-4766-5053-1

Front cover: (left to right) Purple honeycreeper, eastern bluebird, blue-gray tanager,
Jamaican tody, saffron finch, spot-breasted oriole (Luis Marquez/Shutterstock),
and summer tanager (all photographs by Dominic F. Sherony unless noted);
(background) *Archaeopteryx* specimen (Roger De Marfà/Shutterstock)

Printed in the United States of America

*McFarland & Company, Inc., Publishers
Box 611, Jefferson, North Carolina 28640
www.mcfarlandpub.com*

To all the scientists who have worked to provide
an understanding of the world of birds.

Acknowledgments

I want to thank the thousands of scientists and citizens who have contributed so much to the understanding of the world of birds. The many people who have struggled to support conservation efforts also deserve our gratitude. There is no doubt that scientists will continue to gain understanding in avian evolution, taxonomy, and many other important subjects. There is already enough basic knowledge to allow us to appreciate the avian world without resorting to myths. We understand that the future of birds and all nature rests in our hands and in the decisions we make.

Many have provided help in writing this book. I owe a debt of gratitude and a special thanks to Randi Minetor and John Waud for their effort and comments on the draft. There are many people who contributed to this work by providing information or comments on portions of the text. I thank Carol Bland, Evelyn Brister, Jon Dunn, Shaibal Mitra, Frank Morlock, Charles Newton, Helen Nguyen, Chris Norment, Andrea Patterson, Tom Schulenberg, Allan Strong, and Scott Weidensaul.. The staff of the Fairport Public Library, Sarah Standish and Hema Parthasarathi, provided many needed references. Mike Tetlow helped with reference material. Roger White assisted with information systems. Nobu Tamura provided the drawings of prehistoric creatures. Many others have provided photographs or drawings including: David D. Beadle, Nik Borrow, Brad Carlson, Chris Chafer, Gary Chapin, Darren Clark, Kester Clark, Robert Clark, Kathy Dashiell, Kyle Elliott, Patrick Gains, Mélanie Guigueno, Diane Henderson, Gary Kaiser, Lucretia Grosshans, Ed Harper, Ian and D.G. Mackean, Aaron Maizlish, Celeste Morien, Gordon Ramel, Christopher Scotese, Eric VanderWerf, Jeanne Verhulst, Nigel Voaden, and John Waud. Nic Minetor aided in editing photos. Others who were helpful with photographs include: Rob Alicea, Joachim Bertrands, Rob Drummond, Somchai Kanhanasut, and Kayo Roy.

Table of Contents

Part V—Birds in Our World:
Today and In the Future   273

Preface

This work was inspired by a series of lectures I gave on the life and history of birds at Grand Learning in Surprise, Arizona. Although I am an engineer by training, I am also an avid bird watcher and nature photographer and give frequent talks to different birding groups, various clubs, and a few schools about my photographs and experiences. In doing this, I became familiar with the areas covered in this book.

This amazing story takes a broad view of the avian world and is intended to introduce the readers to many subjects and facets about birds: their history, unique features, and lifestyle—subjects with which most birders are unfamiliar, based on my many talks before audiences.

I have aimed to use information from the vast array of technical material available today regarding scientific advances in the study of birds. My hope is that this would not only shed light on the subject, but also might be of interest to the reader. I've included citations to sources in the text. My goal is to familiarize the reader with the many researchers who have provided so much of this new understanding. These researchers have brought clarification to many misconceptions, and also provide surprising insights into a world we barely understand. Since the technical literature can be complex and difficult to understand, I have tried to simplify some of this material. For instance, I use as few technical terms as possible. Nevertheless, some terms might be unfamiliar, so I've provided definitions in a glossary when it would benefit the reader to understand them.

I use the scientific names of orders and families of birds throughout. My intention is to show the connection between modern birds and their earliest forbearers, and to show how biologists have come to understand the relationships of the various orders and families of birds. I have provided a simple table of geological time in the earliest chapter and have referred to it throughout the book.

My aim in this book is to provide a general, overall view for the reader of four main topics:

1. **The long-term history of birds**: Birds go back to the time of dinosaurs, and although their origin is not completely understood, little doubt remains that they stem from these creatures. However, they have had many dead ends and new starts on their way to becoming the birds we know today.
2. **Birds' unique biology**: The discussion of the unique aspects of their biology

focuses on how birds live their lives as compared to other creatures, and not on the more complex, underlying descriptions of this subject.

3. **The amazing diversity of birds**: The variety of living and extinct birds is extensive. Developing an appreciation for this diversity and the lifestyle of so many different kinds of birds will benefit those who never have the opportunity to see some of the various species that might be of interest.

4. **The importance of habitats for birds**: Birds have diversified by adaptation to their environments, and habitats play a central role in this process. There is a great deal of evidence that birds are disappearing across the globe, and in order to reduce their decline, a first priority is conservation of habitats.

Introduction

Human fascination with birds dates back at least 30,000 years, when an image of an owl was created and left in the Chauvet-Point-d'Arc Cave in France [Figure 1.0]. Another archaic image of birds appeared 10,000 to 15,000 years later in the paintings of a bird and bird-headed man in the Lascaux Cave in France. It is clear that early humanity found some sort of mythological connection with birds. Many aboriginal cultures incorporated birds in their beliefs and practices as indicators of time, weather and the seasons; as a resource for hunting, eating, medicine and farming; as domestic pets; and as omens and intermediaries between the gods and humankind (Mynott 2020). Birds are ubiquitous in ancient Greek and Roman literature. Bird wings have been adapted as symbolic representations of mythical animals and spiritual beings like angels and Pegasus, the winged horses. Ancient Egyptians depicted many species of birds in their art, but the falcon, ibis, and vulture had particularly important places in their culture. Almost all aboriginal cultures have associated birds with gods because of their ability to fly, their songs, their changing plumage due to molt, and their mysterious disappearance and reappearance due to migration. No doubt, the fact that birds are colorful, easy to observe, and able to fly has earned them a permanent place in human art and culture.

Birds have a significant place in our lives today. They are a food source: people eat their eggs and meat from both domesticated fowl and game birds. At one time, domesticated carrier pigeons were used to transmit information. In many ways birds are valuable to agricultural practices: for instance, seabird guano makes a fertilizer that is exceptionally high in nitrogen, phosphate, and potassium. The feces of Canada Geese, in moderation, also contribute nutrients to the soil (Paulin and Drake 2003), while at the same time dispersing seeds, providing for the survival of other plants and animals. This dispersal of seeds, nuts, and pollen may be one of the most important benefits of birds in our lives, thus broadening the range of many native plants and trees. Birds also play a significant role in pest control by foraging on insects.

The presence of birds is considered an indicator of environmental quality—in particular, they provide a direct measure of environmental conditions in wetlands, as measured by the number and success of nesting waterbirds; and they have long been employed in coal mines to detect toxic gases.

Birds have a role in the quality of human life as well. Their sheer beauty, diversity, and ubiquity have made them a source of joy for countless people. Their feathers have been and continue to be a source of adornments. Bird song has been a source of inspiration for many centuries, and captive birds are cherished for their aesthetic qualities.

Because of their diversity and the fact that it's relatively easy to study birds, they have also played a role in many important advances in biology in fields such as adaptation, speciation, genetics, and migration.

Although we recognize the beauty and value of birds, the reality is that birds are declining worldwide. The destructive treatment of the environment has had a major impact on all wildlife, and particularly on birds. Edward O. Wilson, an entomologist—a scientist engaged in the study of insects—and an early pioneer in the study of biodiversity, has summarized the major threats to the environ-

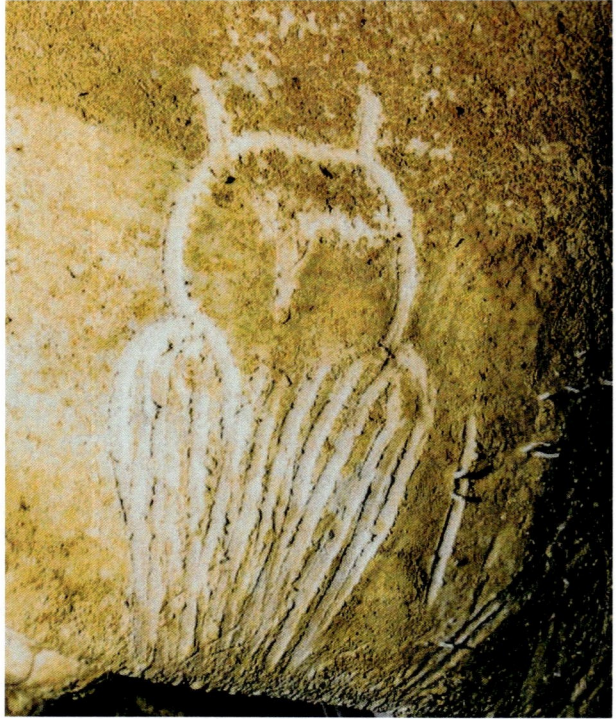

Figure 1.0 Replica of a 30,000-year-old engraving of an owl at the Chauvet Cave, France (photographer: HTO; from the Brno Pavilion, Brno, Moravia).

ment with the acronym HIPPO: habitat destruction, invasive species, pollution, population of humans, and overharvesting by hunting and fishing. For birds, the top of the list is habitat destruction through the spread of agriculture, the decimation of forests, and industrial development. The United Nations Intergovernmental Science-Policy Platform on Biodiversity and Ecosystem Services (IPBES) has estimated that 75 percent of the land-based environment and about 66 percent of the marine environment have been significantly altered by human actions (IPBES 2019).

Other significant, human-induced factors have a direct impact on birds: the decrease in oceanic food supplies, toxic pollution, senseless slaughter and poaching for subsistence and profit, feral cats, and human obstructions such as wind farms, illumination of high rise buildings, and climate change occurring too rapidly for species to adapt.

This book is about those amazing birds all around us, if we take a moment from our busy lives to see them. This is their story, as told by the thousands of observers and researchers who have tried to understand their world. If people have more appreciation of the birds and the environment, perhaps they would be inspired and motivated to conserve the state of the natural world. It is my hope that more people will learn something about these creatures from this book, and will be influenced enough to try to help conserve what is remaining of the avian and natural world.

PART I

BIRDS IN THE DISTANT PAST

Introduction

Birds differ from other animals in so many ways that ancient people hardly knew what to make of them. Their ability to fly, their covering of feathers instead of fur, scales, or a hard shell, and—before the concept of migration became well known—their appearance and disappearance with the change of seasons seemed like great mysteries. With no understanding of the science, people instead shrouded birds' origins in creation myths, believing in birds as gods—or, at least, as emissaries of the gods.

This view began to change in the eighteenth century as the science of paleontology developed, and with it the understanding of fossil forms. Around 1860, the discovery of *Archaeopteryx*, a feathered bird-like fossil, gave the first real hint about the origin of birds. This discovery followed close upon the publication of Darwin's theory of evolution by natural selection: the process through which organisms most suited to specific conditions survive and thrive, while those less suited do not. Over time, scientists including Thomas Huxley studied these remains and recognized the close anatomical resemblance of *Archaeopteryx*, a creature that lived hundreds of millions of years ago, to both birds and dinosaurs. Once Huxley and other scientists began to understand that there was a link between birds and dinosaurs, many more fossils were found that confirmed their hypothesis: Prehistoric birds had existed originally in reptilian-like forms—like dinosaurs—and had gone through extensive transformations to become the birds we now enjoy in our own backyards.

The first two chapters of Part I explore how we came to understand the origins and history of the birds that are part of our lives today. Uncovering avian history has been a long and arduous task for scientists, and the fossil record is still incomplete and often difficult to interpret. What we do know, however, tells us a great deal. In the last 30 years, an astonishing number of fossils have been discovered that have clarified the transitions between dinosaurs and birds. This, combined with the recent development of techniques for genetic analysis of living and even some extinct birds, has brought further insights into birds' complex and fascinating past, and the transitions to the birds we see today.

Geologists express the earth's past in orders—or geological time—by segmenting its history according to rock strata or major geological events. Figure 1.1 provides a simplified diagram of the span of "geological time" in which birds originated and developed.

The highest order of the earth's history is made up of three major eons, very long periods of time spanning hundreds of millions of years.

- **Eons** represent the highest order of the earth's history. Currently we are in the Phanerozoic eon, which started 541 million years ago; scientists group all eons before this into the Precambrian eon. It was during our Phanerozoic eon that plant and animal life developed, became abundant and diversified.
- **Eras** are the next level of geological time, during which important stages took place in macroscopic fossils. The eras in our current Phanerozoic eon include five major extinction events. Two of them are the Permian–Triassic (P–T) event took place at the end of the Paleozoic era, and the Cretaceous-Palogene (K–Pg) event marked the end of the Mesozoic era. We are now in the Cenozoic era.
- **Periods** are a further subdivision of time, determined by particular rock strata and identified by radiometric dating.
- **Epochs**, the last subdivision of time, are characterized by a particular event, a series of events, or a specific development.

The bottom line of the table represents time in Millions of Years Before Present (MYBP), when that period or epoch began. You will see this abbreviation used throughout these chapters.

You may want to put a tab on this page, as you may find yourself returning to this table often as you read this book, to keep track of the timing of important events in the history of birds.

Figure 1.1 shows the timing of the earliest appearance of birds in the Cretaceous period and their presence throughout the Cenozoic Era (shaded below). The Geological Time Scale in Figure 1.1 is expressed in Millions of Years before Present (MYBP). Dates are the beginning of the Epoch or Period. Note that the Permian period is the last period of the Paleozoic era, which began much earlier. By the upper Cretaceous period, birds that are distant relatives of modern birds had made their appearance. Some of these birds survived the end of the Cretaceous period and began a process of diversification in the Paleogene. By the Neogene period, modern families of birds are identifiable.

Figure 1.1 Geological Time Scale

Eon	Phanerozoic				
Era	Paleozoic	Mesozoic			
Period	Permian	Triassic	Jurassic	Cretaceous	
Epoch				Lower	Upper
Begins MYBP	300	250	200	145	100

Eon	Phanerozoic					
Era	Cenozoic					
Period	Paleogene			Neogene		Quaternary
Epoch	Paleocene	Eocene	Oligocene	Miocene	Pliocene	Pleistocene
Begins MYBP	66	56	34	23	5.3	2.6

1

Birds in the Jurassic and Cretaceous Periods

What was the earth like when birds first made their appearance? To understand this, we need to begin at the end of the Permian period, around 260–250 million years before present (MYBP), when dinosaurs were the dominant animals (time periods refer to Figure 1.1).

The term *dinosaur* usually refers to a group of land reptiles that were different from marine reptiles and the flying reptiles known as pterosaurs—which, despite their ability to fly, were not the ancestors of today's birds.

While the general public often uses "dinosaur" to mean any reptile-like animal from prehistoric times, dinosaurs share specific characteristics that other clades—pterosaurs, for example—do not have. (Biologists use the term "clade" to represent a group of organisms that evolved from a common ancestor.) Dinosaurs are descendants of the archosaurs—reptile-like creatures with a variety of different shapes—which emerged by the end of the Permian period, and became the dominant clade in the Triassic period, about 250 MYBP. These include proto-dinosaurs, animals that are not technically dinosaurs but have many of the same characteristics.

The archosaurs included a number of groups of mostly predatory, armor-covered, crocodile-like reptiles, and some plant-eating armored reptiles, some of which had a sail-back, and at least one, *Lotosaurus*, that was toothless; a representation is shown in Figure 1.2. Although we think of all creatures of this period as being large, some of these proto-dinosaurs were small. For example, *Prorotodactylus* was about the size of a modern cat and weighed about 10 pounds (5 kg), based on analysis of its footprints found in Poland (Brusatte *et al.* 2011).

Over millions of years, the archosaurs split into two distinct lineages. One branch evolved into modern crocodiles, while the other began a slow process of diversification that led to the dinosaurs with which we are most familiar today. The earliest evidence indicates that the first of these dinosaurs appeared between 240 MYBP and 230 MYBP (Brusatte 2018, Nesbitt *et al.* 2010).

When the Permian period ended with the worst extinction event in the history of the planet—known as the Permian/Triassic extinction, caused by a series of massive volcanic eruptions in what is now Russia—ninety-five percent of life in the oceans and seventy percent of all life on land was eliminated. Reptilelike creatures that survived in the Permian evolved into dinosaurs that flourished through the

Figure 1.2 The *Lotosaurus* was a heavily built, toothless, sailback archosaur from the early Triassic period that fed by shearing off leaves with its beaked jaws (Nobu Tamura).

Triassic period, and survived another mass extinction event at the end of the Triassic, when more than 50 percent of all species went extinct. Although we don't know precisely what caused this extinction, a number of hypotheses have been posed to explain this massive event, mostly related to changes in climate. We do know that widespread volcanism at the time had caused the oceans to acidify, which led in turn to global warming and decreased oceanic oxygen levels. These volcanic eruptions occurred in the Central Atlantic Magmatic Province, a vast area of igneous rock that formed before the breakup of the supercontinent (more on this in a moment), and that now rests mainly in the North Atlantic Ocean, with some smaller areas found in North and South America.

Whatever the cause of the mass extinction, again some dinosaurs pulled through, and went on to become the dominant animal life form on earth. Their reign continued for 160 million years until another major extinction took place.

After the Triassic period ended and the Jurassic period had begun, all of today's continents were united (or nearly so) as one gigantic landmass, known to us as Pangaea. This gave wildlife the ability to disperse over all of Pangaea. By the end of the Jurassic period, the constant, slow-moving shifting of the tectonic plates produced large rifts in Pangaea, and the landmass began to break up and drift apart, forming the continents we know today. As the continents drifted, they each carried their own complement of the planet's animal life.

With the rifts came new oceans to fill the gaps between continents, as well as large amounts of the greenhouse gas carbon dioxide rising from the rifts. Just as this carbon dioxide contributes to global warming today, it caused the planet to warm back in the early Cretaceous period as well, making sea levels rise to their highest levels in geological history to date. Entire continents flooded, creating inland seas that contributed to a wetter and more humid climate, with lusher vegetation than had previously existed on the planet. In addition, the climate warmed by about 7°C (13°F). This continued into the mid–Cretaceous period, when the planet was rich with life and the continents were free of ice. Insects emerged, and the first large flying animals—pterosaurs of many types and sizes—dominated the skies. The largest member of the group, *Quetzalcoatlus*, first discovered in Texas, had a wingspan of 30 feet (9 m) and weighed 650 lbs. (295 kg). Scientists have struggled to understand how a creature of this physiology could fly at all (Habib 2019).

Seed-bearing coniferous plants became common, and flowering plants appeared, bearing seeds and fruits. In this rich biota, bird-like creatures made their first appearance.

Archaeopteryx: *One Missing Link*

In the summer of 1861, a quarryman whose name has been lost to history found a single fossilized feather in what was then called the Solnhofen Community Quarry in Germany. The quarry had been mined for years for its fine-grained limestone, used for high quality engraving. This fossil feather left impressions on two slabs of limestone, showing that the feather was asymmetrical in that the vanes (the barbs on each side of the shaft) on one side of the shaft were shorter than those on the other side. The fossil also revealed a curved central shaft [Figure 1.3]. All of these characteristics looked like the flight feathers of living birds.

German paleontologist Hermann von Meyer obtained the fossil shortly after it was found, giving him the opportunity to study it closely. He decided that it came from a prehistoric bird, and named the species *Archaeopteryx*, or "ancient wing," and added the species name *lithographica*. Later that summer, he learned that another fossil had been found in Germany, a complete skeleton of another *Archaeopteryx*, and that it had been sold to the London Natural History Museum. He performed a later analysis of this skeleton—now referred to as the London Specimen—and proposed

Figure 1.3 A fossilized feather found in 1861 and examined by Hermann von Meyer, showing the basic structure of the flight feather of a modern bird (Robert Clark).

that the *Archaeopteryx* was the first bird to have been found from the Mesozoic era (Wellnhofer 2009) [Figure 1.4]. By September of 1861, von Meyer had made his analysis known to others.

As is often the case with profound discoveries, however, some disputed his assertion, countering that *Archaeopteryx* was actually a reptile. This disagreement continued into 1868, when anatomist Thomas Huxley first noted the similarities in the fossils of birds and dinosaurs. He compared the structure of *Archaeopteryx* to other reptile fossils and proposed the idea that they might be related.

When Charles Darwin published his book *On the Origin of Species* in November 1859, one of the major criticisms of his theory of evolution had been the lack of transitional species in the fossil record. But now, with Huxley's observations combined with the fossil feather, it became clear that *Archaeopteryx* represented just such a transitional species between a bird and a reptile, one of the first "missing links" to confirm Darwin's concept of evolution.

But was it a bird? *Archaeopteryx* seemed to support Huxley's ideas, but the notion that birds were related to dinosaurs did not gain much support. Over the next century, other analysts also supported Huxley's premise, but it was not until John Ostrom, an American paleontologist, published a series of papers in the 1970s on the similarities of birds and the fossils of theropod dinosaurs—bipedal dinosaurs, usually with small forelimbs—that science paid further attention to this theory.

Figure 1.4 Crow-sized *Archaeopteryx* is a transitional fossil with features of both birds and dinosaurs (Nobu Tamura).

Since then, our understanding of the origin of birds from dinosaurs has accelerated greatly, strengthened by the discovery of many species of small dinosaurs, mostly from Asia. About 700 valid species of non-avian dinosaurs have been catalogued—a large number, but only a small fraction of all the dinosaur species that ever lived. We can certainly expect to see further improvements in our understanding of the evolution of birds from dinosaurs as paleobiologists—scientists who combine the study of fossils with modern biology to understand evolution—use new analyses that reshape our views of life on earth when dinosaurs began.

Since its discovery, thousands of people have studied *Archaeopteryx*. Today there are eleven or twelve body fossils, and the consensus is that *Archaeopteryx* was the earliest bird, dating back to the late Jurassic period—about 150 MYBP, as von Meyer had hypothesized when he first studied the fossil. The specimens range in size from close to the size of a blue jay to some as big as a common raven or large chicken, and likely represent more than one species (O'Connor *et al.* 2011). Although *Archaeopteryx* is very different from modern birds, today archaeologists and ornithologists agree that this ancient birdlike species provides the missing link in birds' evolutionary chain.

Archaeopteryx had some characteristics unique to birds, some unique to dinosaurs, and others shared by both. Like a bird, it was fully feathered except for the head and neck, and the primary feathers on the wing were shaped like those of a modern bird adapted for flight; it had a wishbone (furcula), and its fingers were reduced in size. Because it lacked a complete breastbone (sternum) to anchor the flight muscles (pectorals), scientists did not believe it could fly like modern birds.

Like a dinosaur, *Archaeopteryx* had a very limited sternum, belly ribs (gastralia), a full set of teeth, three clawed fingers on its wrists, and a long, thin, feathered tail. Its breathing system, wishbone, three-toed (tridactyl) foot, fused fifth digit on the hand, and feathers were passed down from the two-legged theropods.

Archaeopteryx was the first known bird-like creature with wings contoured for gliding and rudimentary flight. A recent study on the structure of the wing bones confirmed it could flap its wings in flight (Voeten *et al.* 2018), in part because the bone structure of the wings could withstand the torsion forces of flight. Wings suitable for gliding and flight appeared about 150 MYBP. By the time of *Archaeopteryx,* animals had evolved at least two different types of shoulder joints, one of which would enable powered flight.

The larger *Archaeopteryx* specimen weighed about two pounds; based on recent analysis, the wing feathers were likely black. It probably could become airborne either by climbing trees and gliding down (though some analysts question its ability to climb trees), or by running and gliding, as it had the physical mechanisms for both. Its clawed fingers would allow it to climb and grab prey.

The discovery of *Archaeopteryx* confirmed the existence of a *transitional species,* one that has traits of its ancestral species as well as its descendants. Today there are transitional species for almost every order of animals, and we know that the DNA code links the ancestry of all life on earth.

How Did Birds Evolve from Dinosaurs?

As I noted earlier, modern language tends to lump all prehistoric animals into one big group and call them "dinosaurs." The scientific reality, however, is quite different, with many classifications and subgroups. One of these appears to be the bridge between dinosaurs and birds.

Perhaps it's not surprising that the general public is not familiar with all of these groups, because science continues to struggle to determine perfect classifications and evolution timelines. The current thinking dates back to the nineteenth century, placing dinosaurs into two large lineages: the *Ornithischians* and the *Saurischians*.

The *Ornithischians,* or "bird-hipped" dinosaurs, had a three-part pelvis with the pubis bone pointing backwards, just as it does in our living birds. These dinosaurs include the horned, armored, dome-headed, bipedal grazing dinosaurs, as well as the duck-billed dinosaurs. Ornithischians were primarily herbivores.

The *Saurischians,* or "lizard-hipped" dinosaurs, were characterized by a pelvis pointed forward, as it does in many reptiles. In 1887, archaeologists divided the *Saurischians* into two large orders: **sauropods** and **theropods**.

- **Sauropods** were the largest land animals ever to inhabit the earth, with at least one species reaching a mass of 80 tons and a length of 130 feet. They were quadrupedal with long necks, small heads, lengthy tails, and large, heavy limbs. So far, the fossil record indicates as many as 250 species of sauropods.
- **Theropods** are predominately carnivorous dinosaurs that varied considerably in size, from quite small to very large. *Segisaurus halli*, a small theropod uncovered in Arizona, was about the size of a goose (Carrano *et al.* 2005). The largest and most famous, the *Tyrannosaurs*, began their reign around 160 MYBP. Many species of tyrannosaurs have been identified, and they occur over a span of 100 million years (Hone 2016).

More recently, an analysis of 457 anatomical features from fossils of 73 different dinosaurs (Baron *et al.* 2017) puts less emphasis on the shape of the hip joint, and more on the entire skeleton of these animals. Using this new method of interpretation, archaeologists now date the theropods' origin to the mid–Triassic period (about 225 MYBP). They also reclassified theropods as being more closely related to Ornithischians than Saurischians.

Theropods played a pivotal role in the history of birds. They possessed many of the important characteristics of birds' ancestors: feathers, hollow bones with air sacs that served as extensions of the lungs, and a wishbone, which provides support for bones, muscles, and tendons. Most were carnivores, but some were herbivores, like *Therizinosaurs*, a large family of large and small dinosaurs found in Asia and North America.

Ornithologists have debated the subject of the theropod origin of birds for many years. Bird paleontologist Alan Feduccia (1999), for example, proposed a theory that birds and theropod dinosaurs share an earlier link, and that birds originated earlier in the Triassic period from small arboreal archosaurs, and not from theropod dinosaurs. Research over the past two decades has led to further understanding of their origin. The leading theory today is that birds evolved from theropod dinosaurs sometime after theropods' origin in the mid- to late Triassic period, and that they continued to evolve a number of physical traits that ultimately enabled flight. Most experts agree that this is the best current explanation of both the fossil record and the results of a variety of analyses (Prum 2002, Mayr 2017).

Even though Feduccia's theory has not become the dominant one, birds did inherit a number of anatomical traits from various stages of the archosaurs' vertebrate history. A host of archosaurs' traits in the Mesozoic era eventually showed up in modern birds: the shape of the toes, tail, and leg joints, for example (Makovicky and Zanno 2011).

The case for theropod ancestors appears to be stronger, however, with recent fossil evidence providing unique insight into the evolutionary transition through which theropods gave rise to birds.

Brusatte *et al.* (2014, 2017) traced the evolutionary path from theropod dinosaurs to birds by analyzing the evolutionary development of all known *Coelurosauria*, a subgroup of theropods very closely related to birds. *Coelurosauria* represent the next branch in the avian history. The evolutionary path to *Coelurosauria* is shown in Figure 1.5. Brusatte evaluated 150 different *Coelurosaurs* specimens for 853 anatomical characteristics, and revealed no great evolutional leap from non-birds to birds—but "once the avian body plan was gradually assembled, birds experienced an early burst of rapid anatomical evolution." This team called the comparatively quick emergence of birds "one of the greatest evolutionary transitions in the history of life."

Meanwhile, feathers appeared on a number of dinosaur species, evolving over a period of 100 million years before the appearance of the asymmetric vaned

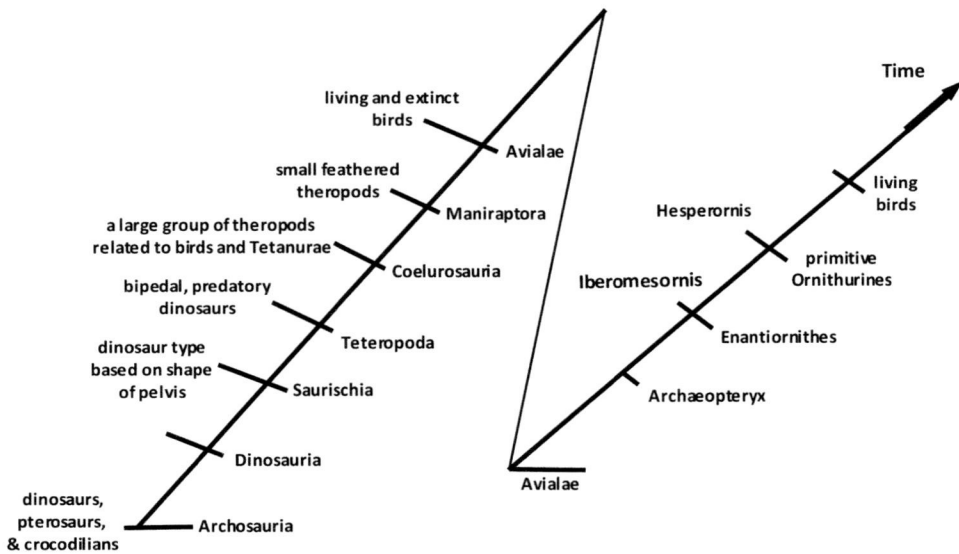

Figure 1.5 Evolutionary tree (phylogeny) of modern birds, beginning with their earliest ancestors (author's collection).

feather, a critical element in a feather's contribution to bird flight. The earliest feathers appeared about 250 MYBP (Zixiao *et al.* 2018) as filament-like structures found on pterosaurs. Feathers probably served as insulation at first—in fact, symmetrical feathers not adapted for flight have been found and dated to at least 10 million years before *Archaeopteryx* (Kaiser and Dyke 2011).

The Maniraptora—Another Link to Birds

Within the theropods, a branch known as the Maniraptora became extremely important in the history of birds.

Maniraptora made its first appearance in the archeological record during the Jurassic period. This two-legged clade of theropods had a long, heavy tail with interlocking parts to balance its own mass, allowing it to walk upright on two legs. This freed its long forelimbs and hands for gripping, for use as support in getting up from the ground, and, when equipped with claws, for slashing.

Some maniraptora were herbivores, while others were carnivores, and some were omnivorous. They were generally small in size—the heaviest weighed about 100 pounds, and most ranged in size from two to 30 pounds (1–14 kg). They had relatively large brain cavities and hollow bones that lessened their weight, so they could outrun their prey and predators. All maniraptora have wishbones, long arms, three-fingered hands, a half-moon shaped wrist bone, and feathers. Notably, the maniraptoras' shoulder joint differed from other theropods, making it possible for them to rotate the arm above the body.

About 30 species of maniraptora have been described by science so far, and these have been sorted into a number of major subgroups (Cau *et al.* 2015). The smaller species had the shorter, lightweight, feathered tails found in later fossils. About 130 MYBP, some with fused vertebrae appeared, resulting in shorter tails. These were followed by specimens with an extended or "keeled" sternum, which supported flight muscles and gave rise to sustained, powered flight. By 100 MYBP, maniraptoras were recognizable as birds.

Some maniraptoras could fly or glide, while others were terrestrial herbivores or carnivores. Their hollow bones likely functioned to assist breathing, as they do in modern birds. They built ground nests and laid eggs with three-layered shells, similar to modern birds but unlike their *Saurischian* cousins, which laid eggs with a single layered structure.

Recent advancements in paleontological work have allowed researchers to determine some basic feather pigmentation of the maniraptoras. Scientists can now analyze melanosomes, microscopic structures in fossil feathers and skin that provide color. Melanosomes contain pigment pouches that are visible as different shapes when viewed using a microscope. Reddish-brown hues result from spherical-shaped pouches, while darker colors come from sausage-shaped ones, and iridescent colors are produced by thinner, longer-shaped pouches (Vinther *et al.* 2016, 2017). Research has shown that the colored, melanin-producing melanosomes in feathers can be identified by chemical analysis (Colleary *et al.* 2015), thus eliminating any confusion with decayed microbes.

Over the last five decades, we have gained an understanding of the avian lineage's progression through time, thanks to a host of findings using new technologies. Biomolecular analysis of living birds (Kaiser and Dyke 2011) and cladistic (measurable characteristics) analysis of many specimens (Makovicky and Zanno 2011) have revealed a great deal of information, but there is much more to learn. The number of orders of small dinosaurs that evolved, and how many co-existed, are subjects about which science continues to expand its knowledge. New discoveries come to the surface almost every year.

Maniraptoran dinosaurs continued into the next recognized phyla in the ascent of birds: the *Avialae*. As this biological group evolved, different branches appeared as they advanced to the birds we recognize. The maniraptorans most closely related to birds mark the beginning of this transition, as shown in Figure 1.5. *Archaeopteryx*, believed to be the first bird species, is considered an early species within *Avialae*—and all living birds (Aves) are part of this phyla as well.

Smaller Dinosaurs: Keys to Discovery

Most of the early dinosaur discoveries were large animals, because large, solid bones are much more likely to be preserved in the ground and discovered by paleontologists. The size of the early sauropods and the enormity and fierceness of *Tyrannosaurus* captivated the public's attention and interest. People came to believe that all dinosaurs were large creatures.

A visit to the dinosaur room of the American Museum of Natural History in New York City bears this out. If you go, you will notice that the mega-dinosaur skeletons captivate the interest of the public, and the display on the much smaller maniraptora gets considerably less attention. This deficit of public interest in smaller dinosaurs makes it even more difficult for people to grasp the diversity of the maniraptora that have come to light in the past few decades—or the fact that the discovery of previously unknown dinosaurs is far from over. Scientists continue to discover more small dinosaurs today—and finding the dinosaur is only the first step in the study of it. Paleontologists continue to revisit and re-analyze poorly understood past finds, improving our understanding of birds' evolution from these prehistoric ancestors.

Today, our understanding of the emergence of avian characteristics in the Mesozoic era is based on at least three separate, co-developing genetic lineages: the small feathered theropods, the *Enantiornithes* (meaning opposite bird), and the primitive toothed *Ornithurines*, the source of modern birds.

During the Cretaceous period of the Mesozoic era, many small dinosaurs co-existed, similar to the way terrestrial mammals do today. The vast majority of these were arboreal and terrestrial maniraptors, along with other small theropods and some mammals. These smaller creatures were widely distributed on earth, but because of their size and the delicateness of their bones, they were not as well represented in the fossil record. Recent finds and more analysis have expanded our understanding of their world, providing further insights into the history of birds.

Finds in China have been particularly important in advancing our understanding of bird evolution. Paleontologists discovered the rich fossil layers known as the Yixian and Jiufotang Formations in northeast China by the late 1940s, but these areas did not receive extensive study until relatively recently. These deposits of weakly laminated fine sediments, sandstone, and shale date from the late Jurassic to the early Cretaceous periods and contain many elements of a Cretaceous ecosystem, including extremely well preserved theropod dinosaurs, birds, mammals, and plants. One of the specimens from China, *Xiaotingia*, is similar in structure to *Archaeopteryx*, although many of the specimens from Jiufotang are markedly different from this earliest bird (Mayr 2017).

For example, one of the smallest dinosaurs ever found—a tiny, non-avian theropod dinosaur that scientists called *Microraptor zhaoianus*, was discovered in China in 2000, and estimated to be sixteen inches long (Xu *et al.* 2000). *Microraptor* is also known as the "four-winged dinosaur," because it had a second set of wings on its hind legs. More detailed specimens have shown that its feet were adapted to climbing.

The archeological discoveries in China have revealed many specimens of *Microraptor* [Figure 1.6]. This non-avian theropod dates from about 125 MYBP, and weighed about 2.2 pounds. In addition to its asymmetric flight feathers on both its wings and long legs, it had a long, narrow, feathered tail. It measured about four feet long, and like *Archaeopteryx*, it had teeth as well as claws on its wrists.

Figure 1.6 **A number of fossils of *Microraptor* found in China show some differences but they all had wing-like feathers on their legs (Nobu Tamura).**

Wind tunnel analysis suggests that *Microraptors* were able to fly or glide, and might even have been able to accomplish powered flight. They used their leg feathers like hind wings, and the long tail for control in reducing speed and in landing (Dyke *et al.* 2013). Recent work analyzing fossilized pigments suggests that the feathers were black and iridescent.

Another maniraptor—*Anchiornis*—was found in China and dated to the Late Jurassic period. Chipmunk-sized and possibly one of the smallest dinosaurs, it predated *Archaeopteryx* and *Microraptor* by millions of years. Many specimens of this maniraptor have been found, allowing us to approximate its color pattern.

Like *Archaeopteryx*, the fully feathered *Anchiornis* had long legs and forelimbs, was about 13 inches long (33 cm) and weighed 0.2 pounds (110 grams) [Figure 1.7]. Its arms and legs were covered with pennaceous feathers (feathers with a vane down the middle, like birds today), similar to those of *Microraptor*. The feathers were tapered near the tip and not aerodynamically shaped for flight. Color pigments preserved in the fine clay indicate that it had an orange crown, black wing feathers with white stripes, and a black face with red spots.

A number of other small dinosaurs contributed to birds' complex ancestry. *Caudipteryx*, a Cretaceous maniraptor that lived around 125 MYBP, was turkey-sized and believed to be an omnivore. It had a small head with a beak that came to a tapered point, and a few small teeth in the front jaw. Covered in feathers, including shorter body feathers, it had claws on its fingers and fan-like longer

Figure 1.7 **A drawing of the small four-winged dinosaur *Anchiornis*, showing colors of its plumage (Nobu Tamura).**

feathers attached to the second finger. Its long tail feathers could spread like a fan.

Paleontologists believe *Caudipteryx* was a swift runner with its long legs, but its short arms were unsuited for flight. Not unlike some modern flightless birds in overall body proportions, it weighed about 20 pounds and was three feet long. The fossil had gizzard stones (gastroliths) where the gizzard would have been; gastroliths are used to grind food in animals lacking teeth, a trait of herbivorous birds.

The clade *Enantiornithes* represents another step along the historical lineage of birds (see Figure 1.5). Very diverse and widely distributed, *Enantiornithes* lived in trees, had bony teeth and a claw on each wing. Some were small, others the size of a crow, and some larger. *Iberomesornis*, depicted in Figure 1.8, was the size of a sparrow. Another very well-preserved specimen of one found in North America, however, was as big as a turkey vulture (Atterholt *et al.* 2018). Many fossils have been found in Eurasia, Australia, and North and South America, indicating that they were a very successful order.

Enantiornithes had teeth, clawed fingers on each wing, and a long, bony tail, but otherwise looked like modern birds. They could sustain flight, but their shoulder joint was the reverse of modern birds' shoulder socket (Hope 2002), leading paleontologists to name them with the Greek word for "opposite birds." By the end of the Cretaceous period, at least some *Enantiornithes* birds had evolved several features found in modern birds (Atterholt *et al.* 2018), including a deeply keeled sternum and changes in the wishbone and feather structure that are closer to the structures we see today. Despite their success, *Enantiornithes* did not survive the

Figure 1.8 A small arboreal Enantiornithe, *Iberomesornis*, of the early Cretaceous, had a wingspan of about 8 inches (20 cm) and a small body (Nobu Tamura).

Cretaceous-Paleogene (K–Pg) extinction event in about 66 MYBP, and disappeared with many other species.

The Next Evolutionary Branch

The primitive and toothed *Ornithurines*, another diverse group, originated in the early Cretaceous period and are more closely related to modern birds than many earlier creatures—in fact, their tails are similar to those of modern birds. Two members of this family include *Hesperornis* and *Ichthyornis* (O'Connor *et al.* 2011).

Hesperornis were large, flightless marine animals of the late Cretaceous period, about 80 MYBP, and are represented by more than 25 species of different sizes, the largest of which was five feet long, represented in Figure 1.9. Most were believed to be flightless; the exception may be one found in England, *Enaliornis*, which weighed about one pound and may have been able to fly (Mayr 2017). *Hesperornis* had lobed feet like a grebe, and legs set back on the body like a loon, so it was not capable of walking on land. In water, *Hesperornis* looked like a cormorant with its long neck and long bill, and a palate to retain fish. Its main difference from modern birds was that it had teeth, while modern birds do not. Its fossils have been found in Kansas, Canada, Russia, and Sweden.

Figure 1.9 *Hesperornis regalis* **was a wide-ranging aquatic bird found around the northern hemisphere (Nobu Tamura).**

Another discovery in China, the earliest known ancestor of modern-looking birds is dated at about 130 MYBP (Wang *et al.* 2015), in the early Cretaceous period. Known from just two specimens, this semi-aquatic wader *Archaeornithura meemannae* was a member of the *Ornithurae* clade. It was about six inches (15 cm) tall, 12 inches (30 cm) long, and stood on bare legs. It had a small feather on the leading edge of the wing called an alula, which increases maneuverability during slow flight. It also had a wishbone, a fan-shaped tail, feathers, and wings with six primary feathers. Paleontologists believe it had the necessary structure to fly.

A seabird called *Ichthyornis*, a ten-inch-long bird about the size of a tern [Figure 1.10], was much like today's birds in bone structure (Mayr 2017), and it could fly because it had a full sternum. First discovered in 1870 in Kansas by Benjamin Franklin Mudge, it was immediately recognized as a bird and named by Othniel Charles Marsh. The beak, like that of *Hesperornis*, had several distinct plates, much like the beak of a modern albatross. It also had teeth in the midsection of the upper and lower mandible, curved backward to allow it to grasp fish (Clarke 2004, Lamb 1997).

Figure 1.10 *Ichthyornis* was a small seabird of the Late Cretaceous from North America with a wingspan similar to that of a medium-sized tern (Nobu Tamura).

Palaeontologists are always making new discoveries from both old and new fossils. Daniel Field and his group reanalyzed fossils found two decades earlier and discovered that misidentified bones were actually part of the bony palate (roof of the mouth) of a bird from the Upper Cretaceous that lived approximately 67 MYBP, and recently named *Janavis finalidens* (Benito et al. 2022). This bird is similar to but larger than the previously mentioned Ichthyornis. However, the characteristics of the palate are unlike those of the Ratites, but are similar to those of more modern birds.

A few other early fossils exhibited some of the characteristics of today's birds: a beak without teeth, a four-chambered heart, a lightweight skeleton with hollow bones connected to the breathing system, and a three-layer egg structure.

Vegavis, an extinct early bird from the late Cretaceous period (68–66 MYBP), was originally found on Vega Island in Antarctica. It was in the order *Anseriformes*, which are related to modern ducks and geese (Clarke *et al.* 2005), and *Vegavis*' bone structure closely resembled today's waterfowl. Fragments of another bird—*Treviornis* from the late Cretaceous period and dated to 70 MYBP, and also of the *Anseriformes* order (Kurochkin *et al.* 2002)—were found in the Gobi Desert of Mongolia.

Finally, the complete, three-dimensionally preserved skull and other bones of a small, quail-like bird were found in northern Europe, and dated to just before the end of the Cretaceous period (Field *et al.* 2020). This long-legged bird's skull size fits between a duck and a quail. It represents a link between modern birds and the ones that existed before the great extinction.

Many fossils of late Cretaceous birds discovered in western North America are little more than fragments, so they cannot reliably be assigned to a family. Specimens from Saskatchewan, Wyoming, Colorado, Montana, and also Argentina cover a range of families including land fowl, waterfowl, loons, and some seabirds, but remarkably, they do not include any arboreal birds (Mayr 2017). One of these closely resembles members of the crane family.

The Cretaceous period left behind many fossils, but most of them are incomplete, making it difficult to understand the diversity of avian forms and how ancient lineages are related to each other and linked to modern birds. Nevertheless, the evidence we have leaves no doubt that some orders of modern birds existed before the radical environmental change that came at the end of the Cretaceous period. Over time, more fossils, molecular data, and analysis will continue to clarify the relationships between these prehistoric species and the birds we see today.

The Great Extinction

The reign of the dinosaurs and many other creatures came to an end at the Cretaceous/Paleogene boundary, in the event known as the K–Pg extinction. (The letter "K" is derived from the German word "kreide," meaning chalk, and is used as an abbreviation for the Cretaceous Period. "Pg" is shorthand for the Paleogene period.) Many animals became extinct, but many others survived, including a variety of birds, mammals, reptiles and amphibians.

The K–Pg extinction affected the entire globe at about 66 MYBP, marking the end of the Cretaceous period and the Mesozoic era, and wiping out all dinosaurs and pterosaurs, as well as other entire classifications, including marine reptiles like plesiosaurs and mosasaurs. At least 17 species of birds went extinct as well (Longrich *et al.* 2011), along with many other creatures (Sakamoto *et al.* 2016). How and why the widely diversified *Enantiornithes* birds became extinct and the *Neornithes* (modern birds) survived remains something of a mystery.

In 1980, Nobel laureate Luis Alvarez proposed that a single large asteroid strike caused this event, based on the discovery of a worldwide layer of iridium found on land and under the sea. Iridium is a rare metallic element on earth, but it is more concentrated in meteorites. Other smaller craters of similar age have been found in the Ukraine, the North Sea, and one off India, however, so the current accepted explanation for the K–Pg event involves one large asteroid or meteor strike, followed by a series of lesser meteorites.

Geophysicists Antonio Camargo and Glen Penfield were the first to produce evidence of this strike when they discovered a crater now known as the Chicxulub impact, 136 miles (219 km) in diameter at the edge of the Yucatan Peninsula in Mexico. The asteroid had a diameter of about six miles, and its impact created an extensive fire storm, burning 70 percent of all trees on Earth (Preston 2019). It darkened the atmosphere with billions of tons of debris and released tons of sulfur, which caused acid rain that scorched the planet's surface. Once the fires subsided, Earth plunged into a period of cold, dropping globally by 7°C (13°F). Based on computer simulations done by Clay Tabor of the National Center for Atmospheric Research (Summer 2017), it appears that no light reached the earth's surface for at least a two-year period. Seventy-six percent of all species on Earth went extinct.

Some in the scientific community disagree that the Chicxulub impact alone caused the end of dinosaurs. They point to the massive volcanic eruption of the Deccan Traps in Central India, which spewed poisonous gases, carbon dioxide, and ash into the atmosphere, perhaps for as long as 30,000 years. Volcanic activity brings about cooling of the earth's atmosphere, which would have been devastating to dinosaurs. The most recent research, however, supports the theory that the extinction event was caused by the Chicxulub impact: A group from Yale University used sediment cores from the sea floor to understand the timing of global temperature shifts, based on marine fossils and climate models spanning several hundred thousand years before and after the extinction. They concluded that the Deccan Traps happened well before the K–Pg extinction event 66 million years ago and, therefore, did not contribute to the mass extinction (Hull *et al.* 2020).

Recent work (Sakamoto *et al.* 2016) suggests that large dinosaurs were in demise long before the end of the Cretaceous period, but maniraptors were not. Based on fossil evidence, maniraptors may have been increasing in number even as the large dinosaurs struggled to survive. Many other ancestors of birds did become extinct as a result of the event, however, including the well-diversified group of *Enantiornithes*. The diagram below [Figure 1.11], created by Kaiser and Dyke (2011), summarizes the current understanding of the evolution of birds at the time of the K–Pg extinction boundary.

The current views on how many orders of birds survived the K–Pg extinction are inconclusive; this subject will be discussed in the next chapter. However, the fossil

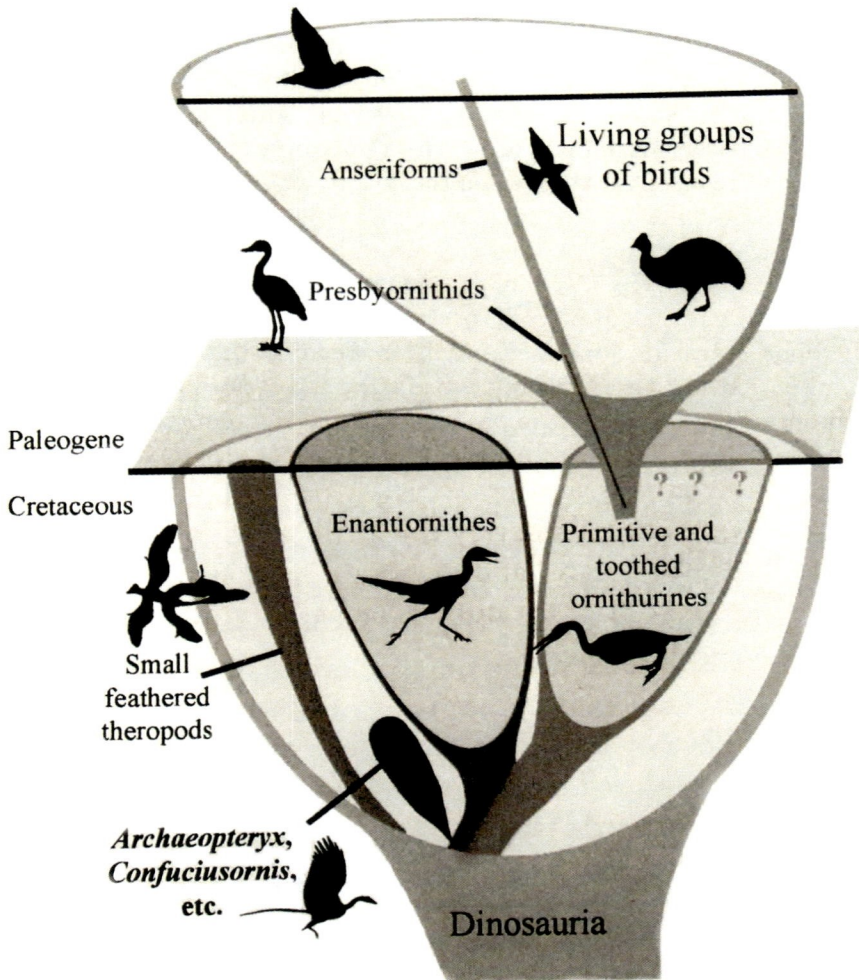

Figure 1.11 Evolution of birds before and after the K–Pg boundary (Gary Kaiser, Dyke and Kaiser 2011).

evidence shows that some early ancestors of modern birds survived. The early members of these families co-existed with the non-avian maniraptora before the extinction event. Phylogenetic analysis based on DNA evidence supports the claim that a major radiation—a blossoming in diversity of early birds—took place in the wake of the mass extinction (Prum *et al.* 2015). Another study based on well-documented fossils shows that several clades related to modern birds diversified around or soon after the K–Pg boundary (Ericson *et al.* 2006).

Summary of the Jurassic and Cretaceous Periods

The Jurassic and Cretaceous periods lasted about 134 million years, and these periods, along with the Triassic, are recognized as the age of dinosaurs. These prehistoric creatures were the dominant life form on Earth during these times. Dinosaurs were distant relatives of many other reptilian life forms that had originated in the Permian period, and became dominant in the Triassic period. During the Jurassic and Cretaceous periods, the continents gradually separated, and the large shallow seas began to disappear. An extinction event took place at the end of the Jurassic period (145 MYBP), and the enormous sauropods and many of the marine reptiles disappeared. The cause of these changes is unknown, but it did open the world to the development of other creatures.

Warm periods and the advent of flowering plants helped to support diversification of life in the Cretaceous period. Insects went through an extensive diversification. Theropod dinosaurs flourished and diversified over the 36 million years of the Cretaceous period, but were likely in demise at the end of this period. Mammals first appeared on Earth in the Permian period and generally remained small creatures until the end of the period. Birds made their first appearance in the fossil record in these two periods, and had significantly diversified and spread around the planet by the end of the Cretaceous period.

The extinction event at the end of the Cretaceous period brought the age of dinosaurs to an end, but some bird families survived the period's catastrophic close. Why did birds survive? In the next chapter, we'll look at what researchers have done to try to answer this question.

2

Birds in the Paleogene Period

Somewhere around 66 MYBP, a meteor struck the earth and caused the Cretaceous-Paleogene (K–Pg) mass extinction event. This killed most of the earlier recognizable life forms, ending the Cretaceous period and beginning the Paleogene period—from about 66 to 23 MYBP. Change was sudden and dramatic: Seventy-five percent of all living species became extinct over an extremely short period of time, in geologic terms.

The K–Pg event wiped out trees for at least 1,000 miles (1,500 km) from the primary meteor strike on the Yucatán Peninsula, and triggered volcanic eruptions and wildfires around the earth, which in turn produced acid rain. In the persistent darkness—the result of the launch of 15,000 million tons of volcanic soot into the atmosphere—the food chain completely collapsed (Tabor *et al.* 2019).

As noted in Chapter 1, extinction events are not uncommon in the geological time scale, and the fact that some creatures survive, reestablish and diversify is not unusual. The K–Pg extinction stands out, however, because it was on such a large scale, and because the darkness disrupted photosynthesis, which curtailed plant growth as well as animals' survival.

Scientists continue to study the question of why dinosaurs and so many other animals were driven to extinction at the K–Pg boundary. Some have formed hypotheses based on comparing fossils from ancient sea beds that predate and postdate the event. In the Hell Creek Formation from northern Montana to the Dakotas, for example, 93 percent of the mammal species went extinct; only very small mammals survived. Unlike mammals, pterosaurs could fly and were diversified, but they did not survive, either. Tree-dwelling birds, the *Enantiornithes*, went extinct even though they had evolved a more advanced skeletal structure, much like the birds that live today, as well as variations in size. Large aquatic birds—various forms of *Hesperonis*—did not survive.

Research by Kaiho *et al.* (2016) proposed that the asteroid's impact forced soot into the air from the "oil-rich area" of the Yucatán Peninsula, where the asteroid made its catastrophic landfall. This soot blocked the sun, resulting in much colder climates than the indigenous animals could tolerate, as well as drought at lower latitudes. Over time, the lack of sunlight affected plant life both on land and under water—the latter happening months or even a year after the impact, as the water cooled to temperatures that would not sustain tropical and subtropical plants usually found in the region.

One theory investigated by the National Center for Atmospheric Research (NCAR) and the University of Colorado–Boulder (Bardeen *et al.* 2017) suggests that the stalling of photosynthesis for even a single year could lead directly to the death of entire species of plant-eating animals. While many large animals may have been wiped out by the meteor's impact and its aftermath, others would starve to death with no plants to eat—but aquatic animals that did not depend on surface plants survived for a time. These perished as well, however, as the darkness dragged on for nearly two years, reducing light levels in lakes and oceans so dramatically that plants could not grow beneath the water's surface.

After hundreds of thousands of years, the entire earth gradually experienced a new beginning, but it did not recover to its original state. The planet's biodiversity changed, and while some life forms recovered within decades, others took millions of years.

Climate simulations by Bardeen *et al.* (2017) tells us that because of the large amount of dust and soot in the air, it took four years just for sunlight to return to at least 10 percent of normal over most of the earth, and about 40 percent of normal at the poles. The earth's temperature plummeted by 29°F (16°C), from an average of about 73°F (23°C) to about 44°F (7°C). Precipitation decreased substantially, and northern hemisphere sea ice expanded to mid-latitudes, much farther south than before. Some areas of the planet were more affected by the event than others: Coastal areas, for example, were better off than inland areas. The equatorial Pacific experienced a dramatic reduction in temperature and precipitation, while the coast of Antarctica barely cooled and precipitation increased only slightly (Tabor *et al.* 2016).

Temperatures remained above freezing in the oceans, coastal areas, and parts of the tropics, but for the first one to two years, the inhibition of photosynthesis significantly delayed the recovery of vegetation. Meanwhile, freezing temperatures persisted in the mid-latitudes for two additional years (Bardeen *et al.* 2017).

Research by Lowery *et al.* in 2018 suggests that small marine organisms first reappeared at the meteor impact crater site in about 30,000 years, while the recovery of the global marine ecosystem in North America took about 300,000 years.

Pollen from before and after the extinction event showed that ferns were the dominant plant in the early recovery. More pollen studies (Vajda and Bercovici 2014) concluded that flowering and seed-bearing plants began recovery a few centuries to a millennia after the event. Donovan *et al.* (2016) studied evidence of insect-feeding damage on fossil leaves, and estimated that recovery of land-based plants to pre-extinction conditions took four million year in South America, and nine million years in western North America.

The animals that managed to survive under these conditions had qualities more adaptable to changing conditions. Animals with small bodies, aquatic lifestyles, night vision and an ability to eat a wide variety of foods including seeds, plants, and insects were more likely to survive (Rosen 2017). Fully feathered birds were much better at surviving cold temperatures than their non-feathered predecessors. Some birds had developed a very efficient breathing mechanism that evolved from their dinosaur lineage, providing them with an advantage in the face of lower oxygen levels. Field *et al.* (2018) determined that only ground-dwelling birds survived, resulting in a predominantly non-arboreal population of birds following the extinction event.

Bird Families That Survived the K–Pg Mass Extinction

The Paleogene period played a pivotal role in the diversification of birds. During the three epochs of its 43-million-year interval of the earth's history, the biology of the earth changed substantially. It ushered in the re-emergence and dominance of flowering plants, the domination of land by mammals, and the domination of the air by birds—and the birds more familiar to us today came to diversify, spread, and populate the earth.

As the effects of the extinction event subsided, birds adapted to the many habitats opened to them. Ornithologists recognize that modern birds (*Neornithes*) became more diverse in the Paleogene period, evolving in the warm cycle of this period and coexisting with mammals and reptiles. The fossil record of the first 10 million years of the Paleogene period, known as the Paleocene epoch, contains more than a dozen families of birds (Lindow 2011), some of which are related to modern birds.

Most of the Paleocene bird species did not survive, their lineages becoming dead ends in this time of diversification. The flightless genus Gastornis, for example, disappeared at the end of the next epoch (the Eocene) as global cooling took hold and ice sheets began to expand from the poles, making it too cold for the ground-dwelling birds that had evolved during a warmer geologic period (Lindow and Dyke 2006). Gastornis, one of the species within the *Diatryma* genus, was a six-foot tall, flightless North American bird with a stout nine-inch beak, depicted in Figure 2.1. It was originally believed to be a predator—dubbed the "terror crane" by modern-day archaeologists—but in more recent studies of the beak structure, scientists believe this bird was a herbivore that probably fed on tough plant material and seeds. The causes of its disappearance are still under investigation by scientists.

The fossil record also tells us that some orders of birds that arose in the Cretaceous period did survive the extinction event. The earliest fossils dating from after the event are primitive, however, and cannot be tied easily to the evolutionary understanding of modern birds (James 2005). These early species were not exactly like those we recognize today, but we can relate them to orders of birds such as *Anseriformes*, the order of waterfowl.

However, fossil evidence supports the belief that many groups of birds arose just a few million years after the mass extinction. For example, a fossil of an ancient species of mousebird was found in New Mexico and dated at 62.5 MYBP, only a few million years after the extinction event (Ksepka *et al.* 2017). It is similar to mousebirds found today in southern Africa.

By the end of the Paleogene period and the beginning of the Neogene period (23 MYBP), fossil birds can be more readily related to modern families, and we can see the appearance of birds we are familiar with today as they evolved from these earlier groups.

© N. Tamura, 2011

Figure 2.1 The fearsome-looking Terror Crane (*Gastornis*) of North America was almost seven feet tall (2 meters) and its estimated weight is over 500 pounds (250 kg) (Nobu Tamura).

Understanding Some Birds' Survival

Fossils continue to be a key element in understanding the families of birds that survived the K–Pg extinction. More recently, scientists have added genetic analysis to their toolkit in determining the factors that led one species to die out after the event, while another one managed to find a way through the dark years.

The fossil record is sparse in the late Cretaceous period, and the quality, age, and incompleteness of fossils make their identification difficult at best. As a result, there are few fossils from this period that are directly connected to modern birds. Exactly how many orders of birds survived from the Cretaceous is not really known, making it a subject of continued scientific interest. Until recently, only representatives of heavy ground-feeding birds (*Galloanseres*) and waterbirds (*Aequornithes*) have been identified from the fossil record before the K–Pg boundary (Mayr 2014). Both of these groups include hundreds of species today.

Based on an analysis of sediments that yield early fossil birds, Dyke *et al.* (2007) have shown that the fossil record of birds across the K–Pg boundary follow a pattern in which aquatic environments dominated the avian fossils of the early Paleogene. They concluded that the ancestors of modern birds lived in these wetland habitats after the extinction event. Fossil evidence appears to demonstrate that in the Mesozoic era (before the K–Pg boundary), birds were evenly divided between land and aquatic species, but in the Paleogene—no doubt because the planet's surface became inhospitable, as described earlier—aquatic species became more plentiful.

Dyke and Gardiner (2011) used a statistical analysis of chronological fossil dates to conclude that four orders of birds, and possibly a fifth, survived the K–Pg event: *Anseriformes* (wildfowl), *Procellariiformes* (tubenose seabird family), *Strigiformes* (owls), *Apodiformes* (swifts) and *Galliformes* (game birds). Lindow (2011), on the other hand, summarized the current molecular and morphological models of the origin and diversification of birds in the Cretaceous period, and concluded that just three of these orders survived that period, as well as the ancestors of the ratites—large, mostly flightless birds of the *Palaeognathae*, one of the two living clades of birds, which today include ostriches and kiwi. Most of these survivors are ground-dwelling species; the earliest fossils do not shed light on perching birds. *Vegavis*, another important fossil dating back to the late Cretaceous period, is recognized as a form of duck, an order that came after the game birds and ratites (Lovette 2016).

The discovery of DNA in the 1950s has provided a greater understanding of the interrelationships of the world's birds. Sibley and Ahlquist (1991) were the first to use a molecular biology technique called DNA-DNA hybridization to determine the genetic similarity between large sets of DNA sequences, particularly between major groups of birds. Since then, this technology has improved substantially as a method to compare the DNA of different bird species. The initial comparisons were based on single segments of DNA, while later work expanded to look at numerous gene markers on the birds' genome to find similarities between taxa. This allowed better resolution and more precise analysis. Most recently, scientists use the entire genome in these analyses to establish information from hundreds of thousands of DNA characters, and use this information to understand how modern species were related to the fossils of the past.

With DNA studies, researchers can extrapolate backward in time by measuring the rate of the genome's change from one species to the next. Prum *et al.* (2015) studied the history of evolution of birds—also called the phylogeny—and their evolutionary relationships based on measured inheritable traits of living birds. These researchers used 390,000 bases of the genomic sequence from 198 species of living birds to understand how birds are related to each other. Through this detailed examination, they concluded that there were five major clades of birds within the clade of all modern birds—known as the Neoaves—that emerged in the Paleogene period. These represent more than 90 percent the modern birds we recognize: (1) the modern nightjars and their relatives, swifts and hummingbirds; (2) the cuckoos, bustards, turacos, pigeons, mesites, and sandgrouse; (3) the cranes and their relatives; (4) a comprehensive waterbird clade related to diving ducks, wading birds, and shorebirds; and (5) a comprehensive land bird clade.

Prum also used 19 fossil taxa to calibrate the emergence of each of the various families of birds by "divergence-time estimation," a process that calculates the development of one species from an ancestor, based on a relatively constant rate of DNA evolution over time (what is known as the "molecular clock"). Using the rates of change of DNA, scientists projected the existence of passerines (perching birds) and other modern birds prior to the K–Pg extinction, but so far, the fossil record provides insufficient evidence to support this claim. These differences of opinion on diversification between DNA evidence and the fossil record are well recognized by ornithologists and molecular scientists.

The fossil record does provide a rough estimation of the timing of the divergence of modern birds, however, and these dates are broadly consistent with the results from molecular analysis (Mayr 2017).

The study of genetics has become an important tool in the search for the origins of species. Evidence at the molecular level indicates that more extensive spread of bird species throughout the continents took place in the Cretaceous period, beyond what the current fossil record supports. This may mean that there are many more ancient bird species yet to be discovered, to provide the clues science requires to understand more fully the evolution of birds from their ancestors.

Avian Diversification in the Paleogene Period

Climate is thought to have played an important role in the adaptation of birds in the Paleogene period. The climate warmed significantly during and at the end of the Paleocene epoch and continued to warm in the early Eocene, reaching its maximum at about 50 MYBP. This temperature rise is referred to as the Paleocene-Eocene temperature maximum.

Little to no ice existed on Earth at the time, and the temperature difference between the equator and poles was much lower than we experience today. Tropical plants lived on all continents across the Arctic. Montana and the Dakotas, for example, were home to tropical rainforest plants such as the magnolia, citrus, fig, pawpaw, and cashew families. At 75° N latitude, Ellesmere Island in the Canadian Arctic hosted subtropical plants, as well as animals like alligators. By the latter part of the Eocene epoch, however, temperatures began to fall; ice formed again, and the Antarctic ice sheet began to appear.

The Paleogene period became an important pivot point in the history of birds' diversification. Scientists Claramunt and Cracraft (2015) performed an analysis that indicates a rapid net diversification in the Paleocene epoch. With increasing temperatures at the end of that epoch, diversification rates decreased, then increased again in the late Eocene and remained high through the end of the Paleogene period. From their results, these macro-evolutionary changes in diversification rates were largely the result of climate change based on ocean temperatures.

During this 43-million-year Paleogene period, new bird species appeared, and others went extinct. Claramunt and Cracraft used complex mathematical analysis of fossils to analyze the rates of diversification and extinction. They selected a wealth of fossils from locations in Europe and the United States: In North America, for example, the Green River Formation in Colorado, Wyoming, and Utah, and the Willwood Formation in Wyoming provide fossils from the Eocene epoch that represent approximately 25 families of birds. In Europe, 55 families of birds are represented from this period (Feduccia 1999, 1995; James 2005). In particular, the Lower Eocene

Fur Formation of Denmark has yielded some of the best preserved fossils relating to major clades of modern birds (Lindow and Dyke 2006).

Some of these fossils were found in three dimensions and with intact soft tissues. These fossils provide the earliest record of modern birds' spread across the continents—an expansion of species and habitat known as "radiation." European and North American animal fossil records have some striking similarities, since they were closely connected in the early Eocene epoch (James 2005).

Using cladistics data and DNA, Claramunt and Cracraft proposed that birds in South America survived the K–Pg extinction event and then started moving to other parts of the world over multiple land bridges, while diversifying during the periods of global cooling. Their work is based on establishing a probability distribution for the age and the analysis of fossil records on a large geographic scale.

Birds used two routes to cover the globe after the K–Pg extinction: first to North America across a Central American land bridge, and then to the Old World; and second, to Australia and New Zealand across Antarctica, which was relatively warm at that time. Figure 2.2 shows a map of the world near the end of the Cretaceous. The continents were close together, sea levels had dropped, and the inland sea in North America was mostly gone.

One More Extinction Event

The end of the Eocene epoch (56–33.9 MYBP) and the beginning of the Oligocene (33.9–23 MYBP) is marked by another large-scale extinction event that affected marine organisms and mammals. This was a time of climatic change

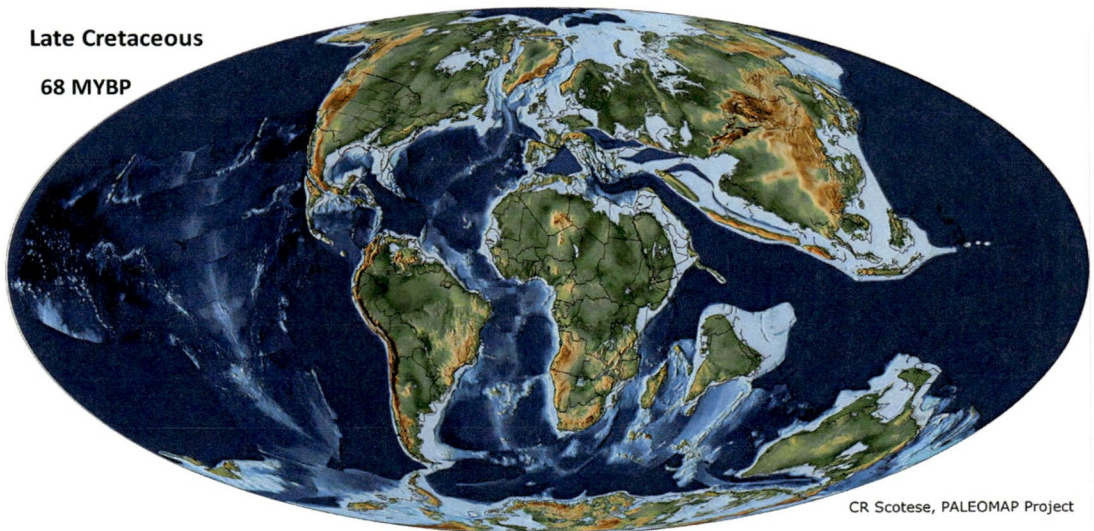

Late Cretaceous

68 MYBP

CR Scotese, PALEOMAP Project

Figure 2.2 At the time of the late Cretaceous, the continents were separated and surrounded by shallow seas, but Australia was still near to Antarctica (C.R. Scotese, PALEOMAP Project).

brought on by cooling, with no currently known links to any volcanic or cata-strophic impact incidents. A large number of European mammals became extinct, and new Asian immigrants appeared. This also became an important factor for the diversity of birds.

Once again, we turn to the only lasting record of birds throughout these geo-logic periods: the fossil remains that have been discovered by archaeologists and studied enough to determine their position in time. The bird fossils of this era are primitive, but we can connect them more specifically to modern bird families and to the earliest records of transitional orders. These primitive fossils include such diverse modern families as ratites, owls, waterfowl, ibises, penguins, game birds and possibly passerines.

Analysis of the fossil record shows that most or all of the diversity seen in mod-ern families of birds evolved in the Oligocene epoch, the last of the Paleogene period (James 2005). Families of birds discovered from the Oligocene epoch include a hum-mingbird described by Mayr (2004) from its wing bones, a tiny barbet-like bird, and songbirds (Mayr 2005a, 2005b). Sources list as many as 130 species or families of birds from the Eocene and Oligocene epochs.

The *Passeriformes* or passerines (perching birds), the largest living order of birds that currently account for more than half of the species on earth, were just beginning to appear in the Eocene epoch. Some examples of passerine birds from this epoch have been discovered among the fossil records. Dating the earliest passer-ine, however, is a contentious question.

Earliest specimens of finch-like passerines have been dated to the early Eocene epoch (Ksepka *et al.* 2019), and passerines were certainly present in the early Oligo-cene epoch. Paleontologists believe they originated in the southern hemisphere and invaded the northern hemisphere (James 2005), but there is also evidence that pas-serines originated in Australia (Low 2017).

The end of the Paleogene and beginning of the Neogene period (23–1.8 MYBP) were marked by a change in marine fossils brought about by continued cooling tem-peratures. Sea levels fell, land bridges appeared, and global temperatures became seasonal. Birds increased their net rate of diversity in the Neogene period (Clara-munt and Cracraft 2015). The avian fossils of this period look very much like mod-ern birds. Apparently even the most modern bird families have been on earth for a very long time.

Summary of the Paleogene Period

Life was severely altered at the end of the Cretaceous period, and it took a long time for the plants, animals, and birds to recover. The fossil record is sparse, but it is sufficient to show that some families of birds survived the extinction event—most likely ground and water birds.

During the Paleogene period, the climate was generally cooler than during the Cretaceous period, with long warm and cold periods. North America and Eurasia separated as part of the continental drift.

The bird fossils from the Paleocene—the first epoch of the Paleogene—are primitive, and few can be related to modern bird orders. By the end of the Paleogene period, mammals had become the dominant vertebrates. Many new species of birds appeared during the period and some became extinct, while others diversified further. In the Eocene epoch, birds became more widely distributed and likely adapted to all the habitats available to them.

The extinction event at the end of the Eocene epoch brought many changes again. Rapid and extensive avian diversification took place during the Oligocene epoch. Most of the modern families of birds made their appearance in the Oligocene and the Neogene, the periods that followed the tumultuous Paleogene.

PART II

THE LIVES OF BIRDS

Introduction

The fossil record and genetics tell us that birds have evolved over a long period of time, but these insights do not explain how their complex history was accomplished. There are many more factors that have led to their success than the environmental issues mentioned in the first two chapters.

Volumes of scientific research have been published on the nature of birds by thousands of ornithologists dedicated to understanding the lives of birds in great detail, through the study of the ecology and behavior of each species. Their insights explain a great deal about how birds have become such a successful class (Aves) of living creatures.

Chapter 3 looks at birds' amazing ability to adapt, a characteristic that has led birds to become extremely diversified in their behavior and habitats: some never leaving the ground, some spending their lives on water, some flying hundreds of miles in a day or thousands every year during migration. This diversity allows them to occupy a wide range of habitats on Earth. We will examine the factors that have influenced this ability to adapt.

Chapter 4 gives an overview of a few of the physical traits that make these creatures uniquely "birds." For example, birds have a specific body structure that sets them apart from mammals, fish, and reptiles; they have varying levels of intelligence based on their species, and they have developed abilities that allow them to succeed in their habitat, whether they live deep in the wilderness or in the middle of human-populated cities.

Chapter 5 looks at the variations in social and breeding lifestyles of birds. Here we explore territoriality, mating rituals, timing and conditions for mating and breeding, pair bonding, nesting, and fledging.

Chapter 6 discusses some basic aspects of bird flight, and what is understood about how birds navigate in migration. Flight is probably one of the most interesting aspects of birds' physiology. Although bats and insects can fly, their flight is not like that of birds. Birds became the model that introduced humans to flight.

Finally, Chapter 7 addresses the post-breeding aspect of birds' lives, vagrancy, and dispersal.

3

Adaptation and Diversity

By the end of the Paleogene period, earth hosted a wide diversity of bird life, as well as many species of plants, animals, and microorganisms. This created the basis for the explosive growth of plant and animal life we see today: Scientists estimate that about 8.7 million species of animals exist on earth now (Mora *et al.* 2011), including many microscopic species. An international research team completed a global vegetation database analysis in 2018, and estimated that there are about 390,000 species of plants on earth (Sexton 2018), though not all of these have been found and described.

Anyone can see that life on earth is diverse, but the fact that living things change over time is not obvious unless we study them in detail. Again, we turn to the fossil record, as it can teach us a great deal about how such changes take place. Fossils show us periods of species development and times when organisms diversify rapidly from an original species—a process scientists call radiation—throughout the three epochs of the Paleogene period, and during the epochs that followed. Scientists have tried for more than a century to understand the mechanism that enables these changes. Some have started with living birds, to figure out the factors that influence their lives.

Adaptation

One such possible mechanism is adaptation. Ornithologist David Snow (1924–2009) and his wife, ornithologist Barbara Snow (1921–2007), spent much of their lives in the tropics of Trinidad, Guyana, and other parts of South and Central America, studying the lives of fruit-eating tropical birds. Their studies focus on how evolutionary paths are dictated by a response to the environment. In his 1976 book, *The Web of Adaptation: Bird Studies in the American Tropics*, David Snow described the social, feeding, and nesting behavior of birds they witnessed through years of observations of tropical birds.

The Snows found that over long periods of time, birds adapt their lifestyles to the environment in which they live. Birds are hatched with inherited traits that are the result of genes, the environment, and the interaction of those genes with the environment. These interrelated traits, collectively referred to by biologists as *phenotypes*, are dictated by the environmental limitations—also referred to as boundaries—imposed on that species. The traits occur as the result of environmental conditions that the species adjusted to over the long term.

The environment imposes a set of boundary conditions for any of its occupants, such as the available food sources and habitat. Competition with other species, defense from predators, and the ability to migrate are also factors in a species' evolution. Over time, a species can adjust its lifestyle to accommodate these boundaries. It may alter the parameters of its nest site selection, the number of nestlings it produces, its plumage, physical size and anatomy, its social behavior, feeding habits, and migratory behavior, so it can protect itself and its young, outpace predators, and succeed in obtaining enough food and other resources to survive and thrive.

To demonstrate the kind of work the Snows undertook, let's focus on one clade, *Catharus* thrush, and a comparison of its lifestyle with another clade—cotingas and manakins, a group of tropical species—to get a sense of the effects of adaptation and how it can differ from one family to the next. Both of these are in the same order (passerines), they are widely distributed, and they are forest species, but they differ in significant ways. Figures 3.1 and 3.2 show how colorful and conspicuous the plumages of some cotingas can be.

Catharus thrushes breed from Alaska through North America, to Central America and mid-way through South America. In North America, most of the common migratory thrush species—hermit, Swainson's, gray-cheeked, Bicknell's, and veery—belong to the *Catharus* family, as do the nightingale-thrushes of Central and South America. The plumage of the orange-billed nightingale thrush, Figure 3.3, is typical of the *Catharus* thrushes (see also Figures 3.4a and 3.4b).

Figure 3.1 The elegant plumage of the Andean Cock-of-the-Rock, a Cotinga of western South America, has a disk-like crest that hides the beak (author's collection).

Figure 3.2 The male Golden-headed Manakin is brightly colored in contrast to the female, who has uniformly muted greenish coloration (Gary Chapin).

There are many species in this clade, but the thrushes' lifestyles are all very similar. They are all about the same size, and they all have muted, cryptic colors, and sexually monomorphic plumages (males and females with same plumage). Most are migratory and have short life spans of less than 10 years. Socially, they are solitary forest-breeders.

Cotingas and manakins, on the other hand, are tropical birds that vary in size from very small to large; they vary in diet, mating practices and nesting strategies. Many are exceptionally beautiful and colorful, but others have more muted plumages.

The feeding habits of these two families are very different. The *Catharus* thrushes are insectivores, spending lots of time searching for food. They are solitary feeders and defend a territory. The cotingas, however, are frugivores (fruit eating) that spend very little time searching for food. They are communal feeders, and do not defend a territory. Because their diet consists of fruit that is abundant in the tropics throughout the year, providing all the nutritious benefits for both adults and immature birds, they can limit the total effort that goes into finding food. Birds and fruiting trees have developed a co-dependent relationship, with the birds spreading the undigestible seeds through defecation, allowing fruit trees to propagate more broadly than they could on their own.

Finally, these two families have very different breeding strategies. The thrushes use vocal attraction to find a mate, and their nests are not well hidden. They have large clutch sizes with high rates of fledgling loss, and both males and females feed the chicks. In contrast, cotingas and manakins have small clutch sizes. Some species

Figure 3.3 The Orange-billed Nightingale Thrush shows the cryptic plumage of the Catharus Thrushes. It breeds in the tropical and subtropical forests of Central and South America (Patrick Gains).

of cotingas and manakins find mates at a lek gathering—a location selected by males, where they perform their displays communally to attending females in hope of attracting a mate.

Like the thrushes, cotingas also have a high rate of fledgling loss, but their nests are well hidden, and only the female feeds the chick. Cotingas have a smaller range and are more diversified than the *Catharus* thrushes, which have large ranges and less diversity.

As mentioned earlier, the thrushes are migratory, while the tropical cotingas are not. Some species migrate to take advantage of the extended periods of light at the higher latitudes, which allow for a blossoming of food sources including plants, insects, and animal life, as well as the expansion of available habitat provided by the temperate forests. Migration imposes many constraints on life expectancy, with high mortality rates for both adults and immature birds. Nonetheless, many bird species on earth today are migratory; for these species, the benefits certainly outweigh the risks.

Although the *Catharus* thrushes, manakins and cotingas are all forest species, they occupy very different types of forests. The *Catharus* thrushes have a lifestyle similar to many other North American passerines, while cotingas and manakins have a lifestyle similar to the birds of paradise of Papua New Guinea.

What can we conclude from this comparison of thrushes to cotingas? The similarities and differences of these bird families are the result of their evolutionary paths in response to their environments. Each group of birds adapted its lifestyle to these physical boundaries through the mechanisms of evolution.

David and Barbara Snow studied the adaptation of cotingas and manakins through painstaking observations over many years. More recently, researchers have used genetic analysis to understand how birds can adapt. Natalie Hofmeister *et al.* (2021) have shown that European starlings, one of the world's most widespread species, have adapted to temperature and precipitation changes across the United States through a small change in their genome. Birds that reside in Arizona showed evidence of adaptation to hot and dry conditions, while those in the Pacific Northwest showed adjustment to cool, wet conditions (Leonard 2021). These changes have occurred in a little over a century.

These are examples of how birds react to their environmental boundaries. In some cases such as the starlings, birds can adapt rapidly. In other cases, adaptation can take much longer.

The Origins of Evolution

Working independently of one another but at the same time, Alfred Russell Wallace and Charles Darwin carried out much of the early work leading to the concept of evolution. Darwin noted how very similar birds were divided into different species, while Wallace proposed that new species come into existence from closely allied species already living successfully. Since the early work of these two scientists, many researchers have studied the subject of diversity of life on islands—a field known as biogeography, which looks at the roles evolution and adaptation play in the distribution of living organisms around the world. This subject is treated in detail by David Quammen in his book *The Song of the Dodo: Island Biogeography in an Age of Extinctions* (1996). The study of all types of living organisms on island communities has led to many of the basic principles of evolutionary changes affecting bird diversity, and an understanding of the causes of colonization and extinction of species.

Years after his initial work, Darwin published his theory of evolution based on the mechanism of natural selection. In this process, living organisms adapt to their environments through accumulated changes that best fit the environmental boundaries. These changes result in a higher likelihood of survival, and are more likely to be passed on to future generations. In his studies of birds, however, Darwin qualified his concept of natural selection. In his later book, *The Descent of Man and the Selection in Relation to Sex* (1871), he proposed that sexual selection—the preferences of animals of one gender for certain characteristics in the other gender—also plays a role in evolution, and sexual and natural selection can be in or out of phase with each other. He postulated that sexual selection could work in opposition to natural selection, which is driven by environmental changes. Darwin's theory of natural selection completely overshadowed this later idea about the role of sexual selection.

Richard Prum, in his book *The Evolution of Beauty* (2017), presents data and examples that support the concept that sexual selection is not always in phase with natural selection. Prum bases his thesis on careful studies of manakins and bowerbirds. He points out that some biologists were not in agreement with Darwin's proposal on the role of sexual selection.

In some cases, the male selects his mate, but in most species of birds the female has free choice in mate selection. Sexual selection is based on visual and physical qualities that give an indication of the chosen mate's ability to provide some benefit. In birds, these qualities include plumage and its coloration, decorative feathers, song, courtship displays, and broader factors such as the quality of the male's territory—all things that indicate the health and fitness of the prospective mate.

Selection based on these traits does not necessarily progress to more elaborate aesthetic qualities, however. For example, Snyder and Creanza (2019) studied the mating systems and songs of 352 songbird species, and found that species that seek multiple mates change songs quickly in the moment, but not in a consistent direction over the long term.

Evolution: The Modern View

When Darwin proposed his theory of evolution by natural selection, he based it on two mechanisms: random variations in a population, and the process of selection that acts on this variation as a gradual response to environmental changes. Today we know that the mechanisms driving evolution are genetic mutations, gene flow between populations, and random genetic drift (favoring one genetic form over another), as well as natural selection (Zimmer and Emlen 2013). Minor (2016) gives us a simple definition of evolution as "changes in the frequency of genes within populations over time."

Natural selection is a mechanism through which populations change over time in response to their environment. It requires variation within the species, inherited traits, and selection over time so the species can adapt to the environmental conditions.

Mutations in regulatory DNA allow species with closely related genes to develop vastly different characteristics (Sackton *et al.* 2019). Parts of the genome regulate when and where proteins are expressed and when genes are turned on and off. For example, recent research has shown that a change in the regulatory DNA of the flightless ratite species (emus, ostriches, kiwis, and others) led these birds to lose the ability to fly. Gene expression—the process through which a gene regulates the appearance of one or several species in a genetic family—also played a role in the Galápagos cormorant becoming flightless (Burga *et al.* 2017), and is an important factor in forming the differences among common, hoary, and greater redpolls (Mason and Taylor 2015).

The creation of hybrids through interbreeding is an important pathway for gene flow between species (Ackermann 2016). It is well understood that interbreeding is a common occurrence in nature. Within the world of birds, we know that many gull species interbreed with one another, as do ducks (virtually every birder has seen a mallard/domestic duck hybrid), and wood warblers including blue-winged and golden-winged warblers hybridize as well, creating the "Lawrence's" and "Brewster's" warbler hybrids. Ornithologists recognize that at least ten percent of all bird species hybridize (Grant and Grant 1992).

Hybridization is a complex subject, because there are different types of genetic changes and environmental factors that can lead to different results (Ottenburghs 2018). In one case, researchers have confirmed that a seedeater species in Argentina has arisen from interbreeding, in which genetic traits among closely related species mixed and matched to create hybrid forms (Turbek *et al.* 2021, Campagna 2021).

The factors responsible for change are always present and operative, but there may be long periods—hundreds of thousands to millions of years—in which the species have adapted to stable boundaries. In these periods of stability, a species can remain in a state of static equilibrium (stasis) until it responds to some instability or opportunity. Paleontologists Stephen Jay Gould and Niles Eldredge call this process of stability "punctuated equilibrium," a process of gradual changes separated by periods of no change.

Punctuated equilibrium has been verified in the fossil record for some types of mollusks, horseshoe crabs, crocodiles, coelacanths (the oldest fish species), and the gingko tree. Extensive studies of fossils covering long time periods have shown only a small number of major evolutionary transitions, and in the time between them, stasis predominates for millions of years (Uyeda *et al.* 2017; Uyeda's study of birds, mammals and other animals was based on metabolic rate and body size). Rapid evolutionary changes can occur over shorter periods of time, but they don't always trend in one direction. Future generations can adapt and later discard these changes.

Diversity of Species and Taxonomy

In the world of birds viewed in the present, we can see that species are the result of gradual changes measured over many thousands to millions of years. Ornithologists arrange birds in order and rank them by their historical appearance. This taxonomic ordering, also referred to as systematics, is important in understanding the evolutionary relationship between species. Families, species and subspecies represent the progressively lower taxonomic ranks. Organisms that are related biologically at the family level share common attributes, but they are also separated from other families by one or more important specific characteristics.

Species is the most pivotal unit of taxonomic order. The species concept is a way of understanding the differences in creatures that result from their basic genetic makeup. A species represents a biological group that can exchange genes or successfully interbreed, but that does not interbreed significantly with other groups. We know that in many but certainly not all cases, species do not regularly interbreed—or if they do, they might produce progeny that are disadvantaged in some way.

For the avian world, the primary concept used to define a species is known as the *biological species concept* (BSC). The BSC is based on the idea that a species is a population of birds that can breed with one another and not with other groups. As simple as this sounds, the BSC presents a number of problems: For example, separated populations might be the same species even though they are geographically isolated (allopatric). They might appear different from each other but have little or no opportunity to interbreed, so they remain isolated even though they have the ability to breed with others who are distant from them. One example of an allopatric species is the white-winged junco, which has a range restricted to the Black Hills of South Dakota. It was once considered a separate species, but today it is understood to be a subspecies of dark-eyed junco. Reproductive isolation does not guarantee that a population is a unique species.

Scientists have proposed many other alternatives to BSC. The *phylogenetic species concept* (PSC) is based on grouping species that are descended from a common ancestor. It seeks to find a combination of derived traits that apply to the smallest set of a population sharing those traits. The traits are assembled by quantitative analysis into a cladistic tree, a diagram of the evolution of a set of species in the same clade, based on either morphological (structural) or molecular DNA data. (The cladistic tree is an attempt to determine the relationships between different organisms, not their specific evolutionary history. It uses the whole genome to allow the DNA analysis to be more precise when applying it to the PSC.)

The problem with both the BSC and the PSC is that it is difficult to establish a criterion for the degree of separation needed to define a species. Ornithologists do not always agree on which populations are species or subspecies; as a consequence, these decisions are evaluated based on the best current information.

A board of ornithologists decides whether or not a group of birds should be treated as a species. In North America, the North American Checklist Committee (NACC) of the American Ornithological Society (AOS) is assigned this task. Decisions are based on published research. The South American Checklist Committee carries out a similar annual process for that continent. In Australia, Bird Life Australia maintains a working list of Australian birds. Africa, Asia and Europe do not have continental-level classification committees, but the International Ornithologists Congress (IOC) has a working group, the Working Group Avian Checklists (WGAC), which works with BirdLife International to address taxonomic and nomenclature issues.

Many countries do not have their own national committee dealing with avian taxonomy. This can lead to some level of confusion for both taxonomy and common names. The same bird can have different common names and scientific names in different countries. This impacts accounting for the number of species on earth.

Three sources maintain their own world checklists of birds: the IOC World Bird List, the Clements Checklist of Birds of the World, and BirdLife International. BirdLife International differs in that it does not recognize the PSC as a criterion. The

Clements Checklist of Birds is updated annually and the IOC updates its list twice annually. All of these checklists are available online.

Other world checklists and national field guides are usually based on the ornithological committees mentioned above. However, field guide authors occasionally have different opinions on certain species, creating further confusion among birders trying to reconcile their life lists.

Defining Subspecies

The lowest level of taxonomic order is the subspecies, a population of birds identifiable with respect to the benchmark for that species. This definition is based on the fact that a population of creatures in a specific breeding range is statistically distinct from other populations, but shares the most basic characteristics of these other populations.

Noted ornithologist Dean Amadon (1949) provided us with the 75 percent subspecies rule, in which 75 percent of the individuals in one area must be separable from 100 percent of an adjacent subspecies. Oklahoma Biological Survey professor Michael Patten (2015) points out the problems of using genetic data to assess subspecies limits, and the need for "a clear and consistent philosophical approach to how genetic data are used to assess subspecies limits."

Opinions vary on how many subspecies are recognized, but let's look at two broadly distributed species with multiple subspecies: yellow warbler and savannah sparrow. Yellow warbler currently has 34 subspecies, broken into four broader groups. Savannah sparrow includes 21 subspecies, sorted into five major groups (Clements *et al.* 2021). Both of these species have ranges that cover all of North America. The debate about which of these subspecies is actually a species is always ongoing within the ornithological community. Indeed, Ipswich sparrow, currently a subspecies of Savannah Sparrow, had at one time been classified as a separate species, and its status is still questioned among ornithologists (McLaren and Horn 2006).

The differences between eastern and western meadowlarks are not readily obvious and are difficult to separate in the field by plumage, but it has been established by direct experiment that the two are, indeed, separate species and can interbreed—although in this case, they produce genetically disadvantaged offspring. Western meadowlark has only two subspecies, and is restricted to Canada and the western and midwestern United States. Eastern meadowlark, on the other hand, has 17 subspecies that range from eastern Canada and the United States to the Caribbean, Mexico, Central America and northern South America. A localized meadowlark species, Chihuahuan meadowlark, has recently been named by the AOS in 2022 as a separate species from eastern meadowlark, and is found in southwest Texas, New Mexico, Arizona, and northern Mexico.

It is clear that broad geographic ranges within these three examples—meadowlarks, yellow warbler, and savannah sparrow—have produced small variations in these widely distributed populations, resulting in genetic changes that are recognizable.

The question of how many bird species there are on earth has been of interest to ornithologists for many years. In 1946, one of the early tabulations of the number of bird species by Ernst Mayr (1946) totaled 8,616 species. The current Clements World Checklist of Birds places the number at about 10,824 (Clements *et al.* 2021). This total included some extinct species. The number of extant species is 10,665. The checklist also lists 20,458 subspecies. Many of these subspecies could be raised to the rank of species with more study.

Ongoing Study and Classification

One might assume that most bird species and subspecies on earth have been well studied, but that is not the case—and that is why there are annual changes in the taxonomy. In a recent publication that looks at species diversity from a genetic viewpoint rather than the BSC, researchers estimate that the actual number of living bird species ranges from 15,000 to 20,000 (Barrowclough *et al.* 2016). We think of species as distinct entities, but the boundary between species and subspecies has some level of uncertainty. There can be a distribution of genes and behaviors within a defined population, and ornithologists try to establish some criteria within which the distribution of these properties provides enough separation to meet the classification of species.

Bird song is a behavioral characteristic that plays an important role in the separation of species. Male birds establish a territory and use song as one way to attract a female. If a statistically significant percentage of females do not respond to the song of a population of males, then the two groups are likely two different species. This lack of acceptance of the song is then a barrier to associative mating.

Allopatric populations are birds of the same species that are separated by physical barriers, so they have no contact in the breeding season. One method to determine if two subspecies are part of the same allopatric species is to use sound playback analysis for both groups, to determine how the males and females behave. Freeman and Montgomery (2017) provide a good example of the results: They used playback analysis on Neotropical passerines of South and Central America that inherit their song genetically. They found that 21 out of 72 pairs of allopatric subspecies did not react to the song of their counterpart. This result provides strong evidence that these subspecies merit classification as independent species as a result of song preferences. Those species that did not react were variable among the populations tested when song differences were low to moderate, but most to all did not react when songs had a high level of difference. Their work also implies that behavioral discrimination needs to be high to support the case for different species. Their results were consistent for both oscines and suboscines (species that acquire their song genetically).

Gray-cheeked Thrush [Figure 3.4a] and Bicknell's Thrush [Figure 3.4b] are an example of two (*Catharus*) thrushes that look very similar but differ by song, habitat type, and wintering ranges. Song is an important factor in their taxonomic classification.

Figure 3.4a Gray-cheeked Thrush is similar in plumage to Bicknell's Thrush (3.5b) and the two species are best separated by song; it breeds all across northern North America in the northern boreal forests (author's collection).

Figure 3.4b Bicknell's Thrush is restricted to coniferous mountaintops in the northeast United States (Darren Clark).

When organisms look very similar, it is difficult to determine if they fit the category of subspecies or species. Steve Howell (2021), in his article "What Isn't a Species," points out all the problems of separating species and subspecies, and how in some past cases these decisions appear to be inconsistent. Ornithologists use the best available information at the time to make these decisions, and have occasionally reversed their previous positions.

More recently, scientists have used whole genome analysis of dominant populations and hybrids to understand the causes of the genetic differences, and link these results to differences in behavior and geographic ranges through field studies. This approach has provided a more detailed understanding, as seen in the study of Baltimore and Bullock's orioles, which hybridize in the central United States. By analyzing the whole genome of these orioles, Jennifer Walsh and her team could explain underlying causes of the differences in these species and hybrids (Heisman 2022, Walsh *et al.* 2020). It is only through significant efforts, such as the case of the orioles, that field ornithologists, biologists, and molecular geneticists can gain sufficient understanding to justify the claim that a subspecies is actually a different species.

Discoveries of New Species

In the past 60 years, the list of bird species of the world has increased significantly. Most of these new species are the result of elevating previously described subspecies to the level of a species by virtue of published analysis. However, between one and five totally new species are found on earth every year. These birds are entirely different from any previously described, and were completely unknown to science. Most of these new species are found in tropical regions that were poorly explored—rare birds with small populations and restricted ranges.

For instance, in 1973 students from the University of Hawai'i found a bird never seen before on the slopes of Haleakala, a mountain on Maui. It was given a Hawaiian name, Po'ouli, and also referred to as the black-faced honeycreeper. Researchers discovered that the population was very small, and after 2004, it was never seen again; the U.S. Fish and Wildlife Service declared it extinct along with seven other Hawaiian bird species.

In Indonesia, a remarkable recent discovery of five new bird species and five new subspecies took place on three islands, Taliabu, Peleng and the Togian group, and the find was announced in *Science* by Frank Rheindt *et al.* (2020). These small islands east of Borneo and Sulawesi (Celebes) are part of the Wallacea, a series of Indonesian islands separated by deep-water straits from the continental shelves of Asia and Australia, where British naturalist Alfred Russel Wallace did his work on the diversity of tropical birds in the 1850s. The mountainous forests of these three particular islands had not been thoroughly explored for bird life. It is almost unprecedented at this time in history that so many different bird species and subspecies were discovered in a small, restricted region.

In South America and throughout the tropics, there are inaccessible valleys between mountain regions that can harbor birds of unique qualities. Birds such as tapaculos, antbirds, and antpittas seen between mountain regions were assumed to be the same species, because they did not differ visually. Detailed analysis later found that they could be divided into different species based on their songs and on genetic analysis (Borgmann 2021).

In rare cases, a unique and totally unknown new species has been found in plain sight. In the Asia Pacific region in 2009, the Cambodian tailorbird, a previously undescribed species, was found in Cambodia's urban capitol of Phnom Penh, and it was recognized as a species in 2013. When someone finds a bird that appears unusual, it takes a lot of analysis to determine if it truly is an additional species, a subspecies, or an unexplained variant.

Today we know that the species concept is complex and difficult to measure. There are many examples of species that are closely related, but they have not been studied in depth because of the limits of our ability to find and understand the differences.

Adaptive Radiation and Convergent Evolution

One of two processes at work in adaptation, *adaptive radiation* occurs when one species is able to secure a foothold in an environment with few competitors. This creates an opportunity for evolution from that common ancestor. Over time, the immigrant population adapts to the different habitats and environmental niches, and evolves into a host of species that stem from the original colonizing source.

Many examples of adaptive radiation can be found among plants, fish, and animals, such as the marsupials of Australia, which originated in North and South America and emigrated to Australia across the Pangaea landmass about 55 million years ago. Here they found plenty of room to diversify into about 250 different species, without much competition for resources from placental mammals.

An extraordinary example are the cichlids of Lake Malawi, a family of fish that has blossomed into 850 species in one large lake in Tanzania. Among families of birds, adaptive radiation has occurred among the Hawaiian honeycreepers, Darwin's finches, Australian honeyeaters, and the vanga, a family of shrike-like birds of Madagascar. In all cases within these families, the birds have variations in size, bill structure, food sources, plumage, and nesting requirements.

Convergent evolution, another adaptation process recognized by biologists, is an outcome in which two or more unrelated lineages adapt to the same ecological niche in different areas. A classic example of convergent evolution involves desert rodents that have developed long hind feet, long tails, and large haunches in order to escape predators by jumping. All major deserts have rodents adapted for jumping, but they are not closely related to each other (Mares 2002).

There are many examples of convergent evolution in birds. Old World and New World vultures are unrelated to each other, the former related to hawks (*Accipitriformes*) and the latter are now in their own order *(Cathartiformes)*, and are regarded as a sister clade to the order of hawks (Jarvis *et al.* 2014). In the southern hemisphere, penguins, diving petrels and petrels are all related and all descendants of early seabirds related to penguins. Their counterparts in the northern hemisphere are the alcids and the gulls, which have their origin in early seabirds related to gulls. Sunbirds in the Old World have a similar lifestyle to hummingbirds in the New World. Old World and New World warblers are not related to each other, but they occupy the same type of habitats and have similar lifestyles. Swifts and swallows are unrelated and are another example of convergent evolution, sharing very similar methods of feeding and roosting.

The earthcreepers of Argentina and western thrashers of North America occupy the same type of habitats, and they have similar size, structure, coloration, and lifestyle. A study of the ovenbird-woodcreeper family has shown both radiation and convergent evolution. Over a long time period, changes in nest building strategy and adaptation to novel habitats may have been a factor in multiple radiations and convergent evolution of these families into more open and bushy environments in South America (Irestedt *et al.* 2009).

Within nature, there are hundreds of examples of convergent evolution between very different living organisms. For example, oilbirds, whales and bats all developed the ability to echo-locate in darkness.

When viewed in the present and looking backward in time, the world of birds presents the strongest evidence that they have diversified through an evolutionary process of natural selection that involves adaptation to their environment, and by sexual selection that has resulted in morphological and genetic changes. The historical record will not uncover the vast majority of past species, because closely related species might not be identifiable from fossils alone. The fossil record is able to preserve only a very small fraction of life. Very few organisms are preserved as fossils, and the light, hollow bones of birds are less likely to survive the ravages of geological time. In addition, some species may be too closely related to separate by bone structure alone.

4

How Birds Are Different

Our familiar birds have had a long history since their earliest ancestors in the Jurassic period. They inherited some of their unique characteristics from their dinosaur origins, but they have also continued to evolve to take full advantage of their ability to fly. Their biology, physiology, skeletal structure, and physical capabilities have all been optimized to enable flight. Flight has given them distinct advantages, allowing them to exploit habitats and distant food sources, to diversify, and to evade predators.

Very few living species of birds are flightless. These have lost the ability to fly because they have no need for it: they have accessible annual food sources at ground level, and they lack predators. Among the flightless birds are the large ratites, species that are part of *Palaeognathae*—ostriches, rheas, emus, cassowaries, and kiwis. These birds are believed to have evolved from flying ancestors, but they adapted to a diet of plants so long ago that they were able to survive by not competing for food with large predators. In this chapter, we will discuss how these ratites differ from the remainder of flying birds (*Neognathae*).

There are twice as many species of birds as mammals and amphibians, and more bird species than reptiles. Birds have succeeded largely as a result of their unique biology. This chapter briefly explores some of the differences in the biology of birds, and how this biology has influenced birds' diversity and ability to fly. Their uniqueness among living organisms is an important factor and should influence our efforts in trying to preserve them. Here we present a basic overview of their qualities: intelligence, body structure, feathers, senses, and the biological features that enabled their success.

The field of avian biology is covered in greater depth in many works written by biologists that have studied birds in great detail. For those who are more interested in further understanding their biology, a few important sources include: *The Cornell Lab of Ornithology Handbook of Bird Biology* (Lovette and Fitzpatrick 2016), *Sturkie's Avian Physiology* (Scanes and Dridi 2021), and *What Is a Bird?* (Williams 2020).

Intelligence

One might assume that because birds are small creatures with small brains, they would similarly have low intelligence. However, research has shown that birds are actually more intelligent than many other animals with larger brains. Jennifer Ackerman (2016), in her book *The Genius of Birds*, has researched the subject of bird intelligence. Gisela Kaplan (2015), in *Bird Minds*, explores avian cognition and how their cognitive processes relate to their behavior.

To measure intelligence, scientists devise puzzles to test cognitive abilities and compare them among various creatures. Mammals and birds are considered highly intelligent compared to reptiles, amphibians, and fish. Measuring the intelligence of birds is not straightforward, because their lives are so different from other creatures, and their brains are not structured like mammals.

One measure of intelligence is based on the encephalization quotient (EQ) metric, a measure of relative brain size with respect to body size. EQ has been used as a proxy for intelligence, and is considered a more refined metric than the more simplistic ratio of brain to body weight. Some species of mammals show patterns of social learning and innovation that correlate with brain size (Laland 2018). These species tend to be more adaptable, are better at problem solving, and copy innovations of other members of their groups. Mammals like chimpanzees, elephants, and dolphins, for instance, have higher intelligence than, cows, cats or panda bears.

How intelligence in birds or any animals evolved over time is unknown, but we know that the relative cranial cavity size of *Archaeopteryx* is smaller than that of modern birds. Birds have higher EQ range than do reptiles and dinosaurs (Hulburt *et al.* 2013). Some species of birds have high EQ values, equivalent to chimpanzees—but, like in mammals, there is a variation in this metric among birds. For instance, hummingbirds and ostriches have lower EQ values than some mammals (Emery 2006).

Birds also show a large variation of intelligence when judged by their problem solving ability. It has been recognized for years that parrots and corvids—which include ravens, crows, and jays—are among the most intelligent birds, ranking high on cognitive tests. Corvids and parrots seem to have superior intelligence compared to other birds, and are more intelligent than apes. Within the orders of birds, however, very few other families have been studied for their cognitive skills. It seems likely that their intelligence would vary considerably from one species to the next, just as they do within the mammals.

Some families of birds rank higher in intelligence even when they have an equal brain-body ratio. This difference may be related to their maturity and independence when hatched. *Precocial* birds, considered older orders of birds in evolutionary history, are hatched fully feathered, and are active immediately after hatching. *Altricial* birds are hatched naked and blind, and are usually nest raised. Relatively speaking, altricial birds have larger brains than do precocial birds, because altricial birds' brain size increases during the longer maturing process. Brain size of some species of birds within the same genus will also vary depending on the territorial environment (Ackerman 2016).

Tests designed around problem solving are inadequate to fully understand bird intelligence, because so many factors influence their cognitive development. For example, each individual bird species lives in a specific, restricted habitat. Many of the larger birds have a long learning process before they fully mature and reach reproductive age, a process of cognitive adaptation to their environment.

Scientists have studied cognitive adaptation in birds by introducing birds into a different setting to see how they adapt. The results show that birds with larger brains are more innovative and respond more successfully to the challenges of a novel environment, illustrated by their response to predators, novel food sources, and different environmental conditions (Sol *et al.* 2005). This result is similar to the ability to adapt that researchers have observed in mammals.

Some species or types of birds are generalists, while others are specialists. Generalists make use of different food sources or different environments to sustain themselves. These include highly adaptable birds such as corvids, caracaras, grackles, great kiskadee, and house sparrow—one of the most adaptable birds on the planet. Specialists, on the other hand, are restricted to either a specific food source or a selected habitat, or they are extremely skilled at one task within their lifestyle. Hummingbirds, accipiters, falcons, spotted owl, and the retiring and secretive tapaculos are all specialist species. Researchers have shown that many species of North American generalists can exploit a wide variety of habitats or food types, but not necessarily both simultaneously (Overington *et al.* 2011).

Another method used to test intelligence involves counting the number of neurons in an animal's cerebral cortex. Neurons connected to the cerebellum are used for motor control and some cognitive functions. Both birds and mammals have these neurons, but the organization of the brains of birds are different from those of mammals. Birds have no cortex; brain activity is centered in an area of the brain referred to as the pallium. Some species of birds have a very high number of neurons connected to their pallium. Corvids and parrots, for example, have a higher degree of connectivity than do pigs, rats, and marmosets. These two bird families also have a higher number of neurons than other species like pigeons. Pigeons can achieve cognitive performances on some tasks the way corvids and parrots can, but not on other tasks, and they require a longer learning time (Güntürkün 2020).

Birds have amazing capabilities of navigation, with the ability to migrate thousands of miles. We know that many long distance migrants return to the same territories in spring to breed, and find the same wintering regions year after year. Non-migratory birds like pigeons are extremely adept at returning to their home range. Many research studies have been able to explain some of the processes that birds utilize in migration, but we have not reached an understanding of all the mechanisms that are involved and how these are interrelated. Recent research has shown that young migratory birds store relatively little spatial information during migration. It is not until they reach their wintering area that their global navigational map is more fully developed (Nemeth *et al.* 2017).

We base our perceptions of how birds use their minds on observations of their behavior. For example, many species of birds use a variety of tools to support their needs. They may use objects to transport water, scoop out insects from holes or tree bark, or to break open seeds, nuts or shellfish. They also use some objects as lures to attract fish. Some species of birds learn by trial and error and by observing other birds, thus teaching and passing skills on to younger generations. They are opportunistic when it comes to exploring for food sources, a skill we can see from their ability to find backyard bird feeders.

Some birds can master complex counting problems. Some species cache hundreds or thousands of seeds or nuts for later use, and remarkably, they remember the locations of all those hiding places, since this larder is necessary for survival. Homing pigeons can remember the route to a loft even after not making the trip there for four years (Collet *et al.* 2021).

Bird songs and calls have specific meanings. Birds that imitate the songs of other birds can have a repertoire of more than 100 songs. Parrots can mimic human speech, but studies with the grey parrot have shown that some are able to associate words with their meanings and form simple sentences. Studies of American crows show that crows can remember the faces of individual people for long periods of time.

Onur Güntürkün (2020) has studied avian cognition and summarized some of the important findings about birds' cognitive abilities. Crows, for example, can solve diverse problems by reasoning about the underlying causal relationships. They plan ahead using mental representations of unseen objects, and make inferences about cause-and-effect relations of observed events. Ravens, another corvid species, demonstrate a high degree of social strategies. They seem to understand what other birds might do, because they deceive potential cache thieves by leading them to places where food is not stored. Tests show that the New Caledonian crows of the Pacific islands have memories that allow them to recall past events and to imagine future events. European magpies are capable of self-recognition in a mirror image, an ability restricted to only a few mammals with larger brains.

Some species of birds are bold and inquisitive, while others are more secretive. Most of the studies on intelligence in birds have been carried out on species that are more inquisitive—so we know more about crows and ravens' cognitive abilities than we do about the skulking rail families, for example, or any of the thousands of bird species that are shy and retiring. The shier birds usually reside in habitats that have difficult access, and they do not venture out to satisfy our curiosity. Rails, some species of North American vireos, and the antbirds and tapaculos of South America are just a few examples of families that are more secretive. Their intellectual capacity is an adaptation to their localized world, focused on cognition, spatial memory, food gathering, social learning and tool use. Tests for cognitive ability are difficult to apply to these types of species.

Sometimes birds will show slow skills at adapting. For example, Skutch (1999) describes the efforts of a pair of trogons to build a nest in a metal structure that cannot be used for this purpose. The birds tried repeatedly and failed each time, and showed no ability to adjust to the conditions. Skutch points out that they eventually succeeded in finding a suitable nest site, and he attributes this to their sheer determination. Rock pigeons, a much more common and familiar species, can remember hundreds of different objects for long periods but cannot solve simple abstract problems (Emery and Clayton 2005).

Another dimension of cognitive ability is the speed of processing. Based on their ability with flight, we know that birds have a remarkably rapid processing speed. Passerines and predatory birds are capable of extremely rapid adjustments in flight in order to maneuver through dense vegetation, to enter a nest hole in flight without perching, or to chase and capture prey. Accipiters and falcons prey on birds seized in flight.

To measure processing rates, research has explored the ability of humans and rock pigeons to switch tasks. Their work has shown that pigeons are on par with humans when a task demands simultaneous processing resources. However, pigeons show faster responses than humans when sub-tasks are separated, requiring fast switches between processes (Letzner *et al.* 2017).

The subject of consciousness in animals is in its infancy. It is difficult to determine if birds feel emotions such as fear, happiness, anger, pain, love, and other states of mind we attribute to humans. Cabanac and Aizawa (2000) show that birds display stress and fear reaction when touched. Their heart rate, breathing, and stress hormone levels change in response, and they recover when released from the threat. To determine the effects of stress on birds in threatening situations, Cockrem (2007) summarizes results of the increases in the hormone corticosterone in birds, including chickens, under stress. The tips of the commercially raised chickens' beaks are cut off to prevent them from pecking at each other. They experience two periods of pain, one instantaneously and another shortly after (Gentle and Wilson 2004). This procedure is painful for birds based on never activity and methods that reduce this stress are recommended (Cheng 2007).

It has been known for many years that most species of birds, especially larger birds, are monogamous through the breeding season, and some species mate for life. When birds lose their mate, they show signs of loss by remaining with the deceased mate for long periods of time, as much as a week in some cases. Other stressors can cause rifts between otherwise peaceable birds: common murres and choughs are social, gregarious birds, but their social structure breaks down when they are confronted with inadequate food to feed their young (Birkhead 2012).

As any bird owner will tell you, pet birds bond with their owners. This is especially true of parrots: when a parrot has bonded with its owner, it will make a fuss if there is a change in its caregiver.

The fact that birds can feel pain and react to stress, that mates engage in preening one another, that they will remain with a deceased mate, and that they form bonds with humans are all indications that birds have emotions. Biologists debate whether these reactions are instinct or genuine feelings, but we have no way of determining the answer at this time. As methods of monitoring brain activity of birds improve, we will be able to better understand how they react emotionally to external stimuli and how these reactions compare to our own.

Body Structure

Birds are part of the vertebrate phylum, which includes all animals with a spinal column: amphibians, reptiles, mammals, birds, and fish. The earliest vertebrates on earth were fish, emerging about 500 MYBP, and they adapted to various environments, and produced different lineages because of this.

Birds and humans are the only true bipedal creatures on earth. As discussed in

Chapter 1, birds evolved from the bipedal theropod dinosaurs. Some apes can walk on two legs, but their pelvis is not designed for prolonged walking. Kangaroos and some small mammals that hop have developed large hind legs that enable forward motion, but this method of propulsion is different from prolonged walking. The ability in birds to both walk and fly results from an amazing set of adaptations to the basic vertebrae of their body structure.

In birds and mammals, the bones are basically the same; their shape and how they function are different. The bones of birds are hollow with internal struts, an adaption that is extremely valuable for flight, because hollow bones reduce weight and improve strength. Hollow bones are lightweight, are structurally strong, and are connected to the respiratory system. Some birds, like soaring birds, have more hollow bones than others, while a few families of birds—penguins and loons in particular—have solid bones.

Some of birds' bones do contain marrow, which generates red blood cells. The bones are made of calcium, which their bodies can add or extract as required to form egg shells or repair stressed limbs (Evans 2016).

The skeletal structure of a bird is far more complicated than is briefly described here, but there are similarities to the mammalian structure. Perhaps the best way to understand the differences in the skeletal structure of a bird and human is to compare them in familiar terms. Figure 4.1 below shows how the same structural bones support the different anatomy of a bird as compared to a human. The differences have allowed birds to deal with the complexity of walking, flying, and landing, while

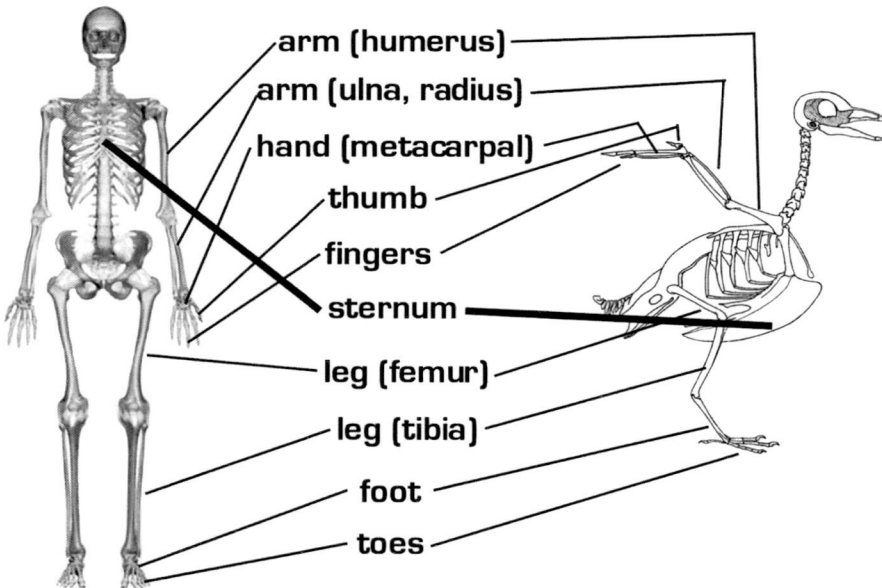

Figure 4.1 Simplified comparison of the skeletal anatomy of birds and humans noting common elements and highlighting the large sternum to anchor flight muscles in birds, and differences between the arms, legs, feet, and hands (D.G. Mackean, www.biology-resources.com).

human anatomy is mostly adapted to the need to support walking on two limbs, and to free our hands for grasping.

The skeletons of birds and humans are different in the skull, neck, sternum, arm, hand and foot. The arm and hand of a bird is homologous (similar) with those of other fauna. The skull accommodates a toothless beak that is much lighter than our mandibles, but as a result, birds cannot chew.

Birds have 13 to 25 vertebrae in the neck, compared to seven in humans and in most quadruped mammals like wolves and horses. These additional vertebrae provide more flexibility to the neck, giving birds the ability to rotate their heads in a much greater angle. Birds can maintain the head in a stable position when moving, especially at high speeds such as flying, landing, or capturing prey.

The remainder of the bird skeleton has also evolved to accommodate the differences associated with flight. A number of bones have been fused—the collar bones, for example, are fused together to form the wish bone. This strengthens the skeleton for flight and acts like a spring, storing and releasing energy as the wings are raised and lowered. The ribs have hooked extensions of bone that overlap and strengthen the rib cage. The vertebrae are fused, making the backbone rigid. The vertebrae of the tail, some of the fingers, and the bones of the braincase are fused as well.

The wing is analogous to the arm in humans, but in birds, the hand has adapted for flight. In primates (apes, monkeys, and humans), the arm and hand are adapted for gripping. In mammals that walk on four legs, arms function as the two front legs and are similar in their structural arrangement to the rear legs. In birds, however, the wing bones—the humerus, radius and ulna—are similar to those in primates, but both wing bones and finger bones support the flight feathers.

Birds have three finger digits instead of four, with one equivalent to the thumb. The other two "fingers" are referred to as digits two and three. The secondary feathers are the inner flight feathers, and are attached to the ulna (arm bone). The inner primary feathers—usually the first six flight feathers—are attached to the metacarpals, the structural bones of the hand; the outer primaries are attached to finger digits two and three, with a single small feather attached to the thumb bone, the ulna (Van Tyne and Berger 1976). The relative lengths of the bones that make up the wing vary considerable among different bird species.

Another major difference related to flight is the sternum. The keel-shaped sternum of a bird is very different from the sternum of a primate, and serves a very different purpose, anchoring the pectoral muscles so they can provide the thrust in the wing's downward motion. Some bird species—some ratites and some other flightless birds—do not have a keel on their sternum; nor did the earliest birds of the Cretaceous period.

A similar set of changes as those in the wing are seen in the foot. When you look at a standing bird like a shorebird or a crane, the toes are on the ground. We would call these the feet, but they are actually the toes. In humans, metatarsals are a set of five bones in the foot, while in birds, these bones are fused into a single long bone, the transmetatarsus, connected to the toes. In many bird species, the foot bone is longer than either of the two leg bones.

Above the bird's foot are the two leg bones, similar to those in mammals. The tibiotarsus is homologous to the tibia in humans. In both birds and mammals, the upper leg bone connected to the hip is called the femur. Quadruped land mammals also walk on their toes, and the foot has a similar arrangement to a bird's foot, but the foot bone is much shorter. It is this arrangement of the foot and leg bones that enables bipedalism in birds, whether the bird walks by striding or hopping.

Even with all these structural adaptations, however, not all birds are skilled walkers. Swifts, trogons, and hummingbirds have very weak legs and cannot walk, or walk poorly, and some passerines (perching birds) also have weak legs. Other birds prefer to walk: grouse, turkeys, and tinamous, for example. Finally, as noted earlier, some birds have strong legs and do not fly: penguins, flightless ducks, ostriches, cassowaries, and others.

Feathers

The Development of Feathers

More than 30 species of feathered dinosaurs have been discovered, mostly from northeast China. One of these, *Dilong paradoxus,* is a theropod dinosaur that is either a Tyrannosaurus or closely related to that family. This animal, dated to about 125 MYBP, was covered with filaments like feathers (protofeathers) that might have served for insulation.

The development of protofeathers first appeared in the fossil record of non-avian coelurosaurian theropods [Figure 1.5], which supports the idea that feathers evolved for an adaptation other than flight (Makovicky and Zanno 2011). The exact reason why feathers first appeared is not fully understood. Most analysts believe feathers first evolved to provide insulation, enabling higher metabolic activity, namely the ability to control body temperature and remain active for longer periods of time. Feathers could also have first appeared for incubating eggs: in 1993, a fossil of a maniraptorian dinosaur was found positioned over a nest with eggs in the same posture used by living birds when brooding (Norell *et al.* 1995).

Maniraptors certainly show feathers with a modern appearance. *Archaeopteryx* unquestionably had asymmetrically veined wing feathers with aerodynamic capabilities, as well as tail and contour feathers. Fragments of early feathers were found in ancient amber recovered in Canada in 2011, dating from about 80 MYBP (McKellar *et al.* 2011); more were discovered in Myanmar in 2015. Some of the feathers archaeologists found were asymmetrical with interlocking elements similar to those of birds, while others were wispy feathers like those found on non-avian dinosaurs. Areas of light and dark suggested patterns and coloration. The samples from Myanmar (Xing *et al.* 2016) are also from the Cretaceous period, dating from 99 MYBP, and are from hatchlings of the sister clade of birds, the Enantiornithes [Figure 1.11]. They show plumage arrangement and microstructure alongside immature skeletal remains.

While they were flightless, many maniraptora had colored flight feathers on the arms, which implies that feathers evolved as a signaling mechanism to attract possible mates or as camouflage to protect the animal from predators, just as they do for birds today. Flight feathers appeared towards the end of the Cretaceous period when dinosaurs were evolving at a slow rate, but birds were evolving more rapidly (Brusatte 2017).

Nature of Feathers

Birds are the only creatures on earth with feathers and, indeed, all of the more than 10,000 species of birds on earth have them. Not only do the feathers provide birds with the ability to fly, but they also aid with insulation and protection from wind, rain and sun, and enable camouflage. Feathers are a useful means of communication through displays for different purposes: attracting a mate, defending a territory, begging for food, incubating eggs, and distracting possible threats.

Waterbirds have 10,000 to 15,000 feathers; tundra swan is at the high end of this scale with about 25,000 feathers. Other birds have from 2,000 to 5,000 (Van Tyne and Berger 1976). The emperor penguin has the most—between 144,000 and 180,000 feathers (Williams *et al.* 2015). Hutt and Ball (1938) show that there is a strong relationship between the weight of passerine birds and the number of feathers per gram of body weight. They show that the number of feathers is not directly related to the surface area of the bird, and that smaller birds like passerines have higher feather densities per surface area to maintain body temperature. In the case of emperor penguins, 80 percent of their feathers are downy plumes needed for their harsh Antarctic environment.

There are many different kinds of feathers, and many have different purposes. We would all recognize the basic types that are common to all birds. The body contour feathers (pennae) cover the bird and include the flight feathers (remiges) and tail feathers (rectices). The flight feathers are aerodynamically contoured; they are stiff and the quills (calamus—lower part of the quill, rachis—upper part of the quill) are attached to the wing bone by ligaments. This attachment is crucial, providing the stiffness that transmits force from the pectoral muscles to the feathers.

Birds have many different kinds of feathers, each of which serves a specific purpose. Beneath the contour feathers are the fluffy, shaftless down feathers (plumulaceous) that provide insulation. The semiplume is a feather with a central shaft; the lower portion is covered with down and the upper portion is part of the contour. Bristles are specialized feathers found on many species of birds, and their function is not always apparent. A third type of feather, the filoplume, is a fine hair-like feather beneath the contour feathers, sparsely distributed over the entire body and sensitive to touch. Some filoplumes around the neck grow beyond the contour feathers and serve a decorative purpose.

The powder down is a very short feather that degrades to a powder at its tip. The waxy white keratin powder spreads over the plumage; it alters the color of the bird and serves as waterproofing. It is visible to some degree in most birds, but is most developed in tinamous, herons, some hawks, and swallow-tailed kite.

The aftershaft is an element of the body contour feather that actually consists of two shafts, an outer one and an inner one. Only two families of birds have aftershafts the same length as the main shaft: cassowaries and emus. Many other species of birds have partial aftershafts, including tinamous, gallinaceous birds (pheasants, and other game birds), herons, hawks, and parrots. Some species have them reduced significantly in size, including woodpeckers and oscine songbirds. Ducks, vultures and owls have a tiny vestigial fringe of aftershafts, and suboscine passerines have lost them entirely (Van Tyne and Berger 1976). The aftershaft is reduced in size in oscine songbirds. (Oscine and suboscine passerines are discussed in Chapter 12.)

Birds care for their feathers by preening, oiling and bathing. The oil gland (preen gland) is located on the upper surface of the tail. Birds collect the oil with their beaks and coat the feathers by running them though the bill. The oil serves as waterproofing, and may have other beneficial effects. Studies of the oil's effect on feather color show that it reduces the overall brightness of white feathers and relative UV reflectance in mallards, but has no effect on UV reflectance in blue tit crown feathers (Biedermann *et al.* 2008). Birds without an oil gland include ratites (cassowary, ostrich, etc.), pigeons and doves, woodpeckers, and North American vultures. The oil gland is most highly developed in waterbirds, though cormorants have preen glands and coat their feathers, but they have less water repellency than other waterbirds. As a result, they are able to reduce their buoyancy for swimming under water, but must hold their wings outstretched to dry after immersion in water.

An unusual adaptation are the breast feathers of the male sandgrouse of Asia. The young are hatched in dry desert habitat, where there is no available water. The adult males dust bathe to reduce their water repellency, and then wade into belly deep water. The feathers have unique properties that allow them to adsorb water. For

Figure 4.2 A Ruffed Grouse taking a dust bath (author's collection).

a two-month period, they carry the adsorbed water to the chicks, which drink the water from these feathers (Crick and Maclean 2003).

As anyone with a bird bath has seen in their backyard, a wide variety of birds and some mammals practice water bathing. Birds partially immerse themselves in shallow water and splash or dip into deeper water. Other means of water bathing include the use of rain and wet vegetation. After bathing, birds will fluff their feathers, and preen to spread oil over their feathers.

All gallinaceous birds (grouse, pheasants, turkeys, etc.), some hawks and owls, birds living in dry habitats, and some passerines practice dust bathing (Van Tyne and Berger 1976). They will create a slight depression or visit an existing dusty area and get the dust entrained in their feathers using body and wing motions. Ruffed Grouse [Figure 4.2] will return to a dusty hollow repeatedly and heap dust over its back and fluff its feathers with dust. After dusting, the birds shake off the dust. Dust bathing is a mechanism of feather maintenance; research has shown that it reduces the amount of feather lipids, thus increasing the insulating properties of the plumage. It has been suggested that dust bathing reduces parasites, but this has not been tested experimentally (Olsson and Keeling 2005).

Molts and Plumages

Birds molt annually to replace their feathers, because feathers degrade over time from bacterial damage, mechanical abrasion and wear, and UV radiation. The sequence and timing of molts in birds have been studied more thoroughly in recent years, and have shown that there are four general patterns that most birds follow. While there are certainly variations and exceptions, birds within a family tend to molt in similar ways. The subject of molt is treated in detail by Steve Howell (2010) and Ron Pittaway (2000) and in the numerous publications by Peter Pyle (1997, 2008).

Birds are often casually referred to as being in winter, summer, or breeding plumage, but these terms are overly simplistic and cannot be applied to many species. The Humphrey-Parks-Howell system is a standard method of referring to the various molt states and the resulting plumages of birds. It can be applied to all birds worldwide, and is used by many bird banders. Understanding molt can help determine the age and even the species of individual birds.

More recently, the Wolfe-Ryder-Pyle (WRP) system provides an advancement in classifying the age of a bird. It uses the Humphrey-Parks terminology, but abandons the calendar-based ageing in favor of a molt cycle-based system (Wolfe *et al.* 2010). It allows for the maximum accuracy of ageing birds based on their molt and plumage state.

The molt processes are referred to in association with the plumage they produce: the prebasic molt leads to the basic plumage, for example, and the prealternate molt leads to the alternate plumage. The plumage traditionally called "winter" or "non-breeding" plumage is the basic plumage, worn for at least some part of the year by all birds. The plumage traditionally called "breeding" plumage is an alternate plumage that gets inserted into the molt cycles of some, but not all birds.

Steve Howell (2003) describes the four most common molt strategies: simple basic, complex basic, simple alternate, and complex alternate. The first set of non-downy feathers for all species is called juvenal plumage, and in passerines is typically the plumage at the time of fledging. In the two complex strategies, species have an additional plumage in their first year, or first cycle, called the formative plumage.

- Some bird families follow the simple basic strategy, with one molt per year: seabirds, vultures, large hawks, falcons, penguins, and other species. These species have only basic plumage in their first year and one molt annually.
- In the complex basic strategy, species have a partial molt in first year (preformative molt), which results in the replacement of some juvenal feathers; this plumage is usually referred to as immature plumage. After first year, these species have one molt annually. Most species follow this strategy: most owls, swallows, nightjars, woodpeckers, hummingbirds, crows and jays, kinglets, chickadees, most wood-warblers.
- Other bird families follow the simple alternate strategy and have two molts per year after their first cycle. Loons, ducks, herons, ibises, gulls, and some alcids follow the simple alternate strategy.
- The complex alternate strategy includes a preformative molt in the first cycle. Sandpipers, terns, some species of gulls, and many species of passerines follow the complex alternate strategy (Howell 2003).

Species in these two alternate categories usually have a complete molt once per year, usually in the fall, and a partial molt, usually in spring. Very few species have two complete molts per year, but examples include bobolinks and Franklin's gull.

Some families of birds like gulls and eagles take a number of years to mature and develop full adult plumage. Their plumages change each year until they reach their definitive adult appearance. During those years, the plumages are referred to by the cycle they are in, such as first cycle, second cycle, etc. Once the adult plumage is attained, the molts and plumage are referred to as definitive, as in "definitive pre-basic molt" and "definitive basic plumage."

In a complete molt, the process of body molt usually begins at the head and proceeds towards the tail. Wing and tail feathers molt symmetrically in pairs for most birds, except for waterbirds. Most partial molts replace body feathers, but not wing or tail feathers. The molt process takes a lot of resources, so birds must supply the nourishment needed to build the feather proteins—so they time their migration to be able to take advantage of important food sources. Some species molt before migration, some during, and others after they have arrived at their destination. For those that molt before or after migration, the molt process can be all at one time, leaving them flightless for a period; or it can be spread out over time.

Color in Birds

Color patterns play a key role in the adaptation of birds to their environment, and for some birds, color patterns are important for their foraging practices. The richness of the colors of male birds is also important in attracting females. Color serves as camouflage for some birds, especially for game birds and for female birds that use open cup nests. Colorants are also important because they can improve the durability of feathers. For instance, melanin pigments give feathers mechanical strength and abrasion resistance (Ralph 1969). Geoffrey Hill (2010) explores the science and dimensions of color in birds in more detail in his book *Bird Coloration.*

One of the most important purposes of coloration is that it allows birds to remain concealed to avoid predators or to capture their own prey. Countershading—a property known as Thayer's Law—makes the parts of any animal exposed to the sky dark, while the parts exposed to the shadow areas are light in tones, with a graduation in tones from darker to lighter. The tonal graduation obscures the shadow area of the animal. This is the principle used to camouflage military hardware. Sparrows, shorebirds and many arboreal forest bird species take advantage of countershading, with dark tones on their dorsal side and light tones on their ventral surfaces. This is also common in seabirds: loons, grebes, alcids, penguins, shearwaters, petrels, and many others. Countershading is not common in larger predatory birds such as hawks, eagles, and vultures, but it does occur in some raptors.

Coloration in birds is an indication of the fitness of an individual, so it is an important factor in sexual selection—but there are other factors that influence bird coloration. Hilty (1994) summarizes the general patterns in bird coloration that are the result of their environments. Birds that feed on fruits and nectar throughout the year are the most colorful of all birds, and in the ones that feed on fruit, the males and females have the same colors and patterns. These traits are found in tropical tanagers, parrots, toucans, motmots and many tropical birds. Birds that feed primarily on insects tend to have muted or dull colorations. Neotropical migrants have a more complex color story.

Birds that spend most of their lives in the tropics have muted plumages in their non-breeding ranges. On their breeding territories in spring, the males are adorned with bright colors, while the females and immatures have muted tones. Ornithologists surmise that these duller colors in the tropics are important for birds' ability to survive, and for their interactions with permanent residents when they transform from a territorial to gregarious lifestyle after breeding. Grassland birds and ground birds have cryptic plumages with dark, muted tones and patterning to avoid detection. There are exceptions to these general rules, of course, such as the Asian pheasants and some very colorful flycatchers.

Color Mechanisms

Fifteen mechanisms in nature give rise to color (Nasssau 2001). Many of these are due to the excitation of atoms at high energy levels, leading to ionic emissions. Four of the color mechanisms, unrelated to thermal causes, are found in biological systems, and three are operative in the coloration of bird feathers. Table 4.1 shows the three mechanisms of color in bird feathers.

Table 4.1 Mechanisms of Color in Bird Feathers

	Mechanism	Physical Effect	What Enables the Colors
1	reflectance	some wavelengths absorbed	dyes and pigments
2	scattering	Rayleigh or Mie scattering	small spherical particle within or on the feather surface
3	iridescence	constructive and destructive interference	thin films on the feather surface made of two materials of high and low refractive index that cause interference of light

Reflectance

The first mechanism, reflectance, involves absorption of light by means of colorants. When light impinges on a colorant, many of the wavelengths of light are absorbed, and the resulting perceived color is made up of the wavelengths that are reflected back towards the observer.

In birds, colorants take the form of organic pigments and dyes. Dyes are usually soluble in the containing media, while pigments are insoluble.

One of these pigments, melanin, is the common pigment in human skin and hair, as well as in the black coloration on the surface of old bananas. In birds, melanin includes two types of pigments: eumelanin produces black or browns, and pheomelanin produces yellow, rufous and red colors.

Carotenoids, another class of color pigments, come in many types and produce the colors of red and yellow. These fat-soluble organic compounds are found in various food sources, including plants and a number of insect larvae. Birds need to ingest these foods in sufficient quantities to acquire the colors we know and recognize. Carotenoids also interact with melanin to produce muted green colors.

Porphyrins are a class of complex, water-soluble molecules that are synthesized in the metabolism to produce bright colors including green, pink, browns, and reds. Porphyrins have been found in some owls, nightjars, bustards, pigeons, gallinaceous species, and turacos (Galvan *et al.* 2016).

Other color pigments are much less common and are restricted to a few species or bird families. Some of these pigments are understood, but some have never been analyzed.

Scattering

The second mechanism of color, the scattering of light within the feather, is the same phenomenon that causes the sky to look blue: the scattering of light by the gaseous molecules in the atmosphere (an effect known as Rayleigh scattering). In birds, blue colors arise from the scattering of light by microscopic bubbles of air within the surface, giving rise to the colors we see in bluebirds, blue jays, and many other species.

The color green can arise from different mechanisms, but the most common is a combination of light scattering plus a yellow pigment. Green can range from muddy olive green in many birds to bright green colors in parrots and parakeets, and in tropical birds like green jay.

A different scattering mechanism known as Mie scattering produces white color in birds, caused by particles embedded in the feather that are larger than the wavelength of illuminating light. (A common example of Mie scattering is the white color of milk.) Very few birds are all white in color. For most birds, white-colored feathers provide highlights, and often serve as a signaling mechanism to other birds. Shades of gray are achieved by mixing melanin pigment with Mie scattering. Gray tones are very common in birds such as gulls, desert birds, and songbirds.

Iridescence

Iridescence, the third color mechanism, accounts for the colors seen in hummingbirds, heads of ducks, starlings, grackles and other birds, and it arises from a process known as interference. Crawford Greenewalt (1960) provided detailed explanations on how the interference colors arise. The color is the result of a thin layer of polymer on the feather surface, consisting of small elliptical structures or particles that encase small air bubbles and are embedded in a polymer matrix. In the case of hummingbirds, multiple layers of these polymers, each about half the wavelength of light, are underlaid with melanin. For instance, red light has a thickness of 0.0000128 inches (325 nm). Variations in the thickness of the films and the overall refractive index create the resulting color. This is a physical phenomenon in which nanostructures on the feather surface create these optical effects (Alù 2022). A well know example of interference color is the graduations in blue and purples when a small film of oil becomes visible at certain angles on the surface of water.

With iridescence mechanism, light enters the layered structure, and certain wavelengths are absorbed by destructive interferences, while others are enhanced and emitted at specific angles with respect to the incoming light. Greenewalt (1960) describes the relationship between the feather surface and the angles of incoming and emitted light. He notes that this emitted light is of very narrow wavelength, thus giving sharp and pure colors. The interference mechanism creates the glittering metallic colors we see when hummingbirds turn their heads to just the right angle. The green backs of hummingbirds and trogons and the greenish or purplish colors of starlings, grackles and ducks are also the result of this mechanism. For starlings, ducks, and some passerines, these thin polymeric films on the feather surface wear over the course of time, and the iridescence eventually ceases to be visible until the feathers are replaced through molt.

Numerous studies have shown that bird feathers reflect ultraviolet (UV) light. Reflections of ultraviolet light cause color shifts, depending on the amount of UV in the incident light. Research on 312 bird species representing 142 families demonstrated that all avian families possess plumages that reflect significant amounts of UV light (Eaton and Lanyon 2003). Another study measured the UV reflectance of 1000 species from nine orders of larger birds, and found all had some UV reflectance (Mullen and Pohland 2008). Among the different species, some feathers have strong UV reflection, some weak reflection, and others have none at all.

Parrots are the only known birds that have fluorescent feathers (Boles 1990). The fluorescent colors glow when UV light is absorbed and reemitted at a longer wavelength. This occurs in many species of parrots; in the case of the budgerigar parakeets, the yellow crown and cheek feathers have pigments that react to UV light. Tests show that both males and females prefer mates with a strong fluorescent response (Arnold *et al.* 2002).

Aberrant Colors

Sometimes birds are born with unusual variations in color—most often with white patches in their plumage, or as completely white birds. The two most common forms of aberrant colors in birds are leucism and albinism.

Leucism is very common, and it appears as large patches of totally white or pale plumage on an otherwise normal-appearing bird. Leucistic birds can be totally white, but their eyes and their fleshy parts, like the legs, will be normal colors. In leucism, the bird lacks sufficient melanin or one or more other pigments that create the appearance of a normal color for that species. Leucism can be genetic or environmental: research on eared grebes that winter on the Great Salt Lake in Utah has shown that leucism can be caused by toxins in the environment (Pyle and McPherson 2017). Leucistic effects can also occur after a molt of otherwise normally colored birds.

Albinism is caused by a genetic mutation and is rare in mammals and birds. In this case, the bird is totally white, and the eyes and fleshy parts are pink. It had been suggested that albinistic birds have short lives due in part to poor vision and susceptibility to predators.

There are many other forms of color aberrations in birds that are the result of pigment abnormalities (Davis 2007) caused by mutations, diet, or age. A very unusual mutation was that of a northern cardinal found in Pennsylvania with one-half male and one-half female plumage. This condition is known as gynandromorphism.

Polymorphisms and Plumages

Other types of color abnormalities are seen in birds, but they should not be confused with polymorphism, or with the annual variations in plumage through molt.

A bird species is polymorphic when it has two or more different appearances (plumages), and birds with these color morphs coexist. This coloration is permanent in any individual and is part of the genetic makeup of the species. It is most common in hawks, but it also occurs in some species of owls, jaegers, cuckoos, herons, geese, nighthawks, gamebirds, and seabirds. Many species of birds can be found in dark or light morphs, and some in rufous (red) or gray morphs. Red-tailed and Swainson's hawks, for example, have light, dark, and rufous morphs and sometimes a combination of these; the parasitic jaeger has intermediate as well as light and dark morphs. Eastern screech-owl has red, gray and brown morphs. Red-footed booby has white and brown morphs.

There is some evidence that polymorphism in raptors is dependent on their environmental niche. Galeotti and Rubolini (2004) found that species of hawks, owls and nighthawks were polymorphic when they occupied a more continuous distribution range, frequented many different habitats, were more migratory, and lived in seasonally alternating conditions. There are no explanations for why polymorphism is found in other birds, but it only occurs in about 3.5 percent of all bird species (Galeotti *et al.* 2003).

Sexual dimorphism among birds refers to species in which the male and female have differences in their plumage coloration and markings, and in some cases, size differences between males and females—especially in raptors. Plumage dimorphism has been correlated with parental cooperation (male and female birds sharing in raising young) and extra-bond paternity, in which otherwise monogamous birds copulate with a second mate of greater size or more attractive plumage, presumably to strengthen the offspring (Owens and Hartley 1998).

Some species of birds have regional differences in plumage colors, and these are usually separated into subspecies. One examples is western gull, with a Pacific northwest subspecies that has paler wings and darker eyes than its counterpart in central and southern California. Great horned owls that are residents of the Arctic region have very pale plumage as compared to those of the more southern part of their range.

Many bird species have plumage color changes within their annual plumage cycle. Ptarmigans change from white plumage in winter to dark, cryptic (camouflage) plumage in summer. Small gulls have black- or dark-hooded heads in breeding, but white heads in winter. Many species of shorebirds have breeding plumage that is generally dark in color, replaced by muted gray plumage in non-breeding season.

Large gulls vary in the darkness of their mantle color depending on their breeding latitude, from lighter-colored wings and back closer to the polar latitudes to dark coloration closer to the equator. Nikko Tinbergen (1959) (see also Nicolson 2018) and others have done extensive studies on the behavior of gulls, and found that the small hooded gulls that have dark hoods in the breeding season, such as black-headed gull or Bonaparte's gull, are much more successful at catching fish with their shallow immersion diving when they are in their basic plumage with all-white heads. In breeding season they do not dive for fish, but feed on insects in grassy fields and pick offal and crustaceans from the surface of the water. This is a case of color shading adapted to different feeding strategies in breeding and non-breeding.

Polymorphism, dimorphism, and molt variations of plumage with the seasons, sex, and maturity all lead to complications and challenges in trying to identify birds visually.

Respiration

The avian respiratory system is one of the most remarkable organs in a bird's physiology, especially from the viewpoint of engineering. Like many aspects of avian physiology, it is adapted to enable flight—a small but incredibly efficient heat-and-mass exchanger.

Birds have lungs that provide for the exchange of oxygen from air to their blood through their lungs (pulmonary system). They also usually have nine air sacs—some birds range from seven to twelve—that exchange air with the lungs, but do not exchange air with the bird's blood. Figure 4.3 shows a schematic of the lungs and air sacs. In this treatment, the five forward air sacs are referred to as the anterior, and the four rear air sacs are posterior. However, the correct names for these are shown in Figure 4.3. The lungs do not expand or contract, but fresh air enters the lungs with both the inhalation and exhalation cycles. The air sacs' volume changes with each respiration.

The air sacs are thin-walled and are connected with air spaces within the bones. They have much larger volume than the lungs and allow for the accumulation or the deficit of air, controlling this with valves.

Unlike mammals, birds have no diaphragm. Air is moved by pressure exerted when the bird flexes the ribs. About one-fifth of the volume of a bird's body is its respiratory system, compared to about one-twentieth for mammals.

Figure 4.4, below, gives an

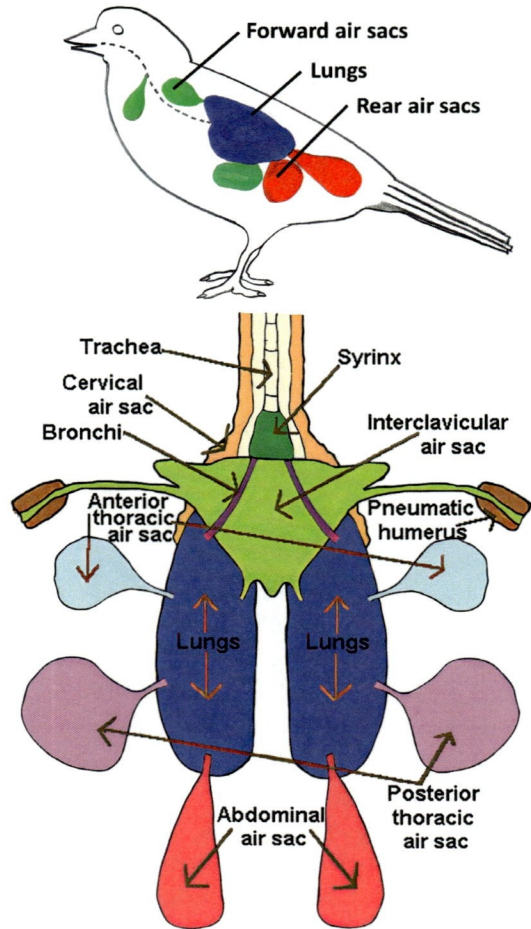

Figure 4.3 Breathing system of birds, showing the placement of the lungs and air sacs (above), and the five forward and four rear air sacs (below) (Gordon Ramel).

approximate state of the lungs and air sacs at the end of each inhalation and exhalation. The breathing cycle includes two full respirations to complete one full period in order to return to the same initial state. The air volume that is circled represents the movement of a single volume of air through the system.

On first inhalation, air enters through the trachea and passes into the posterior air sacs. There it expands the air sacs, and about half of it flows into the lungs, displacing the air there. Thus, at the end of inhalation, the system is full of air. On exhalation, the posterior air sacs contract, pushing air into the lungs, and the anterior air sacs contract and expel air through the trachea, so at the end of exhalation, the lungs are full of air but the air sacs are depleted. During the second inhalation, the process repeats and the air from the first inhalation now exits during the second exhalation. This is an approximate description, since the actual exchange of air is not as

State Changes	Time	Anterior Sacs	Lung	Posterior Sacs
Beginning State of lungs & sacs	0	empty	Full 0	empty
First Inhalation 1. Air fills lung and posterior air sacs		filled	Fill 1	filled
First Exhalation 1.New air flows in from posterior sacs 2. Old air leaves lung through trachea and empties the anterior air sacs		empty	Fill 2	empty
Second Inhalation 1. fresh air enters posterior air sac and lung 2. spent air fills anterior air sac		filled	Fill 3	filled
Second Exhalation 1. Spent air exhausted from anterior air sac and lung. 2 New air enters from posterior air sac		empty	Fill 4	empty

Figure 4.4 **The breathing process in birds has a single period made up of two cycles** (author's collection).

precise as described and not all air is displaced from the lung on each exchange. The important points are that the air flows in one direction, and the lungs are filled on both inhalation and exhalation. Blood enters the lungs transverse to the direction of air current, thus setting up a cross-current flow with the moving air (Scheid and Piiper 1972).

Respiration in birds is a continuous flow with high velocity, compared to mammals, which breathe in a pulsed flow—including a rest after each breath—with low velocity. In addition, the air capillaries in the bird's lungs have a much larger surface area than those in mammals, and the thickness of the lung membrane is much thinner in birds, thus allowing more efficient gas exchange. The respiration of air is asynchronous with the wing beats, and the respiration rate is about one-third that of mammals.

This system of exchange of oxygen is extremely efficient, much more so than the mammalian breathing system. It is estimated that a bird is 33 percent more efficient at extracting oxygen from air at sea level, and 200 percent more efficient at one-mile altitude (Ward and Berner 2011). This system enables flight, since that requires a high energy output, and it enables high altitude flight without suffering hypoxia (oxygen starvation). High altitude flight is common in migrating birds that seek the best wind and temperature conditions.

Of birds that migrate at high altitudes, the bar-headed geese and cranes that cross the Himalayas reach some of the most extreme altitudes. Bar-headed geese routinely cross the Tibetan Plateau at between 16,000 and 19,600 feet (5,000–6,000 m), and some will reach elevations above this—in fact, one was recorded at 23,900 feet (7,290 m) (Hawkes *et al.* 2013). At this elevation, oxygen content in the atmosphere is 40 percent of ground level. Anecdotal reports suggest that these geese fly at even higher elevations than officially recorded. The altitude and the prolonged migration are a significant metabolic cost for these flights (Scott *et al.* 2015).

Ruppell's griffon of Africa has reached an altitude of 37,000 feet (11,300 m), the highest ever recorded for a bird. In 1973, this vulture collided with a commercial aircraft at 37,000 feet over Abijan, Ivory Coast, in western Africa. The altitude was recorded by the pilot shortly after the impact, which damaged one of the aircraft's engines and caused it to shut down (though the plane landed safely). The vulture was identified by the remains of wing feathers collected from the plane (Laybourne 1974). Vultures ascend on updrafts and are not expending the same energy of sustained flight as geese do, even though this bird was able to reach the altitude of commercial jet air traffic.

The avian respiratory system also plays an important role in thermoregulation, serving as the principle method by which birds give off heat generated by the consumption of food. Birds do not perspire as do mammals; while breathing, they take in air at ambient temperature and expel gas at a higher temperature. They can also give off heat by increasing blood flow to their beaks. When not in flight, they can use other mechanisms to reduce heat, such as extending the wings, adjusting their feathers to reduce insulation, and using shade as protection from the sun.

Where did this unusually efficient respiratory system come from? Ward and Berner (2011) discuss the relationship between the avian breathing system and the bird's dinosaur ancestors. A host of characteristics of Saurischian dinosaurs indicate that they had developed the air sac breathing system because they had pneumatized bones, mobile posterior ribs, and other characteristics important to this type of system. It is likely that this high-efficiency system evolved in response to falling oxygen levels at the beginning of the Triassic period. At that time, oxygen levels at sea level had elevated to 30 percent, but began falling to about 17 percent over a 20 million year period. Levels rebounded to 20 percent for a time, then dropped gradually to 13 percent in the Jurassic period and remained low for more than 50 million years until the Lower Cretaceous. To maintain their high activity levels, theropods needed a more efficient breathing system. Consequently, the maniraptors inherited this system from the theropods and passed it on their avian descendants.

Vision

Vision is the most important and dominant sense in birds. They have the largest eyes in the animal kingdom (after normalizing for their size). Their eye is immobile in a fixed socket and requires head movement to adjust the field of view; most birds have 13 to 25 vertebrae in their neck to allow them to turn the head backwards. Some birds—owls in particular—can rotate their head 270 degrees in one direction, compared to humans, who are limited to 90 degrees.

The shape of a bird's eye can be changed very rapidly for distance adjustments through their control of the curvature of the cornea and lens, and the lens is softer than a mammal's to enable rapid adjustments. This ability to adjust also allows birds that swim underwater to see at the same refractive index as water, thereby seeing without distortion. Herons and others that look through the air-water interface have pigment accommodations, correcting their vision to improve their ability to capture prey (Lotem *et al.* 1991).

In birds, the eye also has a third eyelid, the nictitating membrane, which is also found in reptiles, fish, amphibians, and mammals, but is absent in primates. This translucent membrane moves horizontally across the eye. In birds, it serves primarily to clean, protect, and lubricate the eye. The retina is thickened and does not contain blood vessels, allowing a higher density of receptors, rods and cones, and enabling greater acuity (sharpness) than in mammals. Birds have a structure in the eye called the pecten oculi, which provides nutrients to the retina and does not interfere with sight (Kiama *et al.* 2006).

Birds of prey, shrikes, hummingbirds, swallows, and kingfishers have two fovea—focusing points—within the eye compared to one for humans and other creatures (Birkhead 2012). This gives them the highest density of receptors, and they can see with two to three times the acuity of humans, which allows to see clearly for long distances. American kestrels, which routinely hunt dragonflies, can see a dragonfly at 100 yards (90 m). Owls have very large eyes that are tubular shaped with low numbers of color sensors; however, they have a high density of rod cells and other adaptations that give superior vision in poor light. They sacrifice acuity for sensitivity, since they have fewer cone cells that react to color (Birkhead 2012).

In most bird species, the eyes are set on the side of the head, giving them monocular vision: a large field of view but a small forward view. Almost all birds have their eyes placed so they can see forward to at least some degree. As an example, pigeons have monocular vision over an angle of about 300 degrees, and forward binocular vision over a small angle of about 10 degrees. American Woodcock, on the other hand, has monocular vision very close to 360 degrees.

Binocular vision is important for avoidance of obstacles in flight, and for the placement of the bill in capturing of prey, seeds, fruit and nectar. Hawks, falcons, and owls, all of which hunt by sight, have a high degree of binocular vision that allows them to judge the distance of an object. Hawks have a total field of view of about 120 degrees, of which 80 degrees is binocular; owls have an even wider field of view. Shorebirds have vision above them, from in front of the bill to just above their heads (Martin 2009), thus providing a view from the horizon to above their own heads. We can assume this is a way of detecting predators. American bittern has binocular vision on the bottom side of the head for looking down, helping it to spot prey in the water or on the ground. In order to see in the horizontal plane, it must lift its head up so the bill points skyward.

For birds with the eyes placed on the side of the head, such as all passerines, the two eyes are used for different tasks. One eye is better at close range tasks, and the other at long distance sight. This condition is caused by differences in the amount of light the eyes are exposed to in the embryonic state. It is normal that one eye—usually the right eye—has better distance vision. There are many examples of bird behavior that are caused by the sidedness of their vision (Birkhead 2012): domestic fowl, for example, use their right eye for close viewing like feeding, and the left side for distance viewing.

Human eyes have three types of cones that detect the three additive primary colors: red, blue and green. Only primates and birds can see all three primary colors. Most birds (except nocturnal birds) also have a fourth set of cones that detect ultraviolet (UV) wavelengths.

Male and female birds that appear to us to have the same plumage can look different to birds because of these UV reflectance variations between the sexes. Although we do not completely understand exactly what birds see when they view the primary colors plus UV, researchers can determine what birds can see that we cannot. Raptors, particularly kestrels, can detect the urine trails of small mammals from flight because these trails reflect UV. Birds also appear to use UV in the selection of a mate, the identification of their own eggs, and during interactions with their chicks. Ultraviolet also plays an important role in the foraging of herbivorous bird species. This is particularly important to fruit-eating birds, because only mature fruits reflect in the UV (Rajchard 2009).

Hearing

Hearing in birds is the second most important sense organ. Birds hear through ear openings that are covered by feathers and located on each side of the head, behind and below the eye. Some birds have bare heads, so the ear openings are easily seen, but in almost all birds, the ears are covered. For terns, penguins, and other seabirds that submerge their heads, the feathers prevent water from getting into the ears.

Sound in the form of vibrations is transmitted by pressure waves through the outer ear canal to a membrane that protects the middle ear, and then transmits sound by means of a single bone to the inner ear (as opposed to three bones in the human ear), where the auditory signals are converted into sensory signals. In the inner ear, sound causes microscopic hairs to vibrate within the cochlea, and these vibrations send signals to the auditory nerves. The inner ear also encloses a series of three semicircular fluid filled canals, one in each plane that functions to maintain proper balance while perched, such as in windblown conditions, while walking, and in flight.

Birds' ears are most similar to those in reptiles. Mammals' ears differ in a number of respects, such as having three bones to transmit sound. The hairs in the birds' cochlea are replaced on a regular basis, but those in mammals are not and, therefore, decrease in effectiveness with age. Humans, therefore, are prone to hearing loss in the high frequencies.

In all creatures, hearing ability varies with both species type and individuals. Sound is measured in vibrations per second and expressed in hertz (Hz). One Hz is a frequency of one cycle per second, and denotes the pitch of a sound. In humans, average sound sensitivity is between 20 Hz and 20 kilohertz (kHz), but the highest sensitivity is between 1 kHz and 4 kHz. Like humans, the majority of birds are most sensitive in a range of 1 kHz to 4 kHz (Beason 2004).

In general, the full range of hearing in birds is similar to humans, but is dependent on the species. Owls and many other animals can hear much lower frequencies than humans can: below 20 Hz, at a level considered to be "infrasound." The barn owl, for example, has a sensitive hearing range of 0.5 Hz to 10 kHz. Pigeons are also capable of hearing in the infrasound range (0.1 to 20 Hz). Amazingly, there is one species of hummingbird that sings at frequencies in the infrasound range, well beyond the human audible limit. Researchers have shown that the black jacobin, a hummingbird of Brazil, sings its song at 10 to 80 kHz (Olson *et al.* 2018). It is believed that females of this species can hear at that frequency range.

In birds, the length of the cochlea is a reasonable index of the sensitivity to sound. Small birds with short cochlea are more sensitive to high frequencies, and large birds with longer cochlea are more sensitive to lower frequencies (Birkhead 2012). Birds suffer loss of hearing sensitivity with exposure to prolonged loud sounds. Dooling (2002) states that based on auditory tests, when hearing is defined as the softest sound that can be heard at different frequencies, birds on average hear less well than many mammals, including humans. However, birds are able to detect minute noise differences far more rapidly than humans can (Radford 1984).

Anyone who has stalked birds with a camera is well aware of their keen perceptions. They combine signals of sight and sound and respond extremely rapidly if they sense a threat. However, sound quality alone does not explain their alertness. Most birds show poor sensitivity to very low frequencies (less than 20 Hz) because their ears are so close together; however, some bird species will fly in a circle and use the Doppler shifts to determine direction of low frequency sounds (Beason 2004).

As noted earlier, owls and night birds have a higher sensitivity to sound than other birds. In owls, the ears are asymmetrical on the skull, which allows them to pinpoint the direction and distance to prey. Owls' facial disks play a role as well, collecting sound and channeling it to the ears. Some owls, like the great horned, the screech-owls, and many others, have ear tufts—but these should not be confused with actual ears, which are hidden behind feathers. A number of owls and other birds have special skills based on their hunting environment: barn owls can locate prey species in total darkness if they are familiar with the nature and layout of the space within which they are working. Great gray owls can locate rodents from a long distance, even when they are buried beneath a foot of snow. Oilbirds of South America and some species of Asian swifts can navigate by echolocation, emitting a series of clicks as they maneuver in dark caves.

An amazing story of hearing in birds is told by Tim Birkhead (2012), related to him by researchers working with common murres in Newfoundland. Murres breed in colonies on sea cliffs, and the adults and young recognize each other by sounds. For humans, the sounds of adults are the same as young birds. When it is time to depart for the sea, most families depart at the same time; the adults fly off, and the immatures must plummet to the open sea. In the surf within this morass of birds, parent and immatures find each other by sound: the immature must remain with the adult for several weeks at sea; otherwise, it will perish. This demonstrates an amazing ability to recognize subtleties in sound for which we have little appreciation.

Taste and Smell

Most birds locate food sources by sight or, in the case of some predatory birds, by sound. The past perception has been that birds have a very poor sense of smell because most birds have very small olfactory bulbs in their brain relative to their overall brain size. However, research in this field is proving otherwise.

Among bird species, the size of the olfactory bulb varies widely, from 200 to 600 genes within the bulb compared to humans who have about 850, but about half of those in humans are not functional (Verbeurgt *et al.* 2014). Tests on a variety of families of birds show that their brain activity reacts when they are exposed to a variety of scents. Most birds can and do use smell regularly and often for important tasks.

Certain birds have a highly developed sense of smell, including the tubenoses (petrels to albatrosses), many vultures, and the kiwi. This sense began its evolution millions of years ago with these birds' theropod ancestors. Research has verified that theropod dinosaurs had a developed sense of smell based on measurements of the size of their olfactory bulb. Zelenitsky *et al.* (2011) measured the olfactory capabilities in 157 species of non-avian theropods, fossil birds, and living birds. They showed that the relative olfactory bulb size increased during non-avian maniraptorian evolution, remained stable across the non-avian theropod/bird transition, and increased during basal bird and early neornithine evolution.

Some species have an acute sense of smell, and locate food primarily by that means. Kiwis have nostrils at the end of their bill, allowing them to find prey by probing. Turkey vultures can find meat hidden in a forest solely by smell. The work of Kenneth Stager (1964) showed that turkey vultures find dead animals by smell and are able to differentiate the age of the dead carcass, avoiding those that are too putrefied. California condors, however, have a poor sense of smell and rely on their sight and other bird species like ravens and turkey vultures to find food for them.

It has been known that seabirds can be attracted by using chum, a mixture of fish parts and fish oils, placed on the sea surface. Seabirds will respond by seeming to appear from nowhere. Work has shown that dimethyl sulfide, a gas emitted by phytoplankton, is attractive to seabirds and that the albatross can detect these scents from a distance of 12 miles (20 km) (Nevitt *et al.* 1995, Debose and Nevitt 2007). This has been an important tool for attracting birds to pelagic tour boats filled with birders hoping to see these seabirds.

At least some species of birds can identify their mates and chicks by scents. The New Zealand kaka, a parrot, is known to have a strong sense of smell. Anecdotal evidence suggests that parrots are able to smell, but there is no research on this subject.

It was once assumed that birds have little sense of taste because of their apparent indiscriminate diet. However, much more is understood today about their sense of taste, and it is recognized that they can make fine distinctions between different tastes. Birds have relatively few taste buds—50 to 500—depending on the species, compared to about 9,000 in humans (Rowland *et al.* 2015). Human taste buds are on the tongue. In birds, only a few taste buds are on the tongue; most of them are in the salivary ducts and other areas of the mouth. When birds eat, they grasp food between their lower mandible and palate and throw the food backward in a swallowing action. The taste buds are distributed along this path—unlike humans, who have no taste receptors in the throat.

Many birds can taste and reject food laced with unpleasant flavors. It has been shown that when various chemicals are added to water, chickens can discriminate these based on taste (Duncan 1960). Tests with mallards show that they can discriminate between normal peas and those treated with an unpleasant flavor, which they reject. Blue jays learn to reject monarch butterflies based on taste, since the monarchs are toxic to them (Brower 1969).

Hannah Rowland *et al.* (2015) summarize some recent research on the aspects of taste in birds. Fruit-eating birds, orioles, hummingbirds, and other species prefer sweet substances. Hummingbirds can differentiate the types and concentrations of sugar, preferring nectar that is sweeter; they can differentiate sweetness at 1 percent sugar concentrations. Many other bird species are selective of the type of sugar they prefer. On the other hand, sugars are not a part of the diet of chickens, and they are indifferent to sweet foods. Birds can also generally differentiate sweet, bitter and savory meat-like flavor (umami). Omnivorous, meat-eating birds such as crows and ravens often exhibit low thresholds for bitter tastes, because bitterness might indicate poisons. Parrots will eat bitter compounds, but they also consume clay, which neutralizes toxins.

Taste is connected to the sense of smell, but how these two function together in birds is not understood. There does not appear to be much research on the connection of both taste and smell in birds, but we know that some birds rely heavily on both taste and smell.

Touch

Birds are sensitive to touch in a variety of different ways. Touch can play a role in finding food, communicating with other birds, and to sensing temperature. In some cases, birds have high sensitivity to touch, while others have low sensitivity to stimulus.

The bird beak is a sensitive organ. In most shorebirds, the bill tip is soft, pliable and incredibly sensitive to movement and vibrations (Birkhead 2012). This sensitivity allows shorebirds to feel the movement of prey in the ground, such as in soft mud. In ducks, the bill is sensitive to the difference between organic food and gravel and dirt, and they separate the two during feeding. When birds preen each other, they are sensitive to the effects of preening. Woodpeckers chisel wood with their sharp bills closed to protect them. Bristles around the mouth of flycatchers, nightjars, and oilbirds serve a sensory function. A bird's skin sensitivity to temperature is useful in regulating the incubation of eggs.

These are just a few examples of how birds have a sense of touch that results from nerves on their bill and body.

Sensitivity of the bill is the result of tiny pits in the beak that have touch receptors. These receptors are sensitive to pressure and movement. Examination of the long pointed beaks of four extinct birds (*lithornithids*) from the Paleocene epoch show that they had pits in their beaks (du Toit *et al.* 2020), similar to those in today's emu and ostrich. Although emu and ostrich are not known to have sensitivity to touch, these pits are suspected of having been inherited from a common ancestor.

Of birds' many different types of feathers, the filoplumes are fine hair-like feathers sparsely covering the body under their contour feathers, and are sensitive to touch and temperature. Feathers do not have nerves, but they stimulate the nerves that surround the feather attachments. These cells give impulses to adjust the contour plumage for changes in temperature, and the attachments may play a role in making adjustments to the feathers for temperature and pressure in migration.

The legs and feet of birds have very few nerves and little blood circulation. As a consequence, they have low sensitivity to hot and cold temperatures. This allows gulls, waterbirds, and other species to sit on ice and withstand cold water in winter, when many species retreat south of permanent ice during winter.

Bird Sounds

It is wonderful to experience the melodious quality of the dawn chorus of songbirds. Even those who are not preoccupied with birds can appreciate the morning chant of the robin, the ubiquitous calls of the dove, the plaintive song of the chickadee, and the loud sounds of the cardinal. Bird songs are ethereal, ephemeral, and inspiring to those who have an appreciation for the natural world.

Bird songs and calls are rich in variations that few people are aware of or even perceive while listening to birds. We rarely appreciate the innumerable, subtle changes in song even when the song is repeated over and over, or the variations from the same species within their range across the country. In addition, birds have a repertoire of songs and calls that have different meanings and uses for different occasions. Learning to identify species by songs and calls is not easy, but various smartphone apps can be used to identify species by song, most notably Merlin Bird ID, created by the Cornell Laboratory of Ornithology.

In their book *Bird Song: Biological Themes and Variations*, Clive Catchpole and Peter Slater (2008) cover in detail the subject of the complexity, variety and evolutionary changes of bird songs.

Recent work has shown that birds begin to recognize and learn songs before they are hatched. In experiments with superb fairywren eggs, scientists found that embryos' heart rate slowed in response to repeated sounds of their own species, but not to others (Colombelli-Négrel *et al.* 2020). They made measurements of five different species in the field using wild eggs. The embryos of all five species responded to their conspecific song. Overall, the strength of the response was greater in species that are vocal learners than in species that are non-learners. Nevertheless, their results suggest that unhatched birds learn to perceive the sounds of their own species.

Bird song is most highly developed in the passerine family, the largest family of birds on earth. Passerines are divided into oscines (songbirds) and suboscines (birds with less complex songs), and these two groups have differences in how they generate songs; we will go into more detail about this in Chapter 12. The oscines are usually referred to as songbirds, and they learn their songs and add song variations to their repertoire throughout their lives. Hummingbirds also acquire their song by learning, and parrots learn to repeat sounds. But the majority of bird species—sub-oscine passerines, raptors, herons, rails, waterfowl, gamebirds, seabirds, and many other species—make vocal sounds that are not learned, but are inherited genetically. They make a variety of calls that are important communication links, but do not create variations of their main theme.

Song and calls are created in the syrinx, an area where the trachea divides the air stream to the lungs. Vibrations in the air stream are induced by muscle groups surrounding the syrinx. Since there are two air streams, one from each lung, the variations in song can be complex, allowing some species such as parrots to imitate the human voice. The syrinx is located at the lower part of the trachea, which allows for sound resonance.

The syrinx has been found in *Vegavis*, a prehistoric bird of 68–66 MYBP. Computer tomography (CT) scans of a well-preserved fossil specimen suggested that it had two sources of sound in the neck with a large resonating structure (Clark *et al.* 2016).

Species of birds that are closely related to each other will share the same basic song structure and repertoire of songs and calls, but the actual songs and calls will differ for each species. Their repertoire of sounds is an important aid in establishing the interrelationship of different species. The subtle differences in the songs and calls of many species are not always apparent and are best studied for scientific purposes with the use of sonograms.

Bird songs vary in frequency (pitch), starting as low as 30 Hz in the cassowary and about 100 Hz in the drumming of the ruffed grouse. Most songbirds normally sing in the range from 1 to 8 kHz, and their songs have a range of frequencies. The frequency at which birds sing is related to the size of a bird—for example, smaller birds cannot produce lower frequencies, so their songs tend to be higher pitched.

Lower frequency songs travel further than do those of high frequency, because they transfer less energy to the media (air). Many ground-dwelling forest birds use lower frequency songs. Foliage surrounding the bird also reduces the distance the sound travels. As a result, songbirds prefer to sing from the highest perches, where their high-frequency songs can be heard for greater distances. Songbirds can also adjust the frequency of their song over a limited range. Various studies have shown that urban noise causes local birds to change their frequency and bandwidth.

Birds sing different songs to defend a territory, to let other birds know they are there, and to get to know the songs of their neighbors. Some birds use songs as part of their flight displays. It is also thought that birds sing for pleasure. In general, the primary song of the male is used to attract a female and establish a territory. The degree and frequency of songs is dependent on sex hormones.

Some birds will also create variations of their primary song. These variations can result in local dialects, which can vary geographically. Donald Kroodsma (2005) describes the local variations and geographic dialects of five species of American songbirds.

In addition to songs, some species have a repertory of calls. The distinction between a song and a call is not always clear cut, but calls tend to be simpler vocalizations—short sound segments for communication purposes: male-female communication, calls to family groups, and warnings. Calls serve the need for breeding birds to identify each other, and for the immatures to identify their parents. Immatures also use calls to beg for food.

Many species of birds also give flight calls, which are distinct from other types of calls. These calls are mostly heard during nocturnal migration and tend to be of high frequency, giving them a short range. Such calls could be used for coordination and contacting between migrants, and might be a response to atmospheric conditions (Larkin 1982). Birds may produce these calls in other contexts, including while perched and while interacting with fledged young (Farnsworth 2005).

Many birds use warning and threatening calls. The "seep" call is common to many species of small birds and may be used to warn of a predator, switching to "all clear" calls when the threat has passed. Mobbing calls bring attention to predatory birds, signaling to other birds to try to harass the intruder until it leaves the area. Some species have different calls for various social contacts, such as an approaching predator or other aggressive encounters.

Most birds sing during the day. In the breeding season, birds start singing just before sunrise, and other nearby species join in, forming the dawn chorus. Male passerines sing their primary song hundreds of times a day in courtship, and continue through the breeding season. Female birds are very sensitive to very small nuances in song, something so subtle that they are not audible to the human ear and can only be detected with a song spectrum (songram). With very few exceptions, females respond to the males of the same species.

Singing decreases as the day wears on, but some species sing all day long. Later in the breeding season, song becomes less frequent; on wintering grounds, birds will sing very little, but will make their presence known with calls or muted or partial songs.

Some birds sing exclusively at night: most owls, nightjars (whip-poor-wills, nighthawks, and others), and some rails. Passerines including northern mockingbird, hermit thrush, American and European robin, common nightingale, and other thrushes will sing at night. Among non-passerines, herons, shorebirds, and yellow-billed cuckoo will give calls at night.

Throughout the world, songs within certain families of birds are similar to each other. Just a few examples include the warbling-antbirds of Peru, the Empidonax flycatchers, and black-capped and Carolina chickadee. This phenomenon is most common in closely related species occupying the same general region.

In most species of birds, females and males use songs and calls. Benedict and Odom (2017) discuss some of the research on female songs among oscine passerine birds. They show that female song is underrepresented in biological collections compared to male song for most songbird species, because female sounds had not been studied until more recently. In many species in which the male and female have the same plumage (monomorphic), the only way to determine if the male or female is singing is to observe banded birds. Research has shown that in the United States and Canada, more than fifty percent of female passerines are known to sing. Researchers have also found that females of some species have a high diversity of songs and sing frequently, while others only sing occasionally. Female song is more common in tropical and subtropical regions, and is common in species with year-round territories. Females sing to defend a territory or their mate, to communicate with their mate, and to interact with their chicks (Riebel *et al.* 2019).

Males and females of many bird species sing duets. Antiphonal song occurs when the male and female sing alternately, a process practiced by at least four percent of the world's bird species. In North America, species such as red-winged blackbird, whip-poor-will, Canada goose, Virginia rail, Gambel's quail, common grackle, and yellow-throated vireo are known to practice antiphonal singing (Doyle 2018). Tropical wrens, highly secretive birds that live in dense tropical forests, sing duets to find and keep in contact with each other, according to research. In Australia, male and female fairywrens and magpie-larks carry out this practice. It has also been seen in whole families of species in Africa (Thorpe 1973).

There are quite a few species of birds that can accurately copy the song of other birds. Northern mockingbirds can imitate birds and unnatural sounds, such as operating machinery. In North America, blue jay, thrashers, gray catbird and yellow-breasted chat are just a few of the species that imitate the songs of other birds. In Europe, the European starling and sedge and marsh warblers frequently copy other birds. Australian lyrebirds may be the species most well-known for their skill at copying songs of other birds.

Different subspecies that have lived in isolation for long periods of time might find song differences to be a barrier to breeding. This form of separation is called behavioral isolation, manifested as subtle differences in song within a species. There are other forms of behavioral separation, but song is one of the most important and most common. Thus, song is a strong influence on the selection of mates, and it plays an important role in the separation of species (Lovette 2016).

We are most familiar with bird songs and calls from our everyday exposure to birds, especially in the spring. However, there are many species of birds that make sounds for communication purposes by means other than singing. Some species make sounds with their wings while performing display flight routines. Hummingbirds perform looping flights where they combine vocal sounds and feather vibrations in their display. Woodcocks make the most bizarre sounds with their wings in their annual spring display ritual. Snipe produce the "winnowing" sound with their outer tail feathers, and ruffed grouse use beating wing sounds for mate attraction. Manakins use their wings to make snapping sounds as part of their lek displays. This is most highly developed in the club-winged manakin: their sixth primary feather has ridges on the central vane, and the fifth primary has a crooked blade-like tip. They use these primaries to produce sharp raspy sounds (Prum 2017). Grouse and bitterns use their esophagus to make sounds by expelling air, thus creating low frequency booming sounds. The drumming of woodpeckers signal their claim to territory, and the bill snapping of birds like wood stork communicates annoyance or aggression.

Sleep

Sleep allows the brain to recover, preparing the brain for the next day. We know that inadequate sleep causes people to have difficulty with decisions, emotions, and coping with problems; it is also linked to depression and risk-taking.

Humans are not cognizant of their surroundings during sleep. Birds and aquatic mammals, however, function by having one side of their brain asleep at a time so that they can carry on with needed functions. This unihemispheric sleep allows birds to have one hemisphere of their brain go into deep sleep, while the other hemisphere remains awake and alert.

People cannot achieve unihemispheric sleep, so we find it difficult to sleep in a new sleeping location because part of the human brain tries to remain awake until place familiarity is established. This is caused by an imbalance of the short-wave activity on the right and left side of the brain, which will continue until familiarity is reached (Koch 2016). Unihemispheric sleep allows other mammals to get adequate rest, however: dolphins and whales use this slow-wave sleep to control their need to swim and breathe even as they sleep. Researchers have found that crocodiles sleep with one eye open and connected to the awake side of the brain (Kelly *et al.* 2015). This same technique is found in birds as well: roosting birds maintain vigilance as they rest, with one half of the brain active and one eye open as a protective mechanism.

In South America, snakes are the most common predators of birds. Different species of snakes are active during day or night. They attack bird nests for eggs, and are known to target immature and adult birds. Unihemispheric sleep is an important part of the birds' defense. Birds can also fly with only one-half of their brain active. Information from data loggers shows that albatrosses, frigatebirds and swifts fly for weeks on end without landing, a feat made possible by their ability to rest and sleep while still in flight.

The way sleep is regulated in birds is similar to that in mammals. Research on pigeons shows that, like humans, they have two forms of sleep: slow-wave and deep sleep, allowing them to recover from insufficient sleep with short periods of deep sleep (Martinez-Gonzalez *et al.* 2008). Rattenborg *et al.* (2017) have shown that frigatebirds sleep only about 40 minutes a day while on a sustained flight of ten days, taking very brief naps during which the full brain rests up to twelve seconds. When back on land, the frigatebirds will sleep for longer periods. These naps have been recorded in ducks, common swift, and migrating thrush as well. Pectoral sandpipers remain awake and active in the breeding season to maximize the number of possible mates. They maintain high neurobehavioral performance despite greatly reducing their time spent sleeping during a three week period of intense male-male competition for access to fertile females (Lesku *et al.* 2012).

Birds sleep in their nests, in tree cavities and also in communal roosts. Ducks, gulls, seabirds, and waterbirds roost on bodies of water, usually in rafts or flocks. Ducks on the outside of the raft remain vigilant in unihemispheric sleep while those in the center are in deep sleep, and they rotate these positions (Rattenborg *et al.* 1999). Geese and ducks will also roost on protected islands.

Most herons roost in trees or bushes. Shorebirds roost on beaches or on the ground, relying on their camouflaged plumages as protection against predators. Grouse, quails, pheasants, turkeys, and nighthawks all roost either on the ground with cover or in trees, depending on the species. Most passerines roost in dense vegetation near their nest site. Some birds, like bluebirds, chickadees, and woodpeckers, roost in tree cavities. Some songbirds, such as crows and swifts, roost in communal gatherings for mutual protection and warmth. Most owls and hawks roost in hidden locations in trees, and a few species like harriers, snowy owl, and short-eared owl roost on the ground.

Many species of birds, especially waterbirds, sleep with their head pointed backwards and with the head resting on their back, or tucked under their shoulder feathers. In this position, birds can rest their neck muscles and minimize their exposed surface area to conserve heat. It is also possible that in this position, they can reclaim some of the warmth from their breath by exhaling into the annular space under their contour feathers—but this has never been confirmed.

5

Breeding

As with humans and all living creatures, breeding and raising the next generation are the single most important events in life. The onset of breeding for birds is brought about by one or more hormones, triggered by environmental conditions that include the lengthening of the days in spring, warmer temperatures, and food availability. If the timing of initiation of breeding is delayed within the normal breeding cycle, breeding success declines (Verhulst and Nilsson 2008). So it is important for birds to be in the best physical condition when initiating breeding.

The breeding season usually begins when a male establishes a territory and goes through the ritual of attracting a female. For most bird species, males compete and females chooses their mate. Male competition can involve song, ornate plumage, nest construction, preening, feeding, and displays that can be elaborate and complex. In a few species, females compete by virtue of the quality of their plumage and displays, and males select their mates based on their perception of the quality of the female.

Females can display and use song to defend a territory from other females. This practice is well known for jacanas and phalaropes; female jacanas will defend a territory. In rock sparrows, males preferentially mate with more ornamented females (Griggio *et al.* 2005). The work of Caitlin and Servedio (2017) shows that female display is more likely in larger monogamous species.

Most birds breed annually—except the large albatross, which breeds every other year. The timing and conditions when breeding commences vary considerably. We in the northern hemisphere are used to birds breeding during spring and summer, but some species—great horned owl in particular—commence breeding in winter. In the tropics, where there are fairly constant temperatures, timing for breeding varies considerably.

While it takes many years for humans to mature and breed, songbirds, flycatchers, ducks, and others breed within one year of life; they initiate breeding in the spring after the summer they fledged. The house sparrow, one of the most populous species on the planet, can initiate breeding after the age of nine months—thus before other birds, which gives it an advantage in seeking nest sites. Some small birds will not breed until their second year of life because they are still maturing, and it takes first-year males time to find and establish a territory. Some breeding routines are complex, so the birds need to learn how to attract a mate and provide for the hatchlings.

For some species, breeding is delayed for additional time after sexual maturity. Red-tailed hawks mature in their second year, and a few will try to breed at that time. Canada geese will breed in their third year. Jaegers reach sexual maturity in their third or fourth year and breed sometime after that. Bald eagles begin to breed in their fourth or fifth year. Golden eagles reach sexual maturity in their fifth year, so breeding attempts begin then. Large gulls take four years to reach sexual maturity. During this period of maturing, these large birds make the annual migratory trip to the breeding grounds, usually after adults. This time is a period of learning that includes finding a mate and establishing a territory.

Seabirds delay breeding for even longer periods after sexual maturity. These birds spend their early years exploring the vast oceans and developing detailed knowledge of food sources. They do not wander randomly in their long ocean voyages—they become familiar with the ocean in the same way that tropical migrants become familiar with their breeding grounds, wintering grounds and the migratory path between them.

For example, Atlantic puffins reach sexual maturity in three to four years, but do not breed until five to eight years after fledging. Cory's shearwaters reach sexual maturity in their fifth year, but do not breed until about their ninth season (Nicolson 2019). Albatrosses also reach sexual maturity in their fifth year, but do not breed until they are 7 to 15 years old.

Among larger birds, some form pair bonds and mate for life: golden and bald eagles, black vulture, most albatrosses and shearwaters, whooping crane, fulmars, and some swans. Others will mate for a period of years, then change partners. Between 7 and 13 percent of Atlantic puffins change partners annually; for guillemots, the change rate averages about every five years (Nicolson 2019). Research has shown that 11 percent of black turnstones change partners annually (Handel and Gill, Jr 2000). Birds that are concentrated in one area during the nonbreeding season will likely find new partners, but those that return to a specific breeding site are more likely to stay together for a number of years. Jeffery Black (1996) reviewed the subject of pair bonds and partnerships in birds and showed that there is a wide distribution of the frequency of changing partners in birds, from zero to 100 percent depending on the species—and this can be variable within a particular species. Possible causes for dissolving the partnerships involve a better mate to improve reproductive success, or the choice of a better territory. Raising chicks may require both parents to provide food and to cooperate in providing protection, so some birds sustain a pair relationship; they appear to know that experienced adults are better at dealing with these requirements.

Mate selection, referred to as the mating system in birds, is complex and varied, and at least seven different mating systems are known (Johnson and Burley 1997). In almost all small birds, new mates are selected annually. Most small birds are monogamous on an annual basis, but certainly not all. Among all birds, various authors estimate that 80 percent to 90 percent are socially monogamous, meaning that they remain together for the breeding season; but more current research shows that most small birds indulge in extra pair sex. A few species that are known to be polygamous include saltmarsh sparrow, house wren, red-winged blackbird, wild turkey, and bobolink. If both parents participate in raising the chick and defending a territory, they are more likely to be monogamous.

Birds commonly engage in extra pair paternity: the practices of polygamy (multiple partners) and polyandry (females having more than one mate). Males of species that use a lek to choose a mate—greater and lesser prairie chickens, for example— have no social connection to their mate and offspring, and move on to another mate after conception, taking no part in supporting the female or in raising their own chicks. Jacanas and phalaropes practice polyandry, the female laying her eggs and then leaving them to find another partner to repeat the process. The male jacana or phalarope cares for and guards the chicks until maturity.

Some male passerines practice polygamy, and some females practice polyandry. Fourteen of the 291 species of North American passerine birds are regularly polygynous—with the male taking several mates in a season—or promiscuous, with the female taking several mates in succession. Thirteen of these 14 species breed in marshes, prairies, or savannah (Verner and Willson 1966), perhaps because food availability between two males' territories may be great enough to permit a female to rear more young on the better territory. Another mating strategy, found in only a few species (notably Smith's longspur), is polygynandry, where both males and females mate with multiple individuals. In polygynandrous systems, broods may be raised by groups of males, either with or without the help of females.

It has only come to light recently that Adélie penguins engage in homosexual mating, as well as a wide variety of other unusual sexual behaviors (Russell *et al.* 2012). These behaviors were originally observed in 1911 by Murray Levick while studying penguins, but he never published his research notes, so they were not discovered until recently. More recent work has confirmed his findings (Davis 2019). Homosexual necrophilia has also been reported in mallard ducks (Moeliker 2001).

Nesting Practices

Insects and some species of mammals build nests. However, we are very familiar with the fact that nests are most often associated with birds, and almost all bird species build nests. This behavior originated very early in birds' evolution, even before the first birdlike dinosaurs emerged.

Research has shown that about 30 species of larger dinosaurs buried their eggs in nests covered with dirt and vegetation, much like crocodiles and alligators. This was a primitive condition for dinosaurs, but as small, non-avian theropod dinosaurs developed, they laid their eggs on open ground nests. Paleontologists believe this practice was likely widespread in non-avian maniraptora, well before the appearance of birds as an order (Tanaka *et al.* 2015); in fact, fossil evidence for ground nests for maniraptora have also been described (Varricchio *et al.* 1997, *et al.* 1995). Evidence of colonial ground nesting of dinosaurs has been discovered in Montana and as far away as Mongolia. Nest building also may have evolved from theropods placing eggs in natural cavities. The first fossil evidence for a breeding colony of Mesozoic birds from the late Cretaceous was found in Transylvania, Romania (Dykes *et al.* 2012).

The sole purpose of the nest is to provide a safe place for birds to raise their young. Nests are usually partially hidden, and some are extremely well hidden. Some species nest on the ground with little to identify the nest site. Swallows build nests from mud and saliva, some use burrows in the earth, and many species use tree cavities. The vast majority of birds build their nests in trees and shrubs, primarily to avoid ground predators.

In contrast, colonial nesting birds place nests in plain sight, especially those of the larger seabirds. Nests come in all levels of complexity, from simple depressions in sand to complex communal structures

The development of tree nests and the skills to build them is one of the triumphs of the order of birds, and is one of the reasons why they have been so successful for such a long period of the earth's history. Building a tree nest is a complex task that requires some learning. The location for the nest, obtaining the materials, and the process of assembly are challenging tasks. The nest has to be strong enough to withstand wind and rain and be constructed to provide some level of insulation for the eggs. Inexperienced birds will build a nest, and if it is unsuccessful, they will try again.

Very few experimental studies have been carried out to understand the nest building process. One study confirmed that the process of weaving grass strands into a nest by the village weaver depended substantially on gaining experience (Collias and Collias 1962). While some learning takes place as birds gain experience in nest building, birds seem predisposed to build nests.

Some species of modern birds build very simple ground nests, some just lay their eggs in a scraped area, and others use an area on the ground that they line or surround with pebbles. This is especially common among the more primitive species, like the flightless birds, the tinamous, game birds, shorebirds, ducks and geese, and also in some passerines, including some sparrows and larks. Gulls, terns, gannets, and albatross nest on the ground in colonies, relying on cooperation for defense, while larger birds like ostriches, bustards, swans, and others must defend their ground nests on their own.

For most ground nesters, defending the nest against predators must by a primary focus—so they have developed a wide variety of structures to assist with this. Smaller ground nesting birds build nests that are complex and hidden in the grass or foliage. The ovenbird of North America and some of the ovenbirds (*Furnariidae*) of South America build a wide variety of enclosed, oven-like nests—earning them their species' name. Seabirds such as shearwaters, petrels, and alcids build nests in burrows. Burrowing owls build their nests in the ground by digging a new hole or by remodeling a ground squirrel nest.

The most common type of nests are those constructed in trees or bushes. Larger birds such as hawks, eagles, and herons build platform stick nests. Passerines weave their nests on tree branches from grasses, twigs, and other material, making many different types including cup nests, hanging nests, and covered nests. Building these nests normally takes small birds several days to a week, depending on the species and other factors. Some complex hanging nests can take longer, and birds might work on them only a few hours a day.

Owls, woodpeckers, a few duck species, and some passerines build nests in tree cavities. Woodpeckers chisel out their own nest cavities, but small owls, ducks, and passerines use natural cavities or old woodpecker holes. For instance, wood ducks prefer cavities excavated by pileated woodpeckers (Yetter *et al.* 1999) and elf owls use woodpecker holes in saguaro cacti.

Raising the Next Generation

All female birds, except for some female hawks, have one ovary and produce one egg at a time. They cannot afford to be carrying the weight of multiple eggs, because they would not be able to fly. Normally birds lay one egg per day, but that is dependent on the bird species, weather conditions, and other factors. Rapid laying was seen with three prairie warbler females that laid four eggs in three days (Nolan Jr., 1978). The number of eggs per clutch varies, ranging from a high of about a dozen for ducks, possibly more for game birds and other species, to only one for pelagic birds. The number is strongly dependent on the ability of the female to obtain sufficient food for their offspring. If an egg is lost, an adult female might lay a replacement.

Altricial birds including hawks, owls, seabirds, herons, passerines, and others are helpless as hatchlings and rely on the adults for support. Some are naked when they hatch (passerines), and others are covered with down (hawks, owls and herons). Parental care in these species requires providing food and protection against the elements and predators. After fledging (leaving the nest), many immatures remain with the adults for some time as they learn how to support themselves.

Precocial birds such as ducks, shorebirds, and game birds are active after hatching and remain with the adults, but they can feed themselves. Many of these species have large clutch sizes; some ducks and quail can have over a dozen eggs. Shorebirds, however, generally have a clutch of four eggs, and they will reduce clutch size in a year of bad weather or insufficient food. Research shows that parental care of precocial chicks imposes considerable demand on adults of some species (Walters 1984). Since the adults do not provide food, it is assumed that the chicks are self-reliant— but both adults incubate the eggs, and when the chicks hatch, they provide protection. Some species lead their chicks to food sources, while in other species, the fledglings feed while following the adults.

Laying more eggs than can be supported is insurance in case the earliest one or two hatchlings fail. In most cases, the adults cannot supply enough food for all the hatchlings. This is why infanticide, a mechanism of brood reduction, has occurred among 15 of the 28 orders of birds (Sweeney 2008). It has been speculated that infanticide contributes to the fitness and the long term survival of a species, but that has never been verified with data.

A parent might kill its own chick, or an adult bird may kill young from other birds' nest; competition for resources is one possible cause of this practice. Starvation, infanticide by the older siblings, killing by the parents, eviction from the nest, or nest abandonment are all methods for controlling the number of nestlings. Of these, starvation is not uncommon, and has been documented in seabirds that cannot find enough food close enough to their breeding sites to support their one chick.

A few species of large birds lay two or three eggs, but raise only one or maybe two chicks. Examples are some eagles, osprey, kittiwakes, pelicans, some herons, and some boobies. Normally the eggs are laid several days to a week apart, and egg hatching is delayed by the same time gap. This gives the earlier hatchlings more time to grow, and the first one can be quite a bit larger by the time its sibling appears. In many cases, the older hatchling will often kill the younger chick.

Infanticide is common among some communally nesting birds as well, such as large gulls and boobies. The chick will be safe among communal nesting birds if it remains in its own nest and is protected by an adult, unless it is killed by its sibling. Some species of gulls that nest communally, such as Franklin's gull, are quite different in that they will care for orphaned, lost, or abandoned chicks.

In the case of the Nazca booby, the older hatching usually kills the younger bird. Nazca boobies take a few years to mature. The immature males will roost within the breeding colony. If the adult boobies leave their nest unguarded—a common practice—these roaming immature males routinely kill unprotected hatchlings that are too small to defend themselves. This predatory practice is a form of social maltreatment not prompted by the need for food; nor is it known to occur with any other avian species. Research suggests that it is caused by abnormal levels of stress hormones (Grace *et al.* 2011).

Infanticide is only one cause of nestling mortality. The highest overall mortality rate for altricial birds is during the nestling and fledgling periods; it is quite variable, ranging from 40 to 80 percent—and it has been recorded at 100 percent for some seabirds that lack adequate food. The highest mortality rates in the nestling phase are in the first week of fledging (Naef-Daenzer and Gruebler, 2016). If the fledgling survives its first week, the probability of survival improves.

Fledglings do not know enough about surviving. Their inexperience with flight and finding food, diseases caused by parasites, and predators are all factors that contribute to the low survival rate, but predation is the major cause. They rely on hidden sites to avoid most predators, and on frequent adult visits to bring them food or to lead them to food, as well as to develop social interactions. Altricial birds have relatively fast breeding cycles and can raise multiple broods in one season, which helps to sustain the overall population.

Precocial birds face similar challenges with predation, the biggest factor in fledgling mortality. Their overall survival rate ranges from 12 to 18 percent (McGowan *et al.* 2009).

6

Flight and Migration

Bird flight has fascinated people for thousands of years. Leonardo da Vinci wrote a codex on the flight of birds that included more than 500 drawings that showed that he recognized many of the important mechanisms involved in bird flight. The Wright brothers used large, soaring birds as their model for wing design and flight control. Other early pioneers used birds as models for their work on flight.

The mystery of migration is also complex, and has taken many years for researchers to understand the numerous factors involved.

As we discussed earlier in this book, pterosaurs—long extinct flying reptiles— were the first large flying animals. They had a long tenure on earth, evolving a range of sizes from small to very large; some were capable of aerial capture. Pterosaurs had a powerful wing stroke, indicated by their ability to take off from the ground.

Today, bats and birds have the capacity for powered flight—the ability to take off and propel themselves forward—and each has different wing designs, techniques, and capabilities. In both cases (as well as in pterosaurs), the arm and hand provide the power for flight, but each animal uses this in different ways. Bats are the only flying mammal that uses powered flight, using muscles rather than air currents to generate lift and remain airborne. In contrast, powered flight is highly developed in birds, generated by the wing stroke. Birds have a more efficient stroke than do bats, but bats are more maneuverable.

The ability to fly provides a number of significant advantages. It allows for mobility, for rapid escape from predators, and, most important, for access to a much wider variety of large and small habitats that would be inaccessible on foot. This means that bird habitats can include mountains, islands, deserts, aquatic habitats like marshes, tidal estuaries, coastal areas, oceans and lakes, and even specific trees and plants within a forest. Mobility allows birds to follow the seasonal transition of food abundance, many species migrating long distances to areas that provided a rich food supply.

Flight is a gradual adaptation that took many millions of years to accomplish for birds, but the most important enabling changes from dinosaurs to flying birds took place before the K–Pg extinction event. These included the development of feathers, a more efficient breathing system, shoulder joints that allowed wing rotation, and the wishbone, which acts like a spring to store energy, to channel tendons during wing movement, and to strengthen the chest. We noted in Chapter 4 that important skeletal changes in prehistoric birds included shortening the long bony tail, fusing the bones of the fingers and eliminating unnecessary bones, adapting the foot to walk on toes and assist in landing and grasping branches, developing the sternum to anchor breast muscles, replacing the bony jaw and teeth with a light weight keratin bill, and reducing the weight of the skeleton. In addition, other internal changes occurred as birds evolved, driven by the adaptation to fly: weight and complexity of internal organs were reduced, and the pectoral muscles increased in size to power the wing stroke. The distribution of mass and muscles in the body were optimized for the center of gravity, which lies between the wings.

The reproductive organs have also been adapted for flight. Female birds have only one ovary; sex organs are reduced in size for most of the year when not needed, and birds have no urinary bladder. The reproductive organ for almost all birds is internal. For the male, sperm is discharged from the cloaca, the same opening that expels urine and digestive waste. The female accepts the sperm through her cloaca and stores it in a vaginal pouch.

About 3 percent of bird species have a penis. These include ducks, geese, swans and tinamous, and large flightless birds (ratites): cassowary, emu, and ostrich. Prum (2017) discusses this issue and the nature of the avian mating system for those birds that have a penis. The penis of birds was inherited from their dinosaur ancestors, but the exact reason why most male birds do not possess a penis is not well understood. It is possible that the loss of the penis is an adaptation for flight, as virtually all modern birds (*Neoaves*) lack this organ. Researchers at the University of Florida discovered that, in the cases they studied, birds actually do have a penis in the embryonic state, but a gene causes these cells to die off (Herrera *et al.* 2013).

Elements of Bird Flight

Scientists and engineers have a solid understanding of the physics of bird flight. In 1505, in his *Codex on the Flight of Birds,* Leonardo da Vinci showed that he understood the concept of forces acting on a flying bird. In 1738, Bernoulli developed his famous equation for the conservation of energy of a fluid, relating static and dynamic pressure—the equation that relates the lift of a wing to the velocity of air above and below the wing. In the early twentieth century, the Wright brothers spent a great deal of time observing birds in flight. They noticed that birds soared into the wind, and that the air flowing over the curved surface of their wings created lift. They knew that birds change the shape of their wings to turn and maneuver, and used this idea to bend the wings of their flier to allow the plane to bank (roll).

Bird flight involves two different processes, powered flight and gliding flight. In powered flight, the breast muscles attached to the sternum provide the power to move the wing up and down. The primary feathers—those long feathers at the end of the wing—force air down and backwards, thus producing thrust. The wing angles upward with respect to the horizon, a position called the *angle of attack*. The downward thrust of the primaries gives an increase in velocity directed upward. This power generated by the wing motion acts against two forces: gravity pulling downward, and drag pushing against the bird's forward motion. The bird uses its wings to gain lift, and it can use its alula, one or several short feathers attached to the thumb of the hand, to increase the angle of attack. On the upstroke, the primaries are bent to reduce drag. The secondaries—the inner wing feathers—provide lift in response to the increase in velocity caused by the action of the power stroke, increasing lift as it increases velocity. As the bird accelerates from a starting takeoff to steady flight, the power stroke becomes shallower, because less thrust is needed to maintain steady velocity.

Gliding flight relies solely on lift generated over the entire wing by the pressure differences on the upper and lower surfaces of the wing. In gliding, birds rely on lift to remain aloft, but lose altitude unless rising air currents provide sufficient upward force to counter the drag and gravity forces. Soaring birds remain aloft for long distances by gliding from one cell of rising air to another.

The subject of how the earliest feathered birds took to powered flight has been a source of much debate and study. There are two hypotheses: in one, the earliest birds flapped from the ground upward; in the other, they climbed up trees and glided downward.

The first hypothesis holds that early birds had strong enough legs to run and launch themselves into the air and glide, initiating powered flight. In the second hypothesis, early birds are thought of as tree climbers that took advantage of height to glide downward, thus evolving to powered flight. Analysis indicates that the early birds with claws at the hand joint could climb trees (Ostrom 1979).

Kenneth Dial (2003a) proposed a refinement to the ground-up idea by observing that immature gallinaceous (poultry and game) birds use a process of wing-assisted running up an incline to gain height. He has shown that many species of immature birds incapable of flight can ascend even vertical surfaces by using their legs and wings in concert. Flapping the wings is an important part of this process, and provides the connection to a process of flight development. We also know that when birds take off from the ground, they must use their legs as an assist to gain upward momentum, because just flapping will not get them airborne.

Each of these theories has its plusses and minuses. It might be some time before a more definitive answer to the question of the origins of bird flight is accepted by the scientific community.

Flying birds come in a wide variety of weights, from the heaviest great bustard at 46 pounds (21 kg) to the lightest bee hummingbird at 0.07 oz (2 gm). Wing size and shape vary with the type and weight of the bird species. Greater body mass requires larger wing areas to keep the wing loading—the mass of the bird divided by the wing area—within a certain range. The wings of heavier birds account for a greater percentage of their total weight. Birds with low wing loading include gulls, hawks and herons; those with high wing loading include alcids, hummingbirds, game birds and ducks.

Wing loading is an important metric in understanding flight, but wing shape also plays an important role. Wing shape is characterized by three important properties:

- Aspect ratio, the length of the wing divided by the width
- Camber, the curvature in the middle of the wing
- Type, the shape of the wing

These properties are specific to how the wing is used, summarized in Table 6.1. The attributes in this table are general characteristics; in some species, the attributes are more complicated than wing type.

Table 6.1 Wing Type Adapted for Different Flight Styles and Different Types of Birds

Wing Type	Properties	Attributes	Type of Birds
Short & rounded (elliptic)	Low aspect ratio, dependent on species	Rapid acceleration, more maneuverability	Songbirds, doves, ground and game birds
Long and pointed	Moderate aspect ratio, low camber	High speed	Falcons, sandpipers, swallows, terns
Very long and narrow	High aspect ratio	Soaring on localized surface winds, slower speeds	Ocean soaring birds, albatrosses and shearwaters, nightjars, alcids
Long and broad	Moderate aspect ratio, high camber	High lift, soaring on thermals and updrafts	Buteo hawks, eagles, vultures, storks, pelicans

Maneuverability, for example, is an example of an attribute that is more complicated than wing shape alone. Short-winged flycatchers are very maneuverable, while birds like swallows, swifts, and terns are long-winged species that have high maneuverability achieved by different means. These birds have low wing loading with high aspect ratio, which allows a short turning radius. Swallows can achieve high acceleration to gain maneuverability in aerial capture (Warrick 1998). Swifts also rely on speed and changes in their wing shape to make aerial captures. Small terns have low wing loading and use their long tails to make rapid turns, thus gaining maneuverability.

The different wing shapes, body size, and the distribution and development of the birds' muscles are variables that lead to radically different flight styles. Dial (2003b) proposed a model of flight styles based on these variables. Birds that are precocial tend to have higher muscle development in their legs and less in their flight muscles. "Super-precocial" flightless birds like the ostrich have very little wing muscle mass and very heavy leg muscles. Larger precocial birds, like ground-nesting game birds and the duck family, have a high muscular leg development and can fly, but they have low maneuverability. Smaller precocial birds like shorebirds have an equal level of development of flight and leg muscles, and show a higher level of maneuverability—as do some altricial birds, like doves. At the other extreme, small altricial birds including flycatchers, swifts and hummingbirds have highly developed flight muscles and less well-developed leg muscles. Hummingbirds, swifts, and swallows have such weak legs that they sacrifice walking ability.

As wing muscle mass increases proportionately, birds gain maneuverability, which is also related to their wing shape. An important point of Dial's model is the fact that the earliest known birds from the Cretaceous period were precocial birds, and they had well-developed leg muscles for rapid locomotion. This supports the belief that birds first took flight by running to gain lift and to glide.

When a bird is in slow flight, landing, or changing direction in flight, it decreases its speed. Under those conditions, it must avoid going into a stall. One method of avoiding a stall is by elevating the alula, creating a space between it and the wing, which allows these animals to fly at larger angles of attack and lower speeds. The alula serves the same purpose as leading edge slats in aircraft wings (Álvarez *et al.* 2001).

Large soaring birds spread the wing tips of their primaries, creating slots between the feathers. These slots delay stall, reduce overall drag, and increase lift (Wissa *et al.* 2015). Slotting the primaries is also useful in low speed flight to aid maneuvering. Slotted tips allow short wings to have the same induced drag as longer wings without tip slots (Tucker 1993). With inadequate winds for take-off, slotted primaries can provide lift for species with short wings. Figures 6.1a and 6.1b show the slots in a larger and smaller bird.

Studies of a large captive eagle show how soaring birds deal with wind turbulence. The birds collapse their wings in strong gusts, rather than holding them outstretched. This wing-tuck is a direct response to a loss of lift that occurs when a bird flies through a pocket of atmospheric turbulence (Reynolds *et al.* 2014). Mechanical engineers used high speed photography to study small birds in flight in a wind tunnel (Deetjen *et al.* 2017). They observed that while flying, the birds are able to control their flight by changing the shape of the wings, wing twist, and position of the tail on each wingbeat to adjust the angle of attack to compensate for lift and drag. This capability is important in maintaining flight during wind turbulence.

Pelicans, storks, ibises, cormorants and geese fly in a formation in order to draft on the lead birds, synchronizing their wing beats to take advantage of reduced turbulence, which reduces drag. The birds within the formation benefit from drag reduction caused by the vortex coming off the wing of the bird ahead. The energy savings has been estimated to be 20 to 30 percent (Waldron 2014). It has not been

Figure 6.1a Elegant Trogon on take-off showing the slots between flight feathers (Kathy Dashiell).

Figure 6.1b Immature Bald Eagle in flight showing the slots between primary feathers; slots reduce drag and increase lift (author's collection).

possible to estimate or measure if there is any advantage to small birds that fly in large flocks, such as pigeons, starlings, and blackbirds.

The alcids of North America, penguins, and diving petrels of South America are the most adapted to using their wings for propulsion under water. Storer (1960) first proposed that the wing design of birds that swim is a tradeoff between the flight ability in air and the use of wings in water for diving and swimming. For flight in air, the wing area must be great enough to support the bird, which depends on the wing loading—the weight divided by wing area. The alcids have high wing loading and moderate aspect ratios, but they can still fly and swim under water. Common murre has one of the highest wing loading of any bird, and represents the limit of the transition between fly-ability and flightlessness (Thaxter *et al.* 2010). Penguins and the now-extinct great auk have wings reduced in size and shape that function like seal flippers, making them useful for diving and swimming under water. The wings of flightless birds are flat and tapered with a blunt tip, and lack feathers contoured for flight.

The speed of birds in flight differs for takeoff, acceleration, and steady cruising flight. Takeoff speeds are, of course, slower, and larger birds need to get a running start for liftoff, which requires a speed of about 10 mph (18 km/hr). Small birds and those with short wings for acceleration can leap into the air and take off with zero initial speed.

Steady flight has been studied by many researchers using radar measurements of speed, correcting the data for any contribution from wind. Alerstam *et al.* (2007) measured the cruising speed of 138 species of birds, from very large to small, and found that they range from 18 mph (29 km/hr) to 60 mph (97 km/hr). Under steady flight, speed is not strongly dependent on the wing loading, shape, and type. Over land and with normal flight, birds range in speed from 20 to 30 mph (32 to 48 km/hr) (Ehrlich *et al.* 1988), and fly at low altitude, usually less than 500 ft (150 m). Ducks flying over water routinely reach speeds of 40 to 60 mph (64 to 97 km/hr). Predatory birds such as accipiters, jaegers, and falcons will accelerate to higher speeds when chasing avian prey. The peregrine falcon, the fastest living organism on the planet, is well known for its dive (stoop), where it can reach speeds of close to 200 mph (320 km/hr).

Flight speeds also depend on wind speed and direction with respect to the bird. Birds seek wind assist and can routinely increase their ground speed, which allows them to use less effort to cover more distance.

Migration—Life in the Air

Alpine swifts spend a considerable portion of their life aloft. They spend the entire six-month migration period—200 consecutive days—airborne, traveling from their breeding grounds in Switzerland to Western Africa. In one study, researchers attached data loggers to six swifts to record their daily twenty-four hour activity for a year. The data showed that they feed, drink, and sleep on the wing, feeding on aerial plankton (Liechti *et al.* 2013).In another study, Hedenstrom *et al.* (2016) found that common swifts spend ten months airborne.

Some seabirds can also spend a great deal of time aloft. Sooty terns spend four and a half to five years at sea following fledging; tracking studies of adults showed they spend only 3.7% of their time resting on the water, and never land on water during the night (Jaeger *et al.* 2017). The wandering albatross, studied with GPS locators, makes foraging trips of several weeks covering more than 9,000 miles (14,500 Km) (Weimerskirch 2004). This bird is airborne almost all of the time, hunting squid along the way, resting aloft or on the ocean surface if the wind dies down. Wandering albatrosses can embark on wanderings lasting 46 days. Likewise, it is not untypical for frigate birds and shearwaters to cover vast distances when foraging throughout their non-breeding season.

The longest time aloft ever recorded by a migrating bird that does not use soaring flight extensively is the migration flight of the bar-tailed godwits, which leave Alaska and fly over the Pacific Ocean non-stop to New Zealand. Bob Gill and colleagues (2009) fitted seven godwits with implanted satellite tracking devices and followed them over an average non-stop distance of 6,300 miles (10,150 km); the longest flight distance recorded was 7,200 miles. The birds stayed aloft for an average of 7.8 days (range: 6–9.4 days) and averaged 35 mph, consuming 40 percent of their body weight. This record was recently broken by another bar-tailed godwit fitted with a satellite tag that flew a non-stop distance of 8,400 miles (13,400 km) from Alaska to Tasmania in 11 days (Martínez 2023).

Even some small passerines can accomplish long-distance migratory flights over oceans or the Sahara Desert. The blackpoll warbler, a bird weighing 0.4 oz (12 gm), makes an autumn flight from the eastern United States to South America, a distance of 1,900 mi (3,000 km) over water, requiring a potentially nonstop flight of around 72 to 88 hours. Two billion birds cross the Sahara in spring. Geolocators placed on European pied flycatchers measured their non-stop flight time of 40 to 60 hours (Ouwehand and Both 2016) during this crossing.

Long-distance migrants of North America and Eurasia spend as much as one-third of the year migrating. Birds store energy in the form of fat before migration. Two hormones cause the fat buildup, activated by the increase or decrease in the amount of daylight, and storing the fat around the periphery of the abdomen and in their flight muscles. Smaller birds, such as holarctic shorebirds, spend about three weeks of intensive eating to double their weight before they leave their breeding grounds in the northern hemisphere and travel to the southern hemisphere, but they need stopping points to replenish their supply of fat. For example, ruddy turnstones that breed in Siberia winter in Africa. Some reach as far south as Cape of Good Hope, a distance of 8,000 mi (13,000 km). Normally they weigh 3.5 oz (100 gm), but increase their weight to 5.8 oz (165 gm), a 65 percent increase, before departure; they have to make at least three stops to replenish their fat reserves to reach their destination (Alerstam 1980). White-rumped sandpipers, Baird's sandpipers, greater yellowlegs and other shorebirds make annual journeys between the tundra breeding grounds of North America and the wintering grounds of southern South America, a distance of about 8,500 miles (14,000 km) (Harrington 1999). In Eurasia, common greenshanks, wood sandpipers, and green sandpipers, as well as many other shorebirds, hawks, eagles, and some passerines make an annual migration to and from southern Africa. Fat is their fuel for migration, but another important factor is the use of wind assistance. Around the world, these migratory birds need stopover sites to replenish their fuel reserves.

During long flights, the birds burn the fat, and in transoceanic flights where they cannot land, they can burn flight muscles before a landfall. Shorebirds leaving the east coast of the United States for South America will migrate to coastal locations and fatten up before departing, and do the reverse in spring. Humans interfere with this process by harvesting critical food sources such as horseshoe crabs, and this has caused population declines in a number of shorebirds including red knot, which in 2014 was listed as "threatened" under the Endangered Species Act. The elimination of estuaries, marshes, and coastal forests in areas where birds feed has led to a decrease in the habitats and food sources needed by migrating birds.

Many species of land birds, shorebirds, and waterfowl will stop immediately after crossing a large geographic barrier—an ocean, desert, or a large lake. Once over land, birds advance shorter distances at a time than those crossing oceans. The ruby-throated hummingbird, weighing 0.11 oz (3.2 gm), is the smallest bird to cross from the Yucatan to the U.S., but it then proceeds north at about 20 miles per day (m/d) (32 km/d), following the bloom of spring flowers.

Banding data taken of many small passerine species in Europe showed that they traveled an average of 150 m/d (245 km/d) (Hall-Karlsson and Fransson 2008), but some species cover much shorter distances, and others longer. The average over-land distances covered by birds varies from 210 miles per day (m/d) for purple martin to 135 m/d for wood thrush, and to 50 m/d for snow buntings. Blackpoll warblers average 30 m/d (50 km/d) in the United States when traveling north, and increase their distances to 200 m/day (330 km/d) after late May for those traveling to northern Alaska (Zimmerman 1998). Maximum distances covered by passerines range from 120 to 180 m/d (200–300 km/d), while shorebirds cover 240 to 600 m/d (400–1000 km/d) (Newton 2007a). Some birds will migrate to a new stopover every day for a few days, but then spend several days at the next stop. The distances they travel depend on weather conditions, wind direction, and their fuel reserves.

Neotropical migrants in North America annually cover long distances in migration. Most individuals stop frequently during their journey to rest and refuel, so stopover sites are particularly important to help maintain populations of these species. Researchers have developed a framework for identifying the types of stopover sites that are important for conservation, with the goal of reducing mortality among migrating birds. David Mehlman *et al.* (2005) have identified three types of stopover sites that are needed to serve migrant species:

1. Sites adjacent to barriers such as a large body of water, which serve as emergency stopover.
2. Habitats of varying sizes where birds can rest and replenish fat reserves for a brief time period.
3. Sites with abundant resources that allow longer stays.

These types of stopover sites work in conjunction to provide the best support for migrants. Flying at higher elevation and cooler temperatures are more efficient for flight. Soaring birds are usually below half-mile elevation, although they can migrate as high as a mile up (1.6 km). Large waterfowl, like some swans and geese, can reach elevations over 10,000 ft (3 km), and we noted earlier that some geese and cranes migrate as high as 22,000 ft (7 km). Ducks are usually found at low elevations, with some as high as 3,000 ft (900 m), but higher altitudes have been recorded. Passerines migrating over land range from low elevation to about 2,000 ft (600 m), but they can get as high as 11,000 ft (3.3 km) and possibly higher over the ocean (Able 1970). Shorebirds also will migrate as high as 15,000 ft (4.5 km) over the ocean. Bar-tailed godwits have been tracked at over 16,000 ft (5 km), but most were recorded at lower altitudes (Senner *et al.* 2018).

The optimal time of day for migration varies with different species, influenced most frequently by predator avoidance, foraging opportunities and energy usage. Many birds gain a strong advantage by flying at night, because they can maximize their foraging time on the ground and energy usage in the air (Alerstram 2009). Most of the following species migrate in groups at night: owls, flycatchers, thrush, thrashers, catbirds, wood warblers, vireos, kinglets, nuthatches, creepers, wrens, gnatcatchers, cuckoos, warblers, buntings, rails, woodcocks, tanagers, orioles, blackbirds, and bobolinks. Some birds migrate during the day or night: loons, grebes, ducks, geese, swans, shorebirds, swifts, swallows, hummingbirds, auks and murres. Other land birds including woodpeckers, crows, jays, larks, pipits, bluebirds, and finches migrate during the earliest part of the day. Soaring birds such as hawks, cranes, storks, and vultures migrate during the hottest part of the day to take advantage of thermal updrafts.

Migration Strategies and Mechanisms

Ken Able (1999) describes bird migration as a solution to a set of ecological problems. Birds that breed in areas where resources of food, water, cover, and competition are seasonal or variable migrate to regions where these resources are available and predictable. The migration distances are dependent on the birds' resource needs and availability. Insect-feeding birds of the northern hemisphere are obligate annual migrants to the south in their non-breeding season; many travel all the way to the southern hemisphere.

Different populations or subspecies of the same bird species may migrate different distances. Able (1999) points out examples: the white-crowned sparrow in North America and the Eurasian blackcap in Europe are both species that have subspecies populations with different migration destinations. Within a population of birds that are year-round residents, some will change their wintering location in response to resource needs. Some species of birds are nomadic in their search for adequate resources in winter, when resources are not predictable.

Most of the work on migration has involved phenomena of long distance migration, but short distance migration is also important. Some birds migrate from higher elevations to lower elevations, some from drier areas to wetter areas, some from colder to warmer areas, and some do not migrate at all.

Many tropical birds that are thought of as permanent residents actually migrate short distances in order to move to areas where food sources are more plentiful. This is most common among fruit and nectar eating species (Stiles 1988), and it has also been observed in high altitude seed-eating species. During the breeding season in the tropics, fruit-eating birds occupy the higher elevations and move to lower elevations during non-breeding, but not all species carry out this annual cycle. No clear and definitive environmental factors initiate these short-distance migrations (Chaves-Campos 2004), and for many species, there are other possible causes besides food supply. Many case studies have shown that these movements are common in the tropics—and at the same time variable (Hilty 1994). Studies of North

American and Hawaiian birds show the same sort of short distance migrations. Boyle (2017) found that of the species wintering within the USA and Canada, about 30 percent engage in altitudinal migrations. Altitudinal and short-distance migrations are found on every continent.

For long-distance migrants, the onset of bird migration is triggered by the changes to their internal clock. We are all familiar with the cycle of day and night, and we know that we respond biologically to this rhythm. In birds, both biology and the timing of migration are dependent on this internal clock, which has an annual basis as well as a daily basis. Annual molts and the length of daylight prompt birds to begin their migration. Tests done on caged wild birds show that changes in the daylight hours are important in setting this internal clock, and in keeping them in synchronization with their annual cycle. Birds that winter in the southern hemisphere and migrate to the northern hemisphere do not base their dates of departure on the environmental conditions in the north, but on the lengthening days, which tell them that they should find adequate food when they arrive.

Today 19 percent of all species of birds make regular, extended migrations to and from distant breeding locations. The vast majority of these breed in the northern hemisphere and migrate south to the more temperate southern climates to avoid the northern winter (Somveille *et al.* 2013). The primary reasons for this migration-enabled breeding strategy are the extended daylight and food supply in the northern regions, as well as the fact that the majority of land mass on earth is in the northern hemisphere.

Pelagic species spend their lives at sea, except while breeding. Their patterns of migration are much more dependent on the species. The majority are southern hemisphere species, and many will range over the oceans in the region of their breeding locations, while others will range over greater distances. Some that breed in the higher northern or southern latitudes migrate to temperate regions during non-breeding. Most of the larger pelagic species cover long distances during their annual non-breeding cycle. We will discuss them in more detail in Chapter 20.

Some species, particularly those of the far north, normally remain in place through the winter or carry out short distance movements. In some years, they will migrate south in large numbers, usually due to insufficient food supply. This sort of periodic migration is called an irruption, and is seen most often in boreal finches that feed on seeds or nuts, and northern owls that feed on rodents.

Since the phenomena of climate change and global warming have become more severe, the advance of spring has changed in the northern hemisphere. Over the period from 1955 to 2002, the blossoming of spring plants has advanced a little over one day per decade (Schwartz *et al.* 2006). That has changed the timing of the hatching of insects, since their annual cycle is controlled more by the blossoming of plants. Insects emerging on different timetables has led to a mismatch in the arrival of migrant birds and the availability of insect food sources. Some songbirds are able to compensate for this, but most are not (Mayr *et al.* 2017). This mismatch in the synchronization of food sources and arrival of migrants puts the breeding cycle out of phase with food supplies. The outlook is that this phase imbalance will affect the breeding success for some species in a negative way.

The relationship between weather and migration has been studied through correlations with bird movements and weather patterns, and more recently with direct radar measurements, geolocators, and satellite tracking. Birds adapt to meteorological conditions, because weather plays a significant role in migration. They will time their departure with favorable winds and cease migrating under unfavorable conditions. Birds use the wind assistance offered by frontal boundaries between low pressure cells to advance. They will also ride the air flows on the back sides of low and high pressure cells for aerial assistance. Sparks *et al.* (2002) provide a much more detailed explanation of how the factors associated with passing pressure cells—visibility, cloud cover, temperature, and winds—affect the onset of migration.

Radar studies of trans-gulf migration (Gauthreaux 1999) show that birds usually travel at higher altitudes over water than they do over land. The average altitude is about one mile (1,500 m), but this varies depending on the most favorable winds. In spring, the very high altitude winds over the Gulf of Mexico provide favorable tropical surface winds for northbound migrants.

A significant percentage of annual bird fatalities occur during migration, many of them due to human induced causes. In the United States, hundreds of millions of bird are killed by human causes, and most of these impact migratory birds. The most prevalent cause is collisions with human-made structures: buildings, windows, power lines, communication towers, and wind turbines. Collisions with buildings and power lines constitute the highest number of fatalities (Erickson *et al.* 2005). Estimates of collision deaths in the United States annually tell us that between 360 and 990 million birds are killed annually (Loss *et al.* 2014). It is estimated that free-ranging domestic cats kill 1.3–4.0 billion birds annually (Loss *et al.* 2013).

Natural causes of migratory deaths include weather-related mortalities, but there are no global estimates of this problem—only some case studies. A study on Lake Michigan of the effect of a severe rain storm in 1996 found 2,981 dead birds of 114 species on the southwestern shores of that lake (Diehl *et al.* 2014). Two storms in Europe killed hundreds of thousands—some estimates say millions—of swallows and martins, reducing the breeding population from the previous year by 25 percent or more. Newton (2007b) points out that not only storms during transit, but also unseasonably cold weather after arrival brings high fatalities. A study of long-distance raptor migration monitored by satellite showed that the mortality rate during migration accounts for more than half of the raptors' annual mortality (Klaassen *et al.* 2014). Causes of these fatalities between Europe and Africa were only determined for a few incidents, but they included collisions, hunting, and some natural causes.

For passerines, shorebirds and other species that carry out long-distance, trans-oceanic migration, very little is known about the mortalities of these small birds. Storms and depletion of fat reserves will force birds to the water, where they cannot survive. A most unusual discovery revealed that young tiger sharks (*Galeocerdo cuvier*) off the coast of Alabama and Louisiana, where approximately two billion birds undertake the Gulf crossing twice annually, regularly feed on birds that have hit the water. Between 2010 and 2018, a research team looking into the stomach contents of these sharks found that almost 40 percent had ingested small migrant birds

(Drymon *et al.* 2019). Many of the birds were identified by their feather remains. These fatalities demonstrate the risks of long-distance migration for birds of every size and species.

Migratory Map

To migrate to a desired destination, birds must know the direction and distance to the destination. The study of directionality in birds is referred to as *orientation*.

Experiments dating back to the 1950s involved displacing immature European starlings by moving them from Holland to Switzerland, a considerable distance from their normal habitat. The experiments showed the starlings migrated in the correct direction, but made no adjustment for the displacement (distance traveled). Adult birds, however, do make corrections for displacement (Perdeck 1958, 1967).

Migration has been studied for many years in laboratory orientation experiments and from data on retrieval of banded birds, and, in more recent times, with radio transmitters or data loggers carried by birds. Thourp *et al.* (2010) gives a summary of our current understanding of migration based on the most recent information. Many species that migrate as flocks, such as geese, ducks, storks, and some passerines such as finches and blackbirds, learn the route from adults. Other species, left by their parents to find their own way, migrate as individuals: many long distance migrants, some night migrants, and some immatures. Birds have developed amazing skills at accurately migrating to specific destinations using repeatable corridors—the same route from one year to the next.

Once birds have learned the location of a favorable wintering or breeding region, they will return to it with little chance of an error. Their orientation is not fully understood, but it involves the use of sun and star compasses, as well as the ability to sense the magnetic field and to use polarized light. Certain seabirds can navigate by odors, and pigeons can sense very low frequency sounds (Alerstam 1990). Scientists are still trying to fully understand exactly how these different compasses are used in combination, but some are more important than others under certain conditions. For instance, any species of bird that departs near twilight uses polarized sunlight for its initial orientation (Able 1993). The combination of magnetic orientation with sun and star compasses also appears to be very important to long distance migrants. Much work has been done to understand the ability of birds to perceive and use magnetic fields for migration. The status of this technology is summarized by Robert Beason (2005):

> Behavior and electrophysiological studies (of birds) have demonstrated a sensitivity to characteristics of the geomagnetic field that can be used for navigation, both for direction finding (compass) and position finding (map). The avian magnetic compass receptor appears to be a light-dependent, wavelength-sensitive system that functions as a polarity compass (i.e., it distinguishes poleward from equatorward rather than north from south) and is relatively insensitive to changes in magnetic field intensity. The receptor is within the retina and is based on one or more photopigments, perhaps cryptochromes. A second receptor system appears to be based on magnetite and might serve to transduce location information independent of the compass system. This receptor is associated with the ophthalmic (relating to

the eye) branch of the trigeminal nerve and is sensitive to very small (less than 50 nanotesla) changes in the intensity of the magnetic field. In neither case has a neuron that responded to changes in the magnetic field been traced to a structure that can be identified as a receptor. Almost nothing is known about how magnetic information is processed within the brain or how it is combined with other sensory information and used for navigation. These remain areas of future research.

Although the exact mechanism of how birds sense magnetism is not completely understood, there is a connection between their magnetic orientation and polarized light. Songbirds use polarized light cues near the horizon at sunrise and sunset to recalibrate their internal magnetic compass (Muheim *et al.* 2016). This effect has also been shown in zebra finches (Muheim *et al.* 2006), savannah sparrows and white-throated sparrows (Muheim *et al.* 2009). The magnetic calibration takes place in receptor molecules in the avian retina. Research has shown that a quantum mechanical effect of the cryptochrome proteins in the eye can provide magnetic direction (Hiscock *et al.* 2016). Recent experiments with this protein from the retina of European robins show that the protein does respond to magnetic direction (Xu *et al.* 2021).

Studies in the 1930s and since then have shown that at least some shearwaters taken from their nests and transported long distances outside of their normal migration were able to return to their nests relatively quickly without failure, thus showing that they were able to navigate through unknown territory. However, not all bird species use magnetism for navigation. The family of petrels, shearwaters, and albatrosses, collectively known as tubenoses because of the extended nostrils on their upper mandible, do not appear to depend on the earth's magnetic field. When mobile magnets were attached to their heads, eliminating the earth's magnetic field, they had no difficulty finding their way back to their nest sites (Bonadonna *et al.* 2005). This family, the *Procellariiformes*, has an amazingly strong sense of smell and spends many years following the ocean wind currents. Their delayed breeding allows them to learn the ocean habitat and locations of food sources.

Research by Gagliardo *et al.* (2013) with Cory's shearwaters, displaced in the open ocean about 500 miles (800 km) from their nesting grounds in the Azores, subjected the birds to sensory manipulation. Magnetically disturbed shearwaters showed unaltered navigational performance and behaved similarly to unmanipulated control birds, but the shearwaters deprived of their sense of smell were dramatically impaired in orientation and homing. Birds displaced long distances from their breeding grounds and released under cloudy conditions were also disoriented. From this we know that this family uses the sun, winds and odors to navigate, but how these factors are combined is uncertain. Homing pigeons may also use an olfactory map for homeward orientation, since the navigational performance of inexperienced pigeons is disrupted after section of the olfactory nerve (Gagliardo *et al.* 2008).

Other studies with songbirds held in orientation devices open to the sky showed that they have a genetic impulse to migrate in the correct direction, depending on whether it was spring or fall. This demonstrates that orientation impulses are heritable traits (Able 1999). For example, researchers in Europe cross-bred two different populations of Eurasian blackcaps, one that migrated southeast and the other southwest to different wintering quarters. They found that the progeny oriented in an intermediate direction between those of their parents when measured with an orientation device (Helbig 1991).

These examples point to the fact that adaptations in the migratory strategies of birds are attributed to a number of factors, one of the most important of which is genetic variation (Pulido 2007). This is one explanation for why different populations of the same species have different migratory behavior, and it also supports the belief that migratory patterns can be altered within a few decades.

7

Vagrancy and Dispersal

In some cases, migration to and from breeding areas is not the only reason why birds appear in certain habitats and in unexpected places, sometimes way out of range. Vagrancy is a common occurrence among birds, and annually, there are many instances of them found at locations where they are not expected. These are also referred to as *extralimital* records, and birders thrill to chase down these apparently lost birds.

Post-breeding dispersal and irruption are two other mechanisms that bring birds to locations far from where they are expected. As noted briefly in Chapter 6, an irruption is a dramatic, irregular migration of large numbers of birds to areas where they aren't typically found, possibly at a great distance from their normal ranges. The most common cause of an irruption is a lack of food. In the northern hemisphere, we see these irruptions among northern breeding finches like red and white-winged crossbills, common and hoary redpolls, pine and evening grosbeaks, snowy owls, waxwings, and some artic owls. Snowy owls, however, are generally irruptive in years when they produce an overabundance of immatures (Curk *et al.* 2018). Common birds such as black-capped chickadees and red-breasted nuthatches also may wander far from their usual range in search of winter foods.

Irruptions are more unusual than post-breeding dispersal, which is an annual phenomenon among birds. In post-breeding dispersal, immature birds wander after fledging, an important factor in range expansion. Dispersal distances vary with the type of species involved. Finally, some birds are just random vagrants.

To illustrate this, let's take an example. The Hawaiian islands, part of the Emperor Seamounts underwater mountain range, were formed over the volcanic hotspot in the middle of the Pacific Ocean, a location that has been thousands of miles from land for millions of years. As a result, it is unlikely that rosefinches—birds native to Europe and Asia—colonized this island by migration. There had to be some event that caused them to arrive, most likely some type of storm, and it must have brought a sufficient number of these birds to establish a colony that, over millions of years, became the Hawaiian honeycreepers. That initial group of birds was an example of an irruption, and they could only have arrived accidently.

Severe weather conditions can cause a major fallout of displaced migrant birds. Experiments with shearwaters removed thousands of miles from their nests have confirmed that they will return in a very short time, thus demonstrating exceptional migratory skills. However, not every bird meets this migratory challenge successfully. Birds can get blown way off course by storms, get lost in fog, or follow the wrong orientation. They can fail to make course corrections or fight strong head-winds and end up far from where they should be. Most of the navigational methods used by birds don't work well in fog, and birds lose control of their course in storms.

Birds can end up in unexpected places by other causes as well, and these are usually labeled as vagrants. The subject of the occurrence of birds that are rarely found in North America is treated in detail by Howell, Lemington, and Russell in their book, *Rare Birds of North America* (2014).

For example, large ocean-dwelling birds like northern gannets, brown boobies and brown pelicans sometimes end up on the Great Lakes or on large lakes far from the sea by accident, and they have been known to perish in those locations. These are often immature birds that have not had enough experience with migration to find their way back. Coastal hurricanes blow seabirds inland, and many get trapped at locations far from their natural habitats. Terns are strong fliers and invariably find their way back, as do other large seabirds that are still physically fit. In severe storms, however, some birds lose so much body mass that they never recover sufficiently, and perish where they ended up. When birds end up in some unfamiliar location, they may have insufficient food or water and become susceptible to local predators. Cattle egrets, often trapped on the Dry Tortugas, may not find sufficient nutrients to gain strength to move on, and sometimes expire before they are able to leave. Small sea birds like storm-petrels are likely to be killed and eaten by gulls when trapped on large inland lakes if caught during daylight hours (Stenhouse *et al.* 2000; Watanuki 1986).

Most of what is known about the ultimate fate of large birds ending up off course is based on observations. Small birds like passerines routinely end up very far from where they are expected to be found. Birdwatchers the world over get excited about chasing these extralimital species to get a glimpse or a photo. The fate of vagrant passerines is very uncertain; at least some will survive, and these birds will return to the same breeding and wintering locations for many years. In the area of Rochester, New York, for example, a Barrow's goldeneye returned to the same wintering area on Lake Ontario for many years, an unexpected location for this species.

In Arizona, an orchard oriole returned to the same backyard for nine consecutive winters. This species routinely winters from southern Mexico to northern South America. The Sonoran Desert is not suitable to support this species through the winter, but backyard feeders and a manicured subdivision provided all the necessities to sustain this bird. This is one of many examples of how feeders provide lost birds what they need to survive in an unexpected location.

Rarely, some species of South American flycatchers will migrate in the wrong direction. One such example is the fork-tailed flycatcher of South America, which is seen almost annually somewhere in the eastern United States. This is an example of *reverse migration*, an uncommon phenomenon found among first-year birds. Likewise, every year eastern wood warblers are found in the western United States, and those from the west are sighted in the East. These birds usually vanish from view before long, so we can only guess that their eventual fate.

In normally expected migration, birds have some predisposition to the direction of migration and the time of travel, but overshoots are very common. These occur with both spring and fall migrants. For spring migrants, it is unknown if this is caused by tailwinds or just by flying too long. This subject of the various causes of vagrancy and how it applies to those species seen in North America is treated by Howell *et al.* (2014).

However, not all causes of vagrancy are well understood, but recently a team of researchers (Tonelli *et al.* 2023) has found that temporary disturbances in the earth's magnetic field cause a significant percentage of the reported vagrancies. This cause is likely related to the quantum detection of magnetic fields mentioned in the previous chapter.

Within any population of migrant birds, there will always be some individuals that follow a course that deviates from the population average. These variations in direction are rare on a percentage basis, but they result in birds ending up in unexpected locations. If these sites provide the basic requirements, the birds will remain. In at least two examples, bird feeders have led to the establishment of new wintering regions for a migrant species (Able 1999). The Eurasian blackcap established a permanent wintering location in southern England and Ireland over a twenty-five year period. They never wintered in these areas before this time, but feeders became the catalyst that brought this change. The same thing has occurred with the rufous hummingbird of North America: this species typically winters in Mexico, but some fall migrants deviated in a southeast direction and established a population that winters on the Gulf Coast. In these examples, the population of a specific species were predisposed to variations in the normal directionality, and they ended up in locations that eventually served as established sites.

Researchers have noted that Richard's pipit, a small passerine that breeds in Asia and normally winters in Southeast Asia, has established a population that is migrating west instead of south and winters in Europe, thus establishing an additional migratory path. Vagrancy and climate change likely promoted the establishment of this migration route (Dufour *et al.* 2021).

Banding studies of passerine birds show that they return to locations close to where they were hatched and fledged, usually within half a mile (1 km) (Drilling and Thompson 1988). Nevertheless, there are always some that reestablish themselves some distance from their original locations. This sort of dispersal distribution was found for mountain plovers, the majority selecting a site very close to their fledging location and others dispersing as much as 25 miles (40 km) away (Skrade and Dinsmore 2010). Paradis *et al.* (1998) measured the patterns of breeding dispersal of 75 species of land birds and found that dispersal distances were shorter for more abundant species, and further for both migrant species and those living in wet habitats. This process of breeding dispersal is important for range expansion, population biology, and ecology of the species.

From numerous studies, the problem of changes in the breeding range with increasing global temperatures has been well established. Many species like northern cardinal, tufted titmouse, and red-bellied woodpecker have been expanding their range north since the 1940s. This is also true for tropical birds in Mexico: the great kiskadee, which breeds in Texas, has shown this same range expansion.

Extralimital tropical species are sited more frequent in southern United States, which could be a sign of the overall northward evolution of breeding range taking place in North America.

The focus of Part II of this book has been on the adaptation of birds after their early beginning in the Cretaceous period, and the factors that enabled their diversification. The diversity of birds has been centered on their ability to exploit flight as an ecological advantage. This brief discussion of how birds are different, coupled with breeding, migration, and dispersal, represents some of the most important aspects of their lives. These are very complex subjects that have been studied by many scientists; only a very brief summary is offered here. We encourage you to consult other sources to gain further understanding of these subjects. There are many aspects of bird biology and lifestyle that we should take the opportunity to appreciate.

PART III

EXTRAORDINARY BIRDS

Introduction

Before diving into the very extensive field of the diversity of living birds, let's consider a more restricted topic relating to lesser known aspects of the avian world.

The EDGE of Existence program is a global conservation effort to identify the world's most Evolutionarily Distinct and Globally Endangered (EDGE) birds. The program ranks 100 species based on conservation status and how unusual they are compared to other species. Many of the birds that have survived to the present are mentioned in Part III of this book, and are ranked by the EDGE of Existence program.

Extraordinary animals have strange or unusual adaptations that allow them to survive by dominance or by occupying an unusual habitat. Birds can also be unusual based on how they look or behave, if they have very few closely related species, or if they are very rare with restricted populations.

Chapter 8 goes back to the time when birds were a dominant land animal in those geographic areas free of large mammals. Without significant land predators, birds evolved to become very large in size and were the predominant lifeform in their respective habitat. In some cases they were flightless, occupying the ecological niche of large grazers (herbivores).

Many different sources create lists of species they consider unusual birds, mostly based on personal choices. In Part III, birds were selected to highlight the fact that certain species have unique characteristics that we would not suspect.

Chapter 9 is a look at one of the most unusual bird species on earth. Even though it may be related to modern birds, it retains some of the vestiges of ancient bird lineages. Chapter 10 focuses on a few very unusual living species. The lifestyles of these species are atypical when compared to their closest relatives.

8

The Megafauna
of the Avian World

The largest living bird on earth today is the common ostrich, but over their history, many very large birds have emerged in many different categories. The term *megafauna* usually refers to very large animals—those at least as large as a human—but in this context, the term is applied to different orders of birds. Each order has its largest member. Not many of these largest family members are left on the planet; they disappeared through natural or human-caused extinctions. For the most part, our knowledge of them comes from the fossil record.

The heaviest known flying bird was the giant teratorn, similar in shape to North American vultures and found in Argentina from the late Miocene epoch, dated at about 6 MYBP. Its wingspan was 16.7–21 ft (5–6.5 meters) and it weighed 154–159 pounds (70–72 kg). The teratorn genus had other smaller members that ranged into southern California.

Much of the lifestyle of the giant teratorn has been inferred from its anatomy. It had very stout, strong legs, large feet, and a large bill with a hook at the tip. Its leg muscles were likely capable of flight with a running start, or it used wind currents and mountain slopes to become airborne. A soaring bird, it required a large range (Campbell and Tonni 1983). Based on the shape of its bill and its anatomy, it was believed to be a predaceous carnivore and not a scavenger. For comparison, the heaviest extant (living) flying bird, the great bustard, weighs about 46 pounds (21 kg), and the wandering albatross has the largest wingspan of a flying bird at 11.5 feet (3.5 m).

The remains of another large bird, *Pelagornis sandersi,* were discovered in South Carolina in 1983. Knowledge of this bird is based on a single specimen dated to about 25 MYBP, a time when the earth was warmer. It had a wingspan of 20 to 24 feet (6.1 to 7.4 m), the widest wingspan of any bird ever uncovered. It weighed between 50 and 88 pounds (22 to 40 kg). Theoretical calculations confirm that it was an efficient glider (Ksepka 2014). An oceanic species, *P. sandersi* is believed to have lived on cliffs and fed on fish. It probably resembled the albatross, with a large, long bill, and bony teeth not set in sockets.

A number of fossils from Seymour Island, Antarctica demonstrate the early evolution of the giant body sizes of pelagornithids (a prehistoric family of large seabirds) over a time span of ten million years. One from the Eocene epoch was one

111

of the largest flying birds, with an estimated wingspan of 16 to 20 feet (5–6 m), not unlike the one found in South Carolina. This Antarctic bird was a predatory sea bird base on its long jaw, with spikes that resemble teeth (Kloess *et al.* 2020).

Among flying birds, no bird has ever reached the size of the largest known flying pterosaur, the *Quetzalcoatlus northropi*, from the late Cretaceous period of North America; it is estimated to have a wingspan of 30–33 feet (10–11 m), and the most recent estimates of its weight place it over 400 to 650 pounds (180–295 kg) (Witton and Habib 2010).

The largest living birds within different bird families usually occur on islands, because they are usually flightless. In order to survive as a large flightless bird, they must occupy a habitat that lacks predators that can easily overtake them. Sometimes flightless birds can find a way to coexist with predators, as in the case of some flightless ducks. No flightless bird is safe from human predation, however, and almost all large flightless birds have been driven to extinction.

An extinct family of generally large birds of the Paleogene period, the *Phorusrhacidae*, inhabited South America. This family consisted of 14 genera and 18 species. A single species of this family was found in Texas, one of the few South American birds to cross the Panama land bridge that was present 10 to 15 MYBP (Alvarenga *et al.* 2011). Within the family, species varied in size from as small as 3 feet long (1 m) to the largest, the terror bird, which earned this name because of its large size and its enormous beak and head. Its skull was 28 inches (71 cm) long, including a laterally flattened beak of 18 inches (46 cm) in length. It had a long, strong, highly flexible neck, and provided a lethal blow when it struck with downward force. This bird stood 10 feet (3 m) high with the neck fully extended, and it weighed about 550 pounds (250 kg). It likely preyed on small mammals that it probably swallowed whole. It had strong legs that could also be used to attack prey (Bakker 1998).

Two extant species, the seriemas, sole members of the *Cariamidae* family, are long-legged and stand up to 3 feet tall (1 m), and are believed to be related to the *Phorusrhacidae* family. The two species are found on the South American savanna.

Penguins first appeared in the fossil record in the Paleocene epoch, shortly after the K–Pg extinction event (see Chapters 1 and 2). The earliest fossils come from New Zealand, and many fossils have been found in the southern hemisphere; they were flightless and, like all penguins, had a unique shape. From their origins in warmer seas, they adapted to the colder polar seas and diversified.

Fordyce and Ksepka (2012) summarize the many penguin fossils and discuss their development and dispersal. Several were giant penguins that had very long necks and long bills, unlike modern species. The largest specimen ever found, *Kairuku waitaki*, turned up in New Zealand and dated back to 25 MYBP, stood 4 feet 4 inches (1.3 m) tall, and weighed 135 pounds (61 kg). Another large prehistoric penguin, *Icadyptes salasi*, had a long, thin, spear-like bill. The extremely well preserved skeletons indicate that these penguins most likely speared their prey (Ksepka 2009).

The largest bird known to science to have ever lived is the giant Elephant Bird of Madagascar. The skeleton is shown in Figure 8.1. It stood ten feet tall (3.3 m) and weighed between 800 and 1000 pounds (350–500 kg). Its single egg was 13 inches

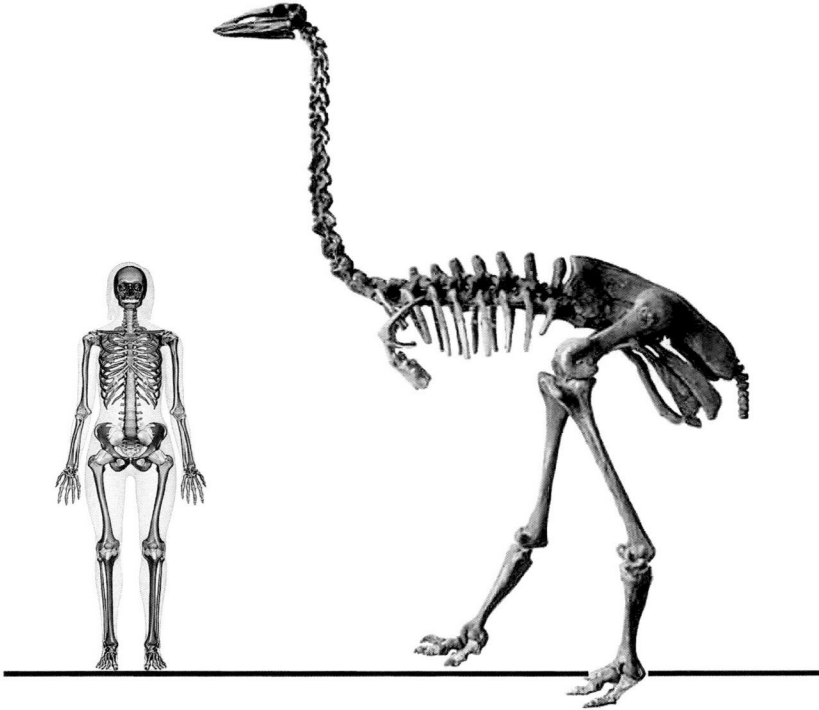

Figure 8.1 The Elephant Bird is believed to be the largest bird that ever lived; its 10-foot height is shown next to a 6-foot human (from: *Quaternary of Madagascar* by Monnier [1913], via Wikimedia Commons).

long (33 cm) and had a volume of more than two gallons (9 liters). Although it looked like an ostrich, it was most closely related to the kiwi (Mitchell *et al.* 2014).

Four species of elephant birds are recognized and placed in the *Aepyornithidae* family, all of Madagascar. This family is a member of a group of ratites, a diverse group of mostly large, flightless birds with a smooth breastbone that lacked a keel on the sternum; thus it had no place to anchor flight muscles. In addition to the ostrich, rhea, emu, cassowary, and kiwi, ratites included the extinct moa and elephant birds. The current thinking is that elephant birds were part of the radiation of ratites that took place in the Cenozoic era after the K–Pg extinction event, from ancestors that must have been flying birds, because the distribution of ratites does not match the pattern of continental drift.

When the *Aepyornithidae* family became extinct is uncertain, but they survived to between 1,000 to 2,000 years ago based on radiocarbon dating of eggshells and bones. These bones also had signs of butchering (Hawkins and Goodman 2003), so they had to exist when the island was first occupied by people. The cause of extinction is not certain, but there is little doubt that hunting pressure and the human harvesting of eggs played a role.

New Zealand was home to the moa, one of the largest families of flightless birds. Moa remains have been found on the south and north island of New Zealand as well as on Stewart, Great Barrier and the D'Urville Islands around New Zealand. What we have known of their taxonomy and biology has been based on skeletons, footprints, egg fragments, coprolites, gizzard stones, and tissue samples that were preserved in caves: muscle tissue, a complete head, a complete leg, a gizzard and other soft tissue. Pieces of skin and parts of the plumage from the upland moa have been found as well. The dark colored feathers with light grayish-white tips are entirely loose, with long aftershafts, like those of emu, kiwi, and others.

Fossil remains show that the moa have a long history in New Zealand, dating back to at least 19–16 MYBP, and almost certainly extending to before the Oligocene epoch (34 MYBP) (Tennyson *et al.* 2010). The modern order diversified about 6 MYBP on the south island of New Zealand, and they populated the north island when these islands were joined.

Moa are most unusual even among flightless birds, because they have no vestiges of wing bones. They were large, tall, long-necked, tailless birds with a small head and beak, resembling an emu or ostrich. Their closest living relatives are the tinamous (Baker *et al.* 2014). The number of different moa species has not been totally clear from fossil evidence, but researchers currently count nine identified species, ranging in size from the giant moa—the largest at 12 feet tall (3.6m) and weighing about 510 pounds (230 kg) (Davies 2003)—to the smallest, the little bush moa and upland moa, which were about 3 feet tall (1 m) and weighed 37–75 pounds (17–34 kg). In all nine species, females were larger than males. The moa were the dominant herbivores on New Zealand; they fed on leaves, tender limbs, seeds and probably small insects. Some, including Mantell's and little bush moa, lived in dense forest, while others occupied upland and mountain zones.

Richard Owen identified moa bone fragments as birds in 1839, but their life history is shrouded in mystery, because there is no information about them as living creatures. It is estimated that they went extinct around 1425 (Holdaway *et al.* 2014). There is some osteological evidence that some species were adorned with crests (Fuller 1988). Researchers believed that they were slow moving. Based on analysis of coprolites, upland moa had a highly diversified vegetarian diet including 67 plant taxa, including nectar-rich flowers of several plants. They grazed in both forest and grassland habitats (Wood *et al.* 2012). The moa probably used their powerful legs and feet with strong claws for scratching or digging up roots.

Prior to the arrival of the Polynesians, the chief predator of the moa was the Haast's eagle. The largest eagle known to have existed, Haast's eagle had short wings, allowing it to hunt in more forested areas, and powerful legs and feet. As with most raptors, the female eagle was larger than the male, with a wingspan of about 8 feet (2.5 m), a total body length estimated to be about 4.5 feet (1.4 m), and a weight of about 32 pounds (14.5 kg), while the male has a wingspan of about 7 feet (2.2 m) and weighed about 23 pounds (10.4 kg) (Brathwaite 1992). Very little is known about this eagle, but with its strong legs and large claws, it must have been able to kill sizable animals.

Based on the dating of campsites, the early Maori settled New Zealand in the late thirteenth century, and they hunted the moa—in fact, more than 300 moa butchering sites have been found. Archeological analysis clearly shows that moa bones were no longer found at campsites after the fourteenth century—an amazingly rapid extinction rate, no doubt brought about because they were easy prey. At the time the Polynesians arrived, the moa population was about 58,000 (Perry *et al.* 2014), and most of them became extinct after about 200 years of human settlement, based on dating of moa remains. Haast's eagle became extinct about the same time, as it lost its main prey.

In 1954, a fossil of a giant owl—the largest owl ever found, along with other avian remains including other owls—was first discovered in Cuba at Pio Domingo Cave, located in the Sierra de Sumidero (Arrendondo 1976). Since then, additional fossils were found in caves in the district of Havana and in other locations in Cuba; at least three almost complete fossil skeletons of this giant owl are known. They date from the late Pleistocene Epoch of the Quaternary period, estimated to be between 12,000 and 126,000 years ago. The Cuban giant owl measured 3.5 feet tall (1.1 meter), and weighed about 20 pounds (9 kg), and its structure resembles that of the burrowing owl. Closely related to modern owls, the Cuban giant owl had long legs and large claws, and it was believed to be powerful enough to kill an animal weighing 70 pounds. It had short wings and a sternum with a vestigial keel, which indicated that it was hardly able to fly. This Cuban giant owl ranged throughout Cuba, including on the Isle of Pines.

As noted earlier, the Hawaiian Islands were isolated for millions of years, since their emergence from the volcanic hotspot in the central Pacific 75 MYBP. There is no doubt that they have always been home to seabirds, but since they were so far removed from any major land mass, they were also a stop for migrating shorebirds and host to occasional land birds blown off course due to storms. Most of these land birds would leave or perish there, but a few species managed to colonize these islands, and they became the dominant herbivores. Members of the duck (*Anatidae*) and the rail (*Rallidae*) families were among the few that colonized this island archipelago.

Since the Hawaiian birds had no natural predators, an entire family of very large ducks and geese evolved. The flightless ducks consisted of four species, called the moa-nalo (meaning "lost fowl"), and they were distributed on different islands within the chain. At least four species of geese emerged, but the only remaining extant species is the Hawaiian goose, or nene, which is related to the Canada goose.

The island of Hawai'i had large areas of savannah on Mauna Loa and Mauna Kea, ideal for geese. Bones of the largest goose ever known were found there, and since it was flightless, it evolved there over time—but within the 0.7 million-year history of the island's origin. This goose, as yet unnamed, was five times the size of nene, reaching 3.9 feet (1.2 m) long and weighing about 28 pounds (13 kg), 1.4 times the weight of the giant Canada goose, which is the largest living goose (Paxinos *et al.* 2002). This extinct largest goose was first discovered in 1982, and more fossil bones surfaced over the next ten years. It had massive leg bones, a huge skull, a deep bill, and disproportionately small wings. Like the nene, it was also related to the Canada goose (Hume and Walters 2012). This bird—along with the flightless ducks, the other geese and all the rails—were driven to extinction by the Polynesians, either as a source of food or by habitat loss.

Europe has its share of avian megafauna as well. On the east coast of Italy at Gargano, paleontologists in 2013 found a number of fossils including those of a large eagle, an owl and a large waterfowl. The fossils were dated to about 5 MYBP, when the area was a series of islands. The large waterfowl, *Garganornis ballmanni*, was estimated to weigh 50 pounds (23 kg) (Pavia *et al.* 2017), and it had wings with spurs—but as it was flightless, researchers assumed it used its wings for defense. Among birds, herbivores were likely to evolve to large size when they did not have mammalian competition.

Perhaps the most famous example of an island giant driven to extinction by man is the dodo, a flightless pigeon once found on the island of Mauritius, east of Madagascar in the Indian Ocean. The dodo weighed 23–47 pounds (10–21 kg) and stood 3.3 feet tall (1 m). The dodo's ancestors were pigeons that island-hopped and settled on this island because of its abundant food, lack of competition from terrestrial mammalian herbivores, and absence of predators. Freed from competition and the energy and size constraints needed for flight, it adapted to a sedentary life. This made the dodo easy prey for sailors that landed there. Dutch sailors recorded the first sighting in 1598, and by 1662, they recorded last verified sighting. This case and many others show just how rapidly flightless birds living on island habitats can be driven to extinction through thoughtless slaughter.

When we look at the history of the order of birds on earth in the fossil and historic record, we find many unique, unusual, and even bizarre creatures that were actually birds. Our focus in this chapter has been on birds that reached megafauna classification within their own families, some in the distant past and some in more recent times. Within every family of birds, there are examples of the largest species, but the size of today's birds is less dramatic than those cited here. In birds, giantism is a special outcome of ideal circumstances, and these conditions are almost impossible to come by in a human-dominated environment. Even today, the common ostrich, the largest extant bird, has survived in Africa because of the preservation of the savannah, its home range, and because it is farmed. However, it was hunted to extinction in the Near and Middle East by the 20th century. The fact that many of the earth's megafauna were driven to extinction by humanity says a lot about the fundamental nature of humankind.

9

Hoatzin

A Window Into the Past

We are all familiar with drawings of dinosaurs, most of which are reconstructions based on their fossils. From these, we imagine what dinosaurs probably looked like—and movies help, recreating them as living creatures and depicting what their lives might have been like. Artists have used these same approaches to show us pterosaurs and early birds.

The hoatzin of South America serves as a template of how early birds looked and behaved. Other species provide examples of distant past birds—the cassowary or the ostrich, for example—but since the hoatzin is a flying bird, it provides us with a better model.

The hoatzin is an unusual tropical bird of South America's Amazonian lowlands, and is the sole member of its family. It inhabits the scrub and trees along streams, swamps, ponds and backwaters, and is one of the planet's most amazing bird species because of its unusual characteristics. Its morphology, physiology, and diet are so distinctive that it suggests a long evolutionary history (Mayr 2017).

Hoatzins, Figure 9.1, are herbivores, weighing about 2.2 pounds, about the size of a large chicken. Green leaves make up 80 percent of their diet, and the remainder consists of flowers and fruits from about a dozen plants. They have strong legs for climbing around on branches to find tender leaves, but appear clumsy and are noisy when moving around. Only a few other species of birds eat leaves, and then only small amounts.

In order to digest this much organic matter, the hoatzin grinds the leaves in a greatly enlarged crop instead of a gizzard, a different process from other birds. The vegetation ferments in the foregut—much like a cow, but unlike other birds, which digest in the hindgut. Food remains in the stomach for one to two days. It feeds in the morning and evening, and spends the remainder of the day roosting (Brooke and Horsfall 2003). A large rubbery callus on the breastbone acts as a support to keep the bird's distended stomach from pushing it out of balance. This unusual method of digestion has another drawback: the hoatzin is known as the "stink bird" because its droppings smell like cow manure.

The bird's appearance lends itself to the comparison to prehistoric birds. The hoatzin has a long neck, exposed blue facial skin, red eyes and long, erect crown feathers. Its scientific name, *Opisthocomus*, refers to its projecting crown feathers

Figure 9.1 Hoatzin spend most of their day roosting in tropical forests (© Jeanne Verhulst).

or "hair." Brownish on the back and chestnut on the breast and underwings, it has short, broad wings in addition to its strong legs suitable for climbing, but its poorly developed pectoral muscles make it a poor flier. It flies short distances, lands awkwardly, and spend its life in the trees near where it was hatched. It rarely ever comes to the ground.

The hoatzin lives in small to medium-sized groups, and builds relatively simple platform stick nests in trees overhanging water, where the female usually lays two eggs. A five-year research project found that 17 percent of hoatzin nests fledged at least one young (Mullner and Linsenmair 2007). Their overall survival rate was typical for tropical passerines. Strahl (1988) reported a survival rate of 27 percent for chicks, and noted predation by monkeys. Birds and snakes also prey on hoatzin young.

The young hatch and leave the nest immediately, and are able to climb around the tree. Remarkably, the young hatch with two claws on each wrist, reminiscent of Cretaceous *Enantiornithes* birds. They use these claws to climb. If the chick falls into the water beneath the nest, it will swim and climb back up to the nest tree. Both adults feed the young digested organic matter from the moment of hatching; while its young are active after hatching, they are not precocial, because they cannot feed themselves. As the chick matures, it sheds the claws on the wrists (Brooke and Horsfall 2003).

The evolutionary lineage of the hoatzin remains unresolved, and even the most recent analyses of comprehensive molecular data do not result in a congruent and strongly supported placement in the avian tree of life (Mayr 2014). In the past, ornithologists have thought the hoatzin to be related to game birds (*Galliformes*), cuckoos (*Cuculiformes*), tinamous and other bird families (Thomas 1996). A recent whole-genome sequencing study (Jarvis *et al.* 2014) places the hoatzin as the sister taxon of a clade of crane-like birds, waders, gulls and auks (*Gruiformes* and *Charadriformes*)—but it is still in its own order, the *Opisthocomiformes*. Richard Prum *et al.* (2015), using genomic sequence data from each of 198 species of living birds, indicated that the hoatzin is the last surviving member of a bird line that branched off in its own direction 64 million years ago, shortly after the extinction event that killed the non-avian dinosaurs.

The earliest and most substantial New World fossil record of a hoatzin is *Hoazinavis lacustris* from the late Oligocene—early Miocene epochs (22–24 MYBP), found in Brazil. Another previously misidentified fossil from Africa indicates that this order of birds was once found on both sides of the Atlantic Ocean (Mayr *et al.* 2011). The fossil from Brazil is represented by wing and pectoral girdle bones only, which closely resemble those of the modern hoatzin. In addition, Mayr (2014) describes a fossil hoatzin from Kenya in the Miocene, and indicates that the taxa was more widely distributed in the distant past.

Although its history is uncertain, the hoatzin has a most unusual lifestyle with its weak ability to fly, its confinement to an arboreal existence, its unusual diet and its unique digestive system, and its young that hatch with claws on their wrists. The hoatzin relies on very little in the way of habitat in order to survive. It does not migrate, it has a very simple diet, and the adults expose themselves to very few risks. Its simplified lifestyle allows the hoatzin to exploit a more common and extensive habitat which, in turn, has improved its chances of success.

This set of characteristics is unlike any bird within the avian lineages. It is not unreasonable to speculate that this species is a window into the restricted life some ancient birds in the Mesozoic era might have lived.

10

Unusual Birds

Within the thousands of species of birds, there are a variety of very unusual species. These birds are truly unique and unusual in many different ways. The fact that birds are feathered sets them apart from all other vertebrates; in addition, birds have adapted to so many habitats and environmental niches that they have evolved in many different directions.

We can appreciate the unusual beauty of birds through the many species that have extravagant plumages with deeply saturated colors and patterns that catch the eye. Hummingbirds, sunbirds, pheasants, trogons, cotingas, toucans, pittas, parrots, kingfishers, tanagers, tropical finches, and many more families of birds have colorful plumages. Artists and photographers have used their talents to help us appreciate the brilliance of birds.

Birds are also unusual in their behaviors. Many species carry out elaborate courtship displays; we can only marvel at their dances. Birds-of-paradise, grouse, manakins, bowerbirds, cranes, albatrosses, shorebirds and some grebes and flamingos are just a few of the families that perform complex rituals.

Birds are very unusual for their ability to navigate, and for their extreme endurance with long distance migration. Some species travel very long distances for their size, and navigate with amazing precision to and from breeding and wintering areas; some even circumscribe the oceans in their annual wandering. Penguins are unusual because they spend most of their lives at sea and have developed a lifestyle more similar to seals than to birds.

There are also many very rare birds on earth, but few of them have been thoroughly studied. Walter Jetz *et al.* (2014) conducted studies of the most evolutionally distinct and the most imperiled birds on earth. Their analysis identified the 465 most genetically evolutionary, distinctive species; some of these have large ranges and are well known, like the ostrich, but others are rare with severely restricted ranges. The Jetz study analyzed the 575 most imperiled birds: those that are on the International Union for the Conservation of Nature (IUCN) Red List, and are endangered or threatened with extinction. Many of these species are rare or imperiled because of distinctive genetic traits that also make them unusual or unique.

With all of these exceptional variations, there are different opinions on which species are somehow unusual among all bird species. Some have evolved to stand out because of their very different behavior, lifestyle, or intelligence. Many (but not all) of these species are monotypic, the sole member of their genus—like the hoatzin discussed in Chapter 9. Focusing on a few extraordinary birds emphasizes the vastness of avian diversity, because it points out the extensiveness of their adaptations to environmental niches.

Plains-Wanderer

In the short space grasslands of eastern Australia lives the rare Plains-Wanderer [Figure 10.1], a small (6 to 7.5 inches in length, or 15–19 cm), retiring, secretive bird that is active during the day or night, but rarely flies. It is almost impossible to see during the day, but it can be heard, and it is possible to approach it at night when the bird remains stationary for extended periods. The only species within its family, the Plains-Wanderer's genetic relationship with other species is not entirely certain. It was first thought to be related to button quails because of its appearance, habitat, and behavior, but later work indicated it was more closely related to plovers—and today researchers believe it to be closely related to the seedsnipes, which are in the shorebird order. A fossil from early Miocene (19–16 MYBP) deposits in New Zealand is thought to be related to both the plains-wanderer and seedsnipe families, based on the features of its postcranial skeleton (De Pietri *et al.* 2015).

Figure 10.1 The Plains-Wanderer of Australia is a small retiring bird in which the female defends a territory. Notice the elongated dark area on the lower portion of the iris (author's collection).

Females defend a territory, and give a repeating low-frequency, long-ranging call that sounds like a moan ("oooom"). Normally, birds cannot determine the origin of sounds at such a low frequency. Until recently, it has been a mystery how plains-wanderers locate each other with this unusual call. Research by Jack Pettigrew, a neurophysiologist, found that the birds' ears are connected through the head, allowing sound to arrive twice at each ear. This acts like a directional microphone, localizing the direction of the sound source (Pettigrew 2017).

The Plains-Wanderers' ears are not their only distinctive trait. Their eye pupils appears to be irregular in shape, elongated and broad in the lower half. This results from dark pigmentation in the lower half of the yellow iris, but the pupil is actually circular. Researchers believed that this benefits their sight for day and night activities. (Pettigrew 2017).

As with phalaropes, the female Plains-Wanderer is larger and more brightly colored than the male. Very little is known about the mating system because there have been no direct observations of the male-female interactions or courtship behavior in the wild (Nugent *et al.* 2022). There is some evidence from captive breeding that lekking, mentioned in Chapter 3, may be a part of their biology (Pauligk 2020). The male incubates the eggs and cares for the young.

These birds have a very restricted and fragmented range in southern Australia. They require dense grasses for nesting and shelter and open spaces for foraging and roosting (Nugent *et al.* 2022), limiting their habitat choices. Surveys in 2010 and 2012 showed a population decline of over 90 percent on Victoria's Northern Plains, due to flooding and other changes in native grasslands. Because of the expansion of agriculture, habitat loss, use of pesticides, and introduced predators, they are now an endangered species with a total population estimated at less than one thousand. When some grasslands were protected from overgrazing by cattle and sheep in the early 2010s, their population partially recovered (Baker-Gabb *et al.* 2016).

Greater Honeyguide

The Greater Honeyguide [Figure 10.2] of Africa is a small passerine that has developed a symbiotic relationship with native people in Ethiopia, Kenya, and Mozambique. In 1588, a Portuguese missionary in Mozambique, João dos Santos (1891), wrote about a peculiar habit that a small bird had of leading men to bees' nests by calling and flying from tree to tree.

The process begins when the person collecting honey makes a specific call. These human calls differ depending on the geographic regions. The bird will respond with a special call distinct from its territorial song; it then flies from tree to tree, leading the human honey gatherer until it finds the hive hidden in a tree cavity (Isack and Reyer 1989). Studies show that it is successful in guiding people to an active hive 75 percent of the time. After people harvest the honey, they lay out wax from the cone, and the birds eat this wax left behind—not the honey. Statistical analysis of the sounds made by natives to induce the birds, the guiding call responses, and the rate of success at finding nests indicate that a wild bird in a

Figure 10.2 The Greater Honeyguide has developed a symbiotic relationship with native peoples (author's collection).

natural setting has adapted to a human signal for mutual benefit (Spottiswoode *et al.* 2016).

Several aspects of this behavior are surprising. First, these birds show no fear of people. Second, the species is a brood parasite that lays its eggs in the nest of a host species, so the immatures do not learn this technique from their host parent. Native people who use the greater honeyguide for guiding indicate that immature birds respond to the prompting call, but do not lead the honey-gatherer. So there must be some learning involved for young birds to develop this skill, most likely from adult birds (Spottiswoode *et al.* 2016).

Symbiotic relationships with animals are well known for other bird species. Oxpeckers and cattle egrets will feed on the backs of cattle to remove insects; myna birds in India eat insects off of antelopes. Black rhinoceroses have very poor vision, so red-billed oxpeckers, which ride on their backs, act as sentries and warn of anything approaching, giving alarm calls to alert the rhino of the danger—and the rhino reacts to avoid the hazard. The birds detect a possible threat at a much further distance than the rhino can, with a 97 percent detection rate (Plotz and Linklater 2010).

An unusual symbiotic relationship exists between an albatross and a fish. The albatross extracts a shrimp-like parasite, *pennella*, from ocean sunfish (*Mola mola*); the fish will gather into a school to attract the albatross.

There is a myth that Egyptian plovers are known to pick food material from inside the mouths of crocodiles, but this has never been adequately documented and is probably not true.

The more important and widespread symbiotic relationship exists between birds and plants. Seed- and fruit-eating birds disperse seeds, and nectar-feeding birds spread pollen.

Glacier Bird

Emperor penguins breed on the ice in Antarctica, but there are no other birds that are known to do so because of the difficulty of providing adequate heat to incubate the eggs. In a rare circumstance, black-legged kittiwakes tried to nest on advancing glacial ice, but their nests failed (Irons 1988).

In 2005, two researchers studying the Quelccaya Ice Cap in the Peruvian Andes

Figure 10.3 The Glacier Finch is known as the ice bird because it nests in cavities in glacial ice (David D. Beadle).

at an elevation of 17,000 feet (5,200 m) found a cup nest that had fallen from the glacier. During further work in the next two years, they found numerous nests of the glacier finch, a small bird about seven inches long (18 cm) (Hardy and Hardy 2008). These finches construct a bulky nest of twigs and grasses in April or May that they place directly on the glacier, usually within an ice crevice or under overhanging ice. The nests are almost one foot (32 cm) wide with an inner cup made of finer materials. The immatures fledge in June or July, before the ice melts when the glacier is retreating.

The Glacier Finch [Figure 10.3] is a high altitude species with a small range primarily in Peru. The Diuca Finch is a lower elevation species found throughout southern South America, and the ranges of these two species do not overlap. The name "diuca" is an aboriginal word for bird. Although the two species are similar in appearance and size, they are not closely related to each other—but each belongs to separate groups of tanagers (Cookson *et al.* 2018).

Emperor Penguins

Emperor penguins are, perhaps, among the most remarkable birds on earth. They are unusual because they have adapted to breed on the Antarctic sea ice during the south polar winter. No other animal occupies Antarctica during the winter, the most forbidding environment on the planet. They survive because of their many adaptations, including their ability to withstand the Antarctic winter storms, to time their feeding cycles to sustain themselves and their chick, to incubate their single egg on the frozen ice shelf, and to recognize their mate and chick after the long absence during their commute to the sea.

The emperor penguin is the largest living penguin. At full weight, a male tops the scale at almost 90 pounds (45 kg) and stands about 4 feet (1.2 m) tall, while females are slightly smaller.

Antarctica has only two seasons: summer and winter. Sea ice builds up around the continent during winter, beginning at the end of the summer in February, and melts during the summer, which begins in October. Unlike the ice shelves, which are made up of glacial ice and can be half a mile (1 km) thick, sea ice is a few yards (meters) thick. The emperors return to this ice and gather in a colony, choosing a location on the ice shelf that allows them to commute to the sea to feed, but is far enough from the edge of the shelf to remain stable during the beginning of summer in October. This gives the new chicks sufficient time to mature before going to sea for the summer.

The breeding cycle begins at the end of summer when temperatures are routinely -4 F (-20 C). Under blizzard conditions, temperatures can reach -40 F (-40 C), and storms bring winds from the interior of the continent (Katabatic winds). Although average wind speeds are only about 13 mph (21 km/hr), winds can reach 100 mph (160 km/hr) during blizzard conditions. Emperor penguins routinely face these conditions, which extract a heavy toll on juvenile birds; even some adults perish. These are the most severe weather conditions on earth. Lindsay McCrae (2019) describes the difficulties these penguins face in winter: during storms, adults form a huddle to conserve heat, remaining in the huddle as long as the storm persists. While in this tight group, birds move continually around and the colony gradually moves, sometimes as much as a mile.

Adults commute to the sea to replenish their food supply and to feed their single chick. After egg laying, the female transfers the egg to the male's brood patch, which protects the egg and the chick after hatching from the cold and ice. The female then leaves; her trek to the ocean can be from about 12 miles (20 km) to as far as 60 miles (100 km) (Ancel *et al.* 1992). The adult will not return to relieve her partner for two months. The male incubates the egg, the chick hatches, and the male makes one feeding before the female returns. During this time, the male will loses half his weight.

There are no geographic boundaries or definable territories within the penguin colony. Adults and chicks are often displaced from one another. Studies with Humboldt penguins show that adults recognize each other and their chick primarily by scent (Coffin *et al.* 2011). Emperor penguins may use calls as well.

Shoebill

Throughout the world, herons and egrets tend to be similar in their appearance and feeding habits—but one of them, the Shoebill, stands out. It is heavily built, with a short neck and an unusually large bill. The Shoebill is a monotypic species in

Figure 10.4 The Shoebill, a unique heron of Africa, has a hooked bill and captures a lungfish, one of its favorite prey (Nik Borrow).

the pelican order, which also includes herons and bitterns and another unusual species, the hamerkop. A rare bird with a very patchy distribution in secluded marshes in Africa from South Sudan to Zambia, it stands four to five feet tall (1.2 to 1.5 m), weighs 9 to 15 pounds (4 to 6.7 kg), and has long toes. It gets its name from its large beak, which is eight to nine inches long (20 to 25 cm) and extraordinarily wide, with sharp edges and a large hook at the end.

Shoebills can stand and walk on aquatic vegetation, and they will remain motionless for long periods of time while stalking their prey. On finding prey, they plunge their massive bill straight down to grab the large prey and jerk upwards, as shown in Figure 10.4. They use their sharp beak to pierce their prey and slice it into pieces. They prey on fish, snakes, and baby crocodiles, and a favorite food is the lungfish (genus *Protopterus*), a large eel-like fish up to 3.3 feet long (1 m) that can weigh more than the Shoebill.

Oilbird

The oilbird of South America [Figure 10.5] is one of only a few bird species that uses echolocation. The only other birds known to use sound waves this way are swiftlets. Although oilbirds have good vision with their small eyes and large pupils, they give loud audible clicks to echolocate in the total darkness of caves, in which they live by day. Unlike bats that can detect an insect in the dark, oilbirds have less sensitivity, but they can navigate successfully in their caves and when foraging.

Figure 10.5 The Oilbirds of South America find fruit in the tropical forest at night by echolocation (author's collection).

They were named oilbirds because the young get about 50 percent heavier than the adults, and were, therefore, harvested by local natives and rendered for their oily fat. These large birds forage at night, and roost and breed on rocky ledges by day. Colonies can be extremely noisy with their harsh, screeching calls. Related to nightjars, they have the general appearance of a nightjar—long and slender with elongated wings and white spots on their brown body plumage. Oilbirds differ from nightjars, however, in that they are colonial breeders and feed mostly on fruit rather than on insects—particularly the fruits of palm and laurel trees, both of which produce fruit in drupes. Oilbirds spend their nights searching for fruit, using their sensitive sense of smell.

Poisonous Birds

Several species of tropical birds have toxic feathers or flesh, the result of ingesting poisonous beetles or plants. The three species of the pitohui of Papua New Guinea, for example, carry the same neurotoxin found in poison-dart frogs. The most well-known of these three is the Hooded Pitohui [Figure 10.6]. A bird bander discovered the pitohui's toxicity accidently when bitten by one of these birds. The poison did not come from the bite itself, but was the result of handling the bird

Figure 10.6 The Hooded Pitohui can spread toxins from its feathers when handled (Nik Borrow).

afterward while nursing the wound. The toxin, batrachotoxin, might be a chemical defense mechanism to ward off predators (Dumbacher *et al.* 1992).

The blue-capped ifrita and the little shrikethrush, both from Papua New Guinea, are also toxic in the same way, and from the same poison found in the pitohui.

The spur-winged goose of Africa and the common quail of Europe, Asia, and Africa consume toxic substances, causing their flesh to be poisonous. The common quail will consume toxic hemlock seeds, producing an illness of muscle tenderness, pain, nausea, and vomiting if someone eats quail flesh (Korkmaz *et al.* 2008). The spur-winged goose ingests a beetle that produces a different toxin than the pitohui— but one fatal to humans who eat the goose's flesh. Surprisingly, these substances are not toxic to these birds, and when they are raised in captivity with diets free of toxins, they do not accumulate these substances.

Parrots of the Night

Parrots are showy, loud birds that are not typically shy. They are diurnal, but two species of parrots are active at night. The kakapo of New Zealand and the night parrot of Australia are both critically endangered. They are both green in color with some dark mottling on their plumage.

The kakapo is unique in several ways. It is the largest living parrot, the males weighing from 3.5 to 8 pounds (1.6–3.6 kg). The females are much smaller, a larger size difference than in most birds. They are flightless but capable of climbing trees, and feed on plants, seeds and fruits that they locate with their acute sense of smell and fine bristles near their beak, which act as feelers and as sensors for walking.

New Zealand was free of land mammals (except for one bat), so kakapos became widespread in a variety of habitats and evolved to occupy the niche of a ground mammal. They are known to be long lived, are lek breeders, and do not breed every year—only when certain fruits are available (Powlesland *et al.* 2006). Their numbers were whipped out by human pressures, reducing the population to only 50 birds in the mid–1990s. Today their population has recovered to about 150 birds through extensive conservation measures.

The night parrot, a small parrot about 9 inches long (23 cm) and related to the Australian ground parrot, was first discovered in Australia in 1845. Through habitat losses due to ranching and probably introduced predators, the population dropped dramatically, and they vanished from 1912 to 1979. Some searchers reported sporadic sightings beginning in about 1979, but it wasn't until two were found dead in Queensland in 1990 and 2006 that an effort began to find any remaining birds, and to do something to conserve them. In 2013, John Young took photos of a live wild bird for, perhaps, the first time ever (Bush Heritage 2018).

Very little is known about this rare ground-dwelling nocturnal bird, because its small numbers made it impossible to study. The habitat where it has recently been found is dominated by tall bunch grass in the dry interior of Australia, and most sightings have been in western Queensland.

Lyrebird

The world's most unusual passerines are the lyrebirds of Australia—two species, the superb and Albert's lyrebirds, with the superb being slightly larger [Figure 10.7]. These large, long-legged ground birds have long tails, and ornithologists originally thought they were related to pheasants. There are no songbirds remotely similar to the lyrebirds in terms of their size and behavior.

The Superb Lyrebird is 40 inches long (1 m), with a body about the size of a chicken, weighing about 2 pounds (1 kg). Its long tail is made up of two strikingly marked outer tail feathers and a dozen wiry filament-like feathers. When the male displays to the female, he raises the tail feathers over his head and fans them in the shape of a lyre. Lyrebirds nest on the ground, and spend most of their life on the ground. Although they are weak fliers with short wings, they can use short flights to ascend and roost in trees.

Both species display and call from their elevated platform with metallic and varied sounds, but they also include sounds of other birds and natural sounds in their repertoire. They are among the world's great mimics, with the ability to imitate the sounds of at least 20 different resident bird species, as well as natural sounds such as frogs and other animals. When attracting a female, the males will sing a varied song and dance in concert to the song they are singing, by matching subsets of songs from their vocal repertoire with different combinations of tail, wing and leg movements (Mulder and Hall 2013). When raised in captivity, they can reproduce the sound of human voices and machines, and will remember these sounds for long periods of time (Taylor 2014).

Figure 10.7 The Superb Lyrebird is one of the largest living passerines. In addition to being an excellent mimic, it makes striking displays with its long tail (author's collection).

The lyrebird is known to be one of the oldest living passerine songbirds. The Australian Museum has a fossil of a lyrebird from 15 MYBP (Boles 2002).

Ruff

Depending on the species, birds practice all kinds of mating patterns, including some birds that do not establish pair bonds, but only come together for a short time to mate. Birds have many ways of attracting a mate: calls, songs, displays, establishing a territory, and lekking, where males gather together and display to females, and females select their mate.

The ruff is a very unusual lekking species, in that there are three types of male ruffs that differ in mating strategy, plumage, and slightly in size. The ruff mating system is the result of an evolutionary process in which a rare combination of events many millions of years ago initiated multiple genetic changes (Lamichhaney *et al.* 2016).

Male ruffs are Palearctic shorebirds that grow long showy neck feathers in

Figure 10.8a Male ruff (independents) with black collar in display posture (Bruce Hallett).

Figure 10.8b Male ruffs (independents) can also have white, buff, or reddish collars (Bruce Hallett).

advance of their breeding season. Ruff plumage includes three basic colors: white, black and rufous, with variations of these colors, as shown in Figures 10.8a and 10.8b. These males, called independents, represent 85 percent of the male ruff population.

When they reach their breeding range in northern Europe, the males and females gather at lekking sites, where the males perform elaborate displays, raising their neck feathers as though they are a ruff collar, spreading their wings, and circling on the ground for position. This frenetic activity involves many birds. Among these are two other kinds of male: the satellites and the faders.

The satellites look like the independent males, but they do not display at the lek. Satellites wait for mating opportunities when the female sits in readiness, and will mount the female before one of the independent males has a chance.

Faders make up only one percent of all males. Most unusual because they look exactly like a female, faders do not perform displays. Like the satellites, they use the frenetic activity of the lek to capitalize on mating opportunities when females are ready to mate.

Amazingly, despite their small percentage among the ruffs, the satellite and fader morphs are the result of the dominant genes (allels). The mating system of the ruffs has been called the most remarkable in the animal kingdom (Lamichhaney *et al.* 2016).

Secretarybird and the Seriemas

The bird kingdom offers many examples of convergent evolution—two species that evolved from different backgrounds to occupy the same environmental niche. One interesting example includes the seriemas and the secretarybird.

The seriemas of South America [Figure 10.9a] and the secretarybird of Africa,

Figure 10.9a The Secretarybird of Africa is a raptor of the African grasslands that has crane-like legs (author's collection).

Figure 10.9b The Red-legged Seriema of South America and the Secretarybird are examples of convergent evolution (Bruce Hallett).

[Figure 10.9b] are similar in structure and lifestyle. They are both predatory birds of grasslands, and have very long legs and long tails; they are predominately ground birds. The seriemas are weak fliers, unlike the secretarybird, which perform soaring aerial mating displays. Both species are predators, taking small animals and snakes.

The seriemas are represented by two similar species, the red-legged and black-legged. Relic species with a long evolutionary history in South America, they are the last remaining members of their order, the *Cariamiformes*, and their predecessors included the terror birds (Mayr 2017a). Today they are considered a sister taxon to falcons and parrots (Prum *et al.* 2015). The two-foot-high (60 cm) seriemas feed on insects, small fruit, tree gum, snakes, reptiles, small birds and mammals. They kill small animals by grabbing them with their bill and bashing or throwing them against the ground or rocks.

The secretarybird, at four feet tall (1.2 m), is twice as large as the red-legged seriema, and has stronger legs. Secretarybirds take snakes and other small mammals. They kill their prey by thrusting a leg and beating it with their strong feet and toes. Once they flush a snake, they will use short flights to follow and harass the snake until they can strike it; then with repeated strikes, they make the kill and swallow the entire snake. They can apply a force of five times their weight while striking, using extremely accurate and quick kicks to kill venomous snakes (Portugal *et al.* 2016). The strike duration has been measured at 15 milliseconds, ten times faster than the blink of an eye, with a strike force measured at about 36 pounds (16 kg). Although the secretarybird looks like a crane, it is a raptor and placed in the order of birds that include hawks—but there are no hawks that kill their prey in this manner.

Magpie Goose

The Magpie Goose of Australia is only weakly related to true geese. The last remaining member of a family of waterbirds that went extinct 20 to 40 MYBP (Miller 2018), the magpie goose has been called a living fossil. While it is an Australian bird, an early partial skeleton from the Eocene epoch was found in the London Clay Formation in England (Mayr 2017b), suggesting that the family was much more widespread eons ago. Its closest relatives are the screamers of South America.

Like the hoatzin, the Magpie Goose [Figure 10.10] is unique in its physical characteristics when compared with more recently evolved waterfowl. It has an elongated trachea for loud honking, partially webbed feet, a beak with poorly developed lamellae (thin layers of bone), and a gradual molt with no flightless period. Its long toes allow it to perch on small branches, so it can nest in trees, but it normally nests on the ground. Magpie geese are colonial wetland breeders—some males breed with two females that lay eggs in the same nest. The geese feed on land, and also consume vegetation they dig up with their long toes in shallow water. They are the only waterfowl that feed their chicks by passing food bill-to-bill.

Figure 10.10 The Magpie Goose of Australia is a distant relative of modern geese and ducks (author's collection).

Ostrich

Everyone knows what an ostrich looks like, a long-legged, gangly bird with a long neck and a mythical reputation for putting its head in the sand. Many people have eaten ostrich meat, which is farmed in many parts of the world. The ostrich is a member of the ratites, an ancient group of birds whose origins date back before the great extinction of 66 MYBP. Researchers believe that the ostrich, as well as the other ratites, descended from flying birds that originated in the north and invaded the southern hemisphere (Maderspacher 2016). They lost the ability to fly, and became large herbivores in habitats that lacked sizable predators. The ostrich is the largest ratite to survive to the modern era.

Standing ten feet (3.1 m) tall and weighing up to 400 pounds (180 kg), the male ostrich is the world's largest living bird. An omnivorous bird, the majority of its diet is plant material including leaves, roots, grasses, and flowers. The all-black males are larger than the females. Like emus and rheas, their feathers are soft, and the plumes are harvested as fashion items and for feather dusters; native peoples have uses for all parts of this animal.

The ostrich has only two toes, both very large and pointing forward, and very strong legs. These adaptations allow it to run as fast as 43 mph (70 km/hr). When running at high speed, it uses its large, feathery wings as stabilizers when making turns. Both males and females carry out elaborate displays with their wings. The structure of their wings suggests that flightlessness was an earlier adaptation at a time when they were not threatened by ground predators.

In prehistoric times, the ostrich had a much wider distribution then today, living in Europe, central Russia and China (Mayr 2017c). Today they are confined to the open savannahs of Africa. It is amazing that such a large, flightless bird has survived with all the mammalian predators of Africa, while the once-abundant subspecies of Arabia and Syria became extinct in 1966 through human hunting.

The Reach of Adaptation

The fifteen highlighted species show just how extensively birds have adapted to their environment when boundaries are broad enough to provide opportunities for them. The featured examples included very diverse behaviors: a bird whose ears are connected through the skull so it can accurately hear very low frequency sounds and sense their direction; birds that establish symbiotic relationships with other living vertebrates; a small bird that nests on glacial ice, and a penguin that is the only living vertebrate to breed on Antarctica in winter; a rare heron that uses its sharp beak like a raptor to kill its prey; birds that consume toxic substances, parrots that are active at night, and a parrot that has a lifestyle like a squirrel; a passerine that looks like a pheasant and behaves like a grouse; two unrelated species that kill prey by bashing them with their long legs; a goose that is very different from all other geese, a shorebird with unusual mating practices, and a large flightless bird that survives among a host of large mammalian predators.

These are only some of the extreme example that show how birds have evolved in many directions, and exploited habitats that one would not have thought possible.

PART IV

An Overview of Birds and Their Habitats

Introduction

Part IV of this work is focused on selected families of birds and their habitats. Chapter 11 gives an overview of the 41 orders of living birds and the types of birds within these orders. Many species from different orders occupy the same basic type of habitat, and yet other species have colonized a variety of habitats. More specialized families of birds can reside in more restricted habitats. Part IV explores some of the different types of birds and the habitats they occupy.

Birds' ability to adapt to various habitats has allowed them to diversify. They have exploited every habitat on earth, and have divided these into smaller segments where different species coexist. Birds further divide habitats by environmental factors such as food sources, time of day, height above the ground, depth of the ground that can be probed, cover, and many other differences that can be exploited for their basic needs. Examples of the major habitat types they occupy include tropical and temperate forests, open woodlands, conifer forests, mountains, scrublands, arctic tundra, deserts, oceans, swamps, wetlands, and grasslands. If we want to protect the diversity of birds, protection of habitats is the first priority.

Chapters 12, 13 and 14 deal with the largest order of birds, the Passeriformes. These insect-eating birds are the largest group of forest species. Flycatchers are highlighted as an important topic related to this environment in Chapter 13. The Hawaiian honeycreepers, described in Chapter 14, provide us with an example of adaptive radiation and diversification to take best advantage of multiple environments.

Chapters 15–18 are concerned with habits or food sources that are important for both passerines and non-passerines. Grassland birds are discussed in Chapter 15, while nectar-feeding birds, described in Chapter 16, are specialists adapted to flowering plants. Species adapted to wetlands are the focus of Chapter 17, and finally, fruit- and seed-eating birds are the subject of Chapter 18.

Chapters 19–22 are related to non-passerine species and their habitats. Trogons are an example of tropical and subtropical forest birds. They have a long history on earth, based on fossil records that stretches back to 50 MYBP. Oceanic birds, covered in Chapter 20, have adapted to occupy one of the most challenging habitats on earth. Shorebirds, discussed in Chapter 21, are members of a very diverse order that use a variety of different habitats, depending on the family. This order is particularly noteworthy because it can be exceptional in many different ways and have many different families, second only to the order of passerines. Raptors, detailed in Chapter 22, are primarily birds of open country, but some types make use of other environments.

11

The Classification of Birds

We are all aware of the fact that there are many different species of birds. Backyard birdwatchers whose yards are set up to attract birds can record as many as 15 species or more in a day. In spring when birds are migrating, bird watchers can find as many as 50 to 100 species on a day of searching, and avid birders can find many more on a "big day" quest. The variety of bird species that we can enjoy without much effort is really amazing—and certainly, birds that are more secretive or inhabit distant and inaccessible places require much more effort to see, making sightings of them even more desirable for enthusiastic birders.

In their book *Ornithology*, published in 1676, Francis Willughby and John Ray made one of the earliest studies of birds, in which they described the birds known at that time in Britain and Europe. The study of birds and their relationship to one another has changed considerably since then, with emphasis on taxonomy, the science of classifying living organisms to provide an accurate and complete listing of all living and extinct life on earth. Today taxonomists are constantly gaining further understanding of how birds are related to each other, and this has brought about many revisions in avian taxonomy.

Birds are biologically classified in the phylum of *Cordata*, which are organisms with a spinal column. Under *Cordata*, birds are in the class known as *Aves*, differentiating them from other classes that have vertebrae: fish, mammals, amphibians and reptiles. From there, birds are sorted into the other levels of classification: order, family, genus, and species.

Within the avian class, there are currently 41 orders of birds, which include 248 extant families [Table 11.1], based on the Clements Checklist of Birds (Clements *et al.* 2021). If you think of evolution of life as a tree with branches, then the "orders" are the main branches stemming from the trunk of living organisms that represents all birds.

The International Ornithological Congress (IOC) currently recognizes 40 orders. All orders of birds have the Latin suffix "...*forms*" which literally means "shape." All birds within an order share some important anatomical distinction that sets them apart from all other birds. For instance, the *Anseriformes* contain all the ducks, geese and swans, and these species share a certain type of bill and a type of webbed foot. Another example are the *Procellariiformes,* which represent all seabirds that have the nostril enclosed in a tube above the bill—they are sometimes referred to as tubenoses. Table 11.1 lists the major groups of birds within that order.

The arrangement of the orders of birds in Table 11.1 is based on their appearance in time, with the older lineages at the top of the list, and those that appeared later in time lower on the list. This is a taxonomy of living birds, and does not include extinct fossil birds. The first five orders of bird form an infraclass (subgroup) called the *Palaeognathae,* and are grouped together by their primitive skull anatomy and other features. They include the four orders of flightless birds (ratites), and the tinamous, which can fly and have a keeled sternum.

All other orders of birds belong to the infraclass *Neognathae,* and they include 99 percent of all living species. They differ from the *Palaeognathae* in specific skeletal traits. The next two orders, *Anseriformes* (ducks, etc.) and *Galliformes* (game birds, etc.), were most likely present during the Mesozoic era alongside the other dinosaurs, based on the fossil evidence.

James Peters began to publish a checklist of birds of the world in 1931. This monumental work of 15 volumes was completed by other authors in 1962 and listed 34 orders of birds (Peters 1931). He made use of suborders, families, and subfamilies to capture the taxonomic differences among the genera of birds.

There are a number of reasons why the systematics have changed and grown to 41 orders. The orders within the *Neognathae* have changed positions and been expanded as research made their origins and relationships clearer. New orders were introduced by splitting former orders based on genetics. For instance, falcons and New World vultures were formally part of the order that included hawks and eagles, but now they are assigned to a separate order. Some very unusual bird species are not clearly related to other birds and are placed in a separate order until further work might determine a better placement—the *Opisthocomiformes* (hoatzin) and *Eurypygiformes* (sunbittern and kagu) are examples.

Table 11.1 Orders, Families, Genera and Species of Birds

	Order	*Families*	*Genera*	*Species*	*Types of families*
1	*Struthioniformes*	1	1	2	ostriches
2	*Rheiformes*	1	1	2	rheas
3	*Tinamiformes*	1	10	46	tinamous
4	*Casuariiformes*	1	1	4	cassowaries, Emu
5	*Apterygiformes*	1	1	5	kiwis
6	*Anseriformes*	3	56	178	ducks, geese, waterfowl; Magpie-goose, screamers
7	*Galliformes*	5	86	298	pheasants, grouse and allies, quails, guineafowl, guans, chachalacas, curassows, megapodes (mound builders)
8	*Phoenicopteriformes*	1	1	6	flamingos
9	*Podicipediformes*	1	6	22	grebes
10	*Columbiformes*	1	49	348	pigeons, doves
11	*Mesitornithiformes*	1	1	3	mesites (flightless birds endemic to Madagascar)
12	*Pterocliformes*	1	2	16	sandgrouse
13	*Otidiformes*	1	8	26	bustards
14	*Musophagiformes*	1	5	23	turacos, plantain-eaters, go-away-birds

	Order	Families	Genera	Species	Types of families
15	Cuculiformes	1	33	147	cuckoos
16	Caprimulgiformes	8	157	599	hummingbirds, treeswifts, swifts, Owlet-nightjars, oilbird, potoos, nightjars and allies, frogmouths
17	Opisthocomiformes	1	1	1	Hoatzin
18	Gruiformes	6	51	192	flufftails, rails, gallinules, coots, finfoots, trumpeters, limpkin, cranes
19	Charadriformes	19	89	383	gulls, terns, skimmers, auks, murres, puffins, skuas, jaegers, pratincoles, coursers, buttonquaill, sandpiper and allies, jacanas, painted-snipes, seedsnipes, plovers, lapwings, oystercatchers, ibisbill, stilts, avocets, thicknees, sheathbills
20	Eurypygiformes	1	2	2	sunbittern, kagu
21	Phaethonitiformes	1	1	3	tropicbirds
22	Graviformes	1	1	5	loons
23	Sphenisciformes	1	6	18	penguins
24	Procellariformes	4	26	138	shearwaters, petrels, storm-petrels, albatrosses
25	Ciconiiformes	1	6	19	storks
26	Suliformes	4	12	59	cormorants and shags, anhingas, boobies, gannets, frigatebirds
27	Pelecaniformes	5	34	114	ibises, spoonbills, herons, egrets, bitterns, Hamerkop, Shoebill, pelicans
28	Cathartiformes	1	5	7	New World vultures
29	Accipitriformes	3	71	252	eagles, hawks, kites, Old World vultures; osprey; secretarybird
30	Strigiformes	2	28	243	owls, barn-owls
31	Coliiformes	1	2	6	mousebirds
32	Leptosomiformes	1	1	1	cuckoo-rollers
33	Trogoniformes	1	6	43	trogons
34	Bucerotiformes	4	18	72	hornbills, ground-hornbills, wood-hoopoes, scimitarbills, hoopoes
35	Coraciformes	6	34	183	rollers, ground-rollers, bee-eaters, kingfishers, motmots, todies
36	Galbuliformes	2	15	54	puffbirds, jacamars
37	Piciformes	7	54	377	barbets (African, Asian and New World), toucan-barbets, toucans, honeyguides, woodpeckers
38	Cariamiformes	1	2	2	seriemas
39	Falconiformes	1	11	65	falcons and caracaras
40	Psittaciformes	4	98	390	parrots, parakeets
41	Passeriformes	142	1336	6470	passerines
	Total	**249**	**2328**	**10824**	

The next level in taxonomy within the order is family. Taxonomists also use the concept of suborders when an order has many families, and some of these families can be grouped by some common characteristic. Suborder is not a fundamental level of taxonomic classification. The "sub" classifications are auxiliary or intermediate taxonomic rankings.

We can see in Table 11.1 that there are, for instance, three families within the order of waterbirds (*Anseriformes*) and four families in the tubenoses (*Procellariiformes*). All family names end in the Latin suffix "idea" which is a Latin noun that indicates paternity or resemblance. This type of analysis of clades organizes birds by their earliest common ancestor.

The designations of family imply progressively closer relationships of individual species. The divisions of families are usually determined by a consensus of taxonomists. Some families can have very many genera and species, and are sometimes broken into subfamilies, which are defined by the Latin suffix "inae." The subfamily is a collection of closely related species within a family.

Some families have very few species, and a few have only one species and are regarded as a monotypic family. Ornithologists around the world do not always agree on which families of birds are monotypic. Excluding extinct species, the number ranges from 20 to about 40, depending on the world checklist. The Clements Checklist of Birds (Clements *et al.* 2021) records 37 species that are unique to their family. Species of a monotypic family tend to be very unusual for some specific reason—examples include the osprey, oilbird, and the limpkin. In some cases, unusual species might be placed in a monotypic family because their relationship to other species within the order is unclear.

The scientific name of each species is its binomial name, which consists of the genus and the species names expressed in Latin. If they are sourced from other languages, they are expressed in a Latin form. The genus refers to a grouping of closely related species. The species name is usually chosen by a naturalist who publishes the first description of this species. Many of these names involve a descriptive characteristic of the species like their color, size, or behavior, or the name of a person the naturalist wishes to honor.

Taxonomists can use different approaches to determine which species belong in a particular genus. Species might be related based on physical and biological characteristics that have been determined by cladistic analysis. Another method is to group species by a common ancestor, based on genetic analysis. A great deal of ongoing study determines the taxonomic relationship of species, resulting in many changes in the genera over time. In some cases, work has shown that species that appeared to be very similar have turned out to be more distantly related than previously thought. Research will continue in this field, and we can expect continued changes in the future.

The concept of a monotypic genus refers to a bird in a family with other species, but in its own genus. Species of monotypic genus are more numerous in large families. For instance, there are 347 species of hummingbirds, and 40 of these are of a monotypic genus; these include such well-known hummingbirds as sword-billed hummingbird, giant hummingbird, bearded mountaineer, and marvelous spatuletail. Monotypic genera represent some unique or very unusual characteristics and could have appeared earlier or later than other taxa, or they may have adapted to some unusual environment compared to their more distant cousins.

As mentioned in Chapter 3, the separation of species and subspecies in ornithology has been based on two different criteria: the biological species concept (BSC) and the phylogenetic species concept (PSC). Even the methods used to apply these criteria have changed over time, but the use of genetic information, song, behavior, and the environment has further improved this process of establishing the rank of species.

The subspecies name is added to the binomial name for those species with two or more recognizable taxa. Field guides usually omit subspecies names, but they can be important in understanding the range of a bird species and how the species has responded to the environment. Sometimes subspecies are elevated to the rank of a species (splitting), or conversely, reduced from the rank of species to subspecies (lumping). Analysis of the American Ornithological Society checklist of birds of North America has shown that 74 percent of North American bird species recognized today have not had a change by splitting or lumping in the past 130 years. Approximately 20 percent of the splits or lumps revert to their previous state, or are changed again at some later date based on further evidence (Vaidya *et al.* 2018). One example is Thayer's gull, which was at one time a subspecies of herring gull, was elevated to the level of a species, and since has been corrected as a subspecies of Iceland gull. Some species of birds have many subspecies, and any of these might one day be elevated to species rank, sharing the genus name.

Table 11.1 also lists the number of the families and the type of birds within that family, except for passerines, because there are too many to list. Among non-passerines, the families include the generic classes of birds we know well, such as duck, swift, hummingbird, and owl. In some cases, orders of birds include families that appear to be very different, such as the *Charadriformes* with its 19 families that include gulls, shorebirds, and sheathbills. This table gives an overview of the classification of birds, and a sense of their extensive diversity. This represents a comprehensive overview of the class of birds.

We begin to know and refer to birds by their common names as we become familiar with them. Field guides and internet sources provide both common and scientific names. The intent of the scientific name is to avoid confusion, because common names can vary by country and region. In order to understand birds at a higher level, it is important to become familiar with the genera of species you might encounter frequently, because birds of a common genus are closely related to each other and can have similarities in habitat types, behavior, and general appearance.

This introduction to the classification of birds seen in the orders, families, genera, and species is important to help readers understand the full scope of their diversity. Many families of birds are unknown to the general public. In the past fifty years, many studies by dedicated researchers have explored the behavior and ecology of birds at the family and species level. These studies have extended from the more populated areas of North America and Europe to the tropical regions of the world, where birds are more diverse. They provide us with an opportunity to learn about their lives and history of birds.

An account of all the families of birds is beyond the scope of this work, but a brief account of selected bird families and their habitats will provide examples of the various orders and families. Other authors have also treated this subject in more detail. The Cornell Laboratory of Ornithology's Birds of the World, available online by subscription, gives a description of all of the world's birds. Perrins (2003) treats birds at the level of the families and gives a synopsis of the life history for approximately 183 families with photographs and range maps. Winkler *et al.* (2015) provides a thorough summary of the life history and biology of all the families of birds. Tudge (2018) covers many subjects relating to the natural history of birds.

12

Perching Birds
and Forest Habitats

The Earth's Forests

Forests represent a variety of habitat types: tropical, subtropical, temperate, deciduous and coniferous, mixed, boreal, special forests at high elevation, and scrub and thorn forests. A greater variety of birds occupy forests than any other environment on earth. While passerines (perching birds or songbirds) are the most common birds found in forests, many other avian families depend on forests as well: tinamous, pheasants and partridges, doves and pigeons, parrots, cuckoos, owls, hawks, trogons, many hornbills, barbets, woodpeckers and woodcreepers, pittas, and toucans, just to name a few.

Bird diversity depends strongly on latitudes. In Alaska, between 26 and 80 species breed in an area of 350 miles (560 km) on a side. In the eastern United States, as many as 130 species occupy the same area. And in tropical Panama and Costa Rica, 500 to 600 breeding species inhabit an area of similar size (Terborgh 1992).

The highest diversity of birds in tropical forests is the result of the overall diversity of all forms of life in these habitats. They represent 20 percent of the earth's land area. Thirty percent of all tropical forests are rainforests (also known as humid tropical forests); they account for about two percent of the earth's surface. The remaining tropical forests include dry forest, shrub lands, mixed forests, mangrove, conifer, and moist deciduous forests. Analysis has shown that tropical forests harbor 63 percent of all terrestrial vertebrates. When considering bird species on earth, 72 percent utilize tropical forests (Pillay *et al.* 2022). Rajeev Pillay and his team carried out their analysis based on the range maps of 10,935 avian species, and found that 7,918 either breed in tropical forest or utilize them for wintering.

All forested areas are important for birds, since those wintering in the tropical biome breed in temperate and other types of forests, as do thousands of species that do not utilize the tropical zone.

Of the Earth's 57.1 million square miles (M sq mi) (148 M sq km) of land mass, 15.6 M sq mi (40.3 M sq km) are covered in forest. This represents a decrease of 32 percent since 1800, and the rate of decrease accelerated at the end of the twentieth century. In the 1990s, world forest coverage decreased by approximately 60,000 sq mi (0.15 M sq km) annually (Adams 2012). In the first decade of the twenty-first century, however, worldwide net forest losses averaged about 50,000 sq mi (0.13 M sq km) (Hansen *et al.* 2013), about the size of the state of Mississippi. The rate of loss slowed somewhat as efforts to slow climate change took hold, with many organizations and corporations planting vast areas with trees to offset carbon emissions elsewhere.

A team at the University of Maryland (Hansen *et al.* 2013) analyzed Landsat images of forest losses from 2000 to 2012, and determined that globally, 0.89 M sq mi (2.3 M sq km) of forest were lost during the 12-year study period—and 0.29 M sq mi (0.8 M sq km) of new forest were gained. Over this period, the rate of tropical forest loss increased at 810 square miles (2,100 sq km) annually. The tropics had both the greatest net losses and gains from regrowth and plantations, but the basic problem with plantations and regrowth is that they tend to create forests that contain just a few tree species—so they do not restore the original plant diversity.

Africa and South America have seen the largest net losses of tropical forest area today (Adams 2012). Indonesia and the Philippines are just two of many locations that have lost the majority of their forests: forests have decreased by 70 percent in the Philippines and only six percent of the land remains as prime habitat for wildlife. Unfortunately, these developing nations have large populations and few job opportunities, so they support their economies by exploiting their own natural resources, and most of the logging in these tropical areas is either unregulated or illegal. These governments do occasionally set aside land for conservation, but they rarely enforce their own regulations.

Satellite data implies that the United States and Europe are not losing net forest area at the rate of the tropics—but area coverage does not tell the whole story. Replanting takes place more aggressively in the United States and Europe, but maturity takes a long time, and it lacks the diversity of plants found under natural conditions. Satellite data does not distinguish mature forest from replanted or degraded areas. When forests are clear cut or partially logged, they are often replaced by tree species that do not support the natural resident birds, with corporate interests planting trees that provide specific commercial products. Around the world, agroforestry monocultures of palms for palm oils, eucalyptus, rubber, coffee, pine and spruces may provide benefit for carbon sequestration, but they create deserts for wildlife. The variety of plants within a forest are important for providing various food sources to birds, especially in tropical forests, where birds rely on different food sources provided by different trees over the course of the year.

Planting trees is one of the lowest cost approaches to carbon sequestration, but it must be done properly to be beneficial for wildlife. Planted forests need to contain a diversity of naturally occurring plants. In tropical areas, farms that blend trees and crops are more productive than monocultures, because they are more drought tolerant and require less use of pesticides and fertilizer (Lambert 2021).

Shade-grown coffee is a perfect example of this approach. The coffee grows on farms that maintain a diverse shade cover, provided by a wide variety of trees appropriate to that habitat. This sustains increased numbers and species of birds and pollinating insects, while providing natural control of soil erosion and protection from insects that prey on coffee plants. The coffee benefits, the birds benefit, and the environment gets a boost from the carbon sequestration the trees provide.

Monocultures are not the only problem faced by forests. Human development in the form of roads, agriculture, utilities, and other interruptions to natural areas causes fragmentation—the breaking up of large tracts of forested land into smaller pieces. This allows predators to invade forests more quickly and easily, and it limits the movement of feeding flocks throughout wide areas, many of which are dependent on a mixture of habitats.

Pollution by acid rain also causes degradation. Acid rain develops when sulfur dioxide and nitrous oxide are released into the atmosphere by fossil fuels, dissolving to form sulfuric acid and nitric acid. These fall to the ground in rain and leach aluminum and other minerals from the soil—nutrients that plants and trees require to grow and thrive. Areas of dead or dying trees may have been affected by this destructive phenomenon.

All of these factors have a significant impact on forest species of all types, not just birds. Replanting trees can help, but they must be added in the right ecosystems to restore those habitats. Adding trees to temperate forests, for example, does not replace the diversity provided by tropical forests.

Trees have important environmental benefits besides habitat for wildlife. They improve air quality, reduce pollution, reduce water runoff, and convert carbon dioxide to oxygen. In the residential environment, they provide shade and can reduce cooling costs. Research has shown that agricultural land that blends trees with crops can be more productive than monocultures (Lambert 2021). This process of agroforestry has been successful in a number of cases, but it is rarely practiced in the United States in spite of the fact that it could provide some benefits.

Scrublands

Scrublands are plant communities dominated by shrubs, and they are an underappreciated and overlooked forest type. Desert shrubs, chaparral, sagebrush, and barrens are all forms of scrublands, but the most common form are bushy shrubs in the eastern United States. These habitats may be found in their own patches, or on the edges of early-successional forest. They result from some form of ground disturbance, making their existence ephemeral, or they may continue to grow and revert to forest over time. Scrublands represent a biome that is rich in biodiversity, occurring throughout the world.

Too often, scrublands are thought of as having little value and are often destroyed, especially if they are on private land. They are a breeding habitat for many species of passerines and other birds, however, as well as for small mammals and other creatures like butterflies. In coastal areas, they serve as a pathway for migrant songbirds and raptors by providing cover from predators and an area for feeding. The goal should be to manage and protect scrubland areas, not to treat them as vacant lots.

Within the eastern United States, 41 species of birds breed in this type of habitat (Schlossberg and King 2007). Just a few include many species of wood warblers, various sparrows and towhees, flycatchers, wrens, yellow-breasted chat, snipe, woodcock, and whip-poor-will. Some birds prefer taller vegetation, greater than five feet (1.5m), while others prefer lower vegetation. Over half of these eastern species have experienced a decline in populations.

The western half of the country hosts just as many scrubland species in the wide variety of desert scrublands, as well as those along the Pacific coast.

Many of our natural shrub habitats are being replaced with invasive species. Some of these species are sold as ornamental plants: bush honeysuckle, autumn and Russian olive, burning bush, porcelain berry, Callery (Bradford) pear, kudzu, multiflora rose, and Japanese knotweed are just a few examples of widespread invasive shrubs. They pose a major problem for birds because native moths and butterflies do not feed on their leaves, and almost all passerine birds feed on caterpillars as a major source of protein. Hedgerows clogged with invasive species show 68 percent fewer caterpillar species, 91 percent fewer caterpillars, and 96 percent less caterpillar biomass than native hedgerows (Richard *et al.* 2018). When birds cannot find sufficient food, they are less likely to breed. Research has shown that bird populations can only be sustained if nonnative plants constitute less than 30 percent of plant biomass (Narango *et al.* 2018).

Scrubland habitats have been in a long-term decline in the United States, and are considered at their historical low today (Natural Resources Conservation Service 2012). The amount of early successional forests in the eastern part of the country have been in serious decline since the 1950s (Schlossberg and King 2007). In the early twentieth century, many farms in New England were abandoned as farming shifted westward, and scrublands became widespread. Today these scrublands have been lost to successional forests—the orderly, predictable change in the dominant species from low shrubs to tall trees—or to commercial and residential development. In fact, from 1950 to 2005, early-successional habitat of New England forests dropped from about 25 percent coverage of the area to just 5 percent.

Preservation of scrublands is primarily dependent on private land owners, railroad right-of-ways, powerline easements, and agricultural hedgerows. The problem, however, is that private land is rarely left vacant for a long period before further development takes place. The Natural Resources Conservation Service's (NRCS) Wildlife Habitat Incentives Program (WHIP) offers some assistance for this purpose. We'll discuss this more in Part V of this book.

Passerines

Passerines have been mentioned a number of times because they are such a large order of birds (Table 11.1). The Passeriformes take their name from the Latin word "passerines," meaning sparrow, and are also called perching birds or songbirds. Normally, all birds are separated into the two broad groups by dividing them into passerines and non-passerines. The non-passerines include all other orders of birds: loons and grebes, ducks, hawks, marsh birds, seabirds, shorebirds, gulls and terns, owls, doves, parrots, hummingbirds and others. Compared to non-passerines, passerines are small, vocal, colorful, diurnal land birds. Among non-passerines, there are many tropical species that resemble passerines such as barbets, rollers, honeyguides, and mousebirds, but these are not close relatives.

Passerines are the most highly diversified order of birds, and the most recently evolved. Most passerines are particularly adapted to forest and scrublands. To a lesser extent, they occupy other environments including grasslands, desert, and marshes, but are uncommon in mostly aquatic environments.

A variety of molecular analyses have consistently shown that the closest relatives to the passerines are the parrots, which have a different toe arrangement, shorter and stouter legs, and a different type of bill. Passerines are identified by the arrangement of their toes. They have three toes directed forward and one toe backward; they can perch on plane surfaces and have a special adaptation in the legs and toes that allows the fourth toe to curl for perching on cylindrical surfaces, a toe arrangement called anisodactyl. In parrots the feet are zygodactyl—two toes pointing forward and two backward. Generally, the bills of passerines are straight, and those of parrots are curved. Another close relative of the passerines are the falcons, which are not related to the order of hawks and eagles. These discoveries were unexpected because of the apparent differences between passerines and other orders of birds, but the analyses have been confirmed by a number of different studies (Hackett *et al.* 2008, Mayr 2017).

Origin of Passerines

Exploitation of the forest habitat by birds happened long after the K–Pg extinction event. Prior to the great extinction, the Enantiornithes were a well-diversified order of birds, but some authors believe they were mostly arboreal, which might have led to their demise at the end of the Cretaceous period, when most trees on earth were destroyed.

Passerines are divided into three major suborders. About 6,400 species fall into two of these suborders: the *oscines*, which includes more than 5,000 species in 123 families of songbirds; and *suboscines*, which includes 1,356 species in 17 families (Clements *et al.* 2021), distinguished by differences in the development of their vocal organ (syrinx) that prevents them from developing complex songs like the oscines sing. Finally, a few species of New Zealand wren have their own suborder, *Acanthisitti*.

Paleontologists have struggled to understand the origins of this super-diverse order of birds. Their evolution is difficult to follow because their small size and delicate bones have left relatively few fossils. Their first appearance in the fossil record is uncertain, but there are some fossils that establish that true passerines were present by the Eocene epoch, 56 MYBP to 34 MYBP. Boles (1997) reported the discovery of passerine remains from the Early Eocene in Australia. Other examples include two small birds with finch-like bills that are believed to be early fossil specimens of passerines: one found in Wyoming and dated to 52 MYBP, and the second fossil discovered in Germany, dated to 47 MYBP (Ksepka *et al.* 2019). Both were from the early Eocene, before the Oligocene epoch, and were believed to be seed-eating birds based on their bill shapes. From the late Oligocene (about 23 MYBP) onwards, passerines are very abundant in deposits in the northern hemisphere (Manegold 2009).

Ericson *et al.* (2002) have proposed a model for the spread of passerines around the globe based on geographical, genetic and biological data. Their analysis suggests that suboscines originated on Gondwana, the supercontinent that today includes most of the landmass of the southern hemisphere. These birds spread to Europe and Africa and later reached the Americas. The oscines spread from Australia about 40 MYBP and reached North America much later. Barker *et al.* (2004), using nuclear gene sequences of passerine families, support the Gondwanan origin of passerines, and proposed that multiple waves of passerine dispersal took place from Australasia into Eurasia, Africa, and the New World, commencing as early as the Eocene.

More recent research, based on comprehensive genome-scale DNA sequencing calibrated with fossils of known age, also suggests that the early passerines began to diversify during the Middle Eocene epoch (~47 MYBP) and suboscines and oscines diverged ~44 MYBP on the Australian landmass (Oliveros *et al.* 2019). Manegold (2009) points out all the difficulties in establishing the timing of these two groups, one of which is the lack of fossils, while another is the diversity of interpretations of the few fossils available.

Low (2014) summarizes DNA evidence from various researchers that supports the premise that oscine songbirds originated in Australia. The Australian lyrebirds and the two species of Australian scrub-birds (*Atrichornithidae*) are thought to retain the earliest features of oscine songbirds. Tim Low places these two families as the earliest oscines. Other work (Moyle *et al.* 2016) suggests that the earliest suboscine, the pittas and broadbills, were present in the two early periods of the Cenozoic Era.

Carl Oliveros *et al.* (2019) discusses the complexities of the passerine origin, their split into oscines and suboscines, their spread across the globe, and their continued diversification. These are exceedingly complex subjects, and there will be continued analysis and refinements of these ideas as new information is uncovered. One thing scientists agree on, however, is that they originated tens of millions of years ago on continents south of the equator.

Passerine Diversity

So many species of passerines exist that the normal taxonomic divisions of order, family, genus, and species are insufficient to encompass their diversity and variations. As different kinds of birds become more closely related, the ability to divide them into taxonomic classes proves ever more difficult. Ornithologists make use of suborders, subfamilies, and subspecies to group similar characteristics, and use those subdivisions with passerines.

As noted above, modern passerines are divided into three main suborders: New Zealand wrens (*Acanthisitti*), suboscines (*Tyranni*), and oscines (*Passeri*). The names *Tyranni* and *Passeri* have been suggested as an alternatives for suboscines and oscines.

The New Zealand wrens are represented today by two surviving species, but there were as many as six species in past history. They are placed with the passerines and are considered a sister taxa of all other passerine birds, thus sharing a common ancestor with passerines. Three groups of birds on earth are called wrens: the New World wrens (*Troglodytidae*), the Australian wrens (*Mularidae*) and the New Zealand wrens, but they are not related to each other.

No one knows whether the oscines or suboscines evolved first, but extrapolations based on DNA analysis suggest that they separated very early in their history. More precise estimates would require more detailed analysis with additional relevant information (Barker *et al.* 2004).

How does this difference manifest in living birds? As discussed in earlier chapters, birdsong is produced by vibrations in the air stream coming from the lungs and induced by muscle groups within the syrinx, which cause tension on the air duct membrane. The oscines and suboscines differ in the number of muscle groups within the syrinx. Suboscines have one or perhaps several muscle groups, while oscines have six to nine. Since suboscines have fewer muscles to control their airways, they produce fewer sounds with less musical quality. They lack the neural control present in oscine birds.

In suboscines that have been studied, their songs are not learned, but are acquired by the genetic makeup of these birds. Experiments have shown that if the eggs of one species of suboscine are placed in the nest of another, the nestlings will make the calls and sounds of their genetic parent, not their foster parent. Their DNA has distinct differences from the oscines that lead them to produce only the sounds they are genetically wired to sing.

There is, however, at least one family of suboscines that do learn their song. Bellbirds of Central and South America are large, fruit-eating birds that make an extraordinarily loud, bell-like call that is best described as a loud "bonk." This is the loudest sound produced in the avian world. Research by Donald Kroodsma, *et al.* (2013) showed that some species of bellbirds need to learn their song from their genetic parent.

Passerines occupy almost every habitat available on earth, except in the extremes of Antarctica and on water—with one exception. The only family of passerines that feed underwater are the dippers, thrush-sized birds whose streamlined shape allows them to walk underwater in a fast-flowing stream to search for invertebrates. Their plump size allows them to withstand relatively cold temperatures.

The vast majority of passerines are woodland or scrubland birds, and they have evolved to maximize the advantages of their environment. The adaptation of their feet is ideal for grasping a branch. Many of them are permanent residents of the temperate and tropics zones, while others are migrants, flying thousands of miles each spring and fall from tropical regions in the southern hemisphere to the forests of the north and south temperate zones. Passerines have also colonized grasslands, wetlands, boreal habitats, high mountain meadows, and other habitats, but their greatest diversity resides in forests. Unfortunately, this means that passerines are in worldwide decline because of the losses of woodlands and scrublands to timber harvesting and agriculture.

Suboscines Passerines

Suboscines including cotingas, manakins, antbirds, ovenbirds, and woodcreepers are substantially different in lifestyles, both from each other and from all oscine birds. Most of the suboscines are non-migratory, and the vast majority of them reside in South and Central America. There are no suboscines in Europe, and the only ones in North America are the flycatchers. Most are tropical or subtropical and can be broadly divided into three clades: Old World suboscines; New World *Furnariida* (ovenbirds and antbirds) and *Tyrannida* (flycatchers and other families) (Oliveros *et al.* 2019).

Old World Suboscines (Eurylaimides)

The Old World suboscines are made up of five families:

1. *Eurylaimidae* broadbills of Asia
2. *Calyptomenidae* broadbills of Africa
3. *Asites (Philepittidae)*, an unusual family of small, colorful birds that resemble sunbirds and are found in Madagascar
4. The amazing sapayoa of Panama and Ecuador, a small flycatcher whose closest living relatives are the broadbills of Africa and Asia. Broadbills are not related to other South American suboscines (Dzielski *et al.* 2016). The sapayoa's range and those of its close relatives underscore the complexity of historical bird movements that resulted in such an unusual geographic separation for a single species.
5. *Pittas (Pittidae)*, the largest family in this clade—colorful ground birds found in tropical Africa, India and southern Asia, with three species in Australia. The broadbills are primarily insectivores, but some eat fruit.

New World Suboscines

New World suboscines (*Tyrannides*) include eleven or twelve families, depending on the world checklist source. Some sources group just two families. Recent analysis groups them into two clades: Furnariida and Tyrannida. Research based on genetic analysis found that the earliest branches of the Furnariida are much older than the Tyrannida (Ohlson et al. 2013). The Tyrannida include four families—cotingas, manakins, tityras, and flycatchers. A fifth family, the sharpbills and royal flycatchers, were formally part of the family that included the tityras. The Furnariida include seven families; the largest are the antbirds, antpittas, ovenbirds and woodcreepers, and tapaculos. The birds of the Furnariida clade are very different in appearance, lifestyle, and behavior but they share the tropical forests or forest edges.

Furnariida: Antbirds, Antpittas, Ovenbirds, Woodcreepers, Gnateaters, and Tapaculos

The typical antbirds are another large family of suboscines (*Thamnophilidae*) with 237 species. They are generally monogamous, are dimorphic, defend territories against conspecifics, and territory sizes are dependent on body mass (Duca et al. 2006). Males are black to gray and females are brown to reddish brown. They lack bright colors; many have white highlights or spots on the wing coverts. They range in size from the warbler-like smallest antwrens to the slightly larger antvireos and medium-sized antbirds and antshrikes. The bill size and shape of these groups vary, but they are heavier billed, and some have a hook at the end of the bill. They are most numerous in humid forests and thickets, and they stick to the shadows and avoid sunny places. Most are arboreal except the antthrushes and antpittas. A few are migratory but some wander locally. Their common names are size classifications; they are not actually related to wrens, vireos, or shrikes.

The smaller antvireos and antwrens will seldom forage on the ground and are usually in higher foliage, but below canopy level. They are insectivores and use the same feeding techniques as those of warblers. They are frequently found in mixed flocks which usually contain only a single pair of one species. They remain in mixed flocks only as long as it remains in their own territory (Skutch 1996).

Among them are the bare-faced and "professional" antbirds, about 30 species, that follow army ant columns to feed on the small vertebrates and invertebrates flushed by the ants—but these birds do not consume ants. These professional antbirds have long claws and strong feet and legs for gripping branches as they chase prey while following the ant swarm. The larger antbirds prefer the undergrowth and are rarely ascend much above six feet (2 m).

The antpittas and antthrush are secretive, ground-dwelling antbirds, similar to the pittas of Asia and Australia. Antpittas feed on worms, ants, and invertebrates, while the antthrush takes larger prey such as snails, frogs, and lizards. Antpittas [Figure 12.1] are small with a plump body shape, long legs and short tails, and they have muted tones of brown, rufous, and gray; males and females are alike in plumage. They stay on the forest floor most of the time, occupy areas of thick shrub and dense foliage, and can be difficult to see, although they call frequently. In contrast, antthrushes are larger than the antpittas, and they have heavier bodies and longer tails.

The ovenbirds and woodcreepers comprise a large family of passerines with 306 species in a number of groups: ovenbirds, woodcreepers, miners, earthcreepers, cinclodes, horneros, tit-spinetails, spinetails, canasteros, thornbirds, cacholates, xenops, foliage-gleaners and treehunters, leaftossers, and others. Each of these groups of species has its own special characteristics.

Ovenbirds, for example, get their name because they build spherical, totally enclosed, oven-like nests made from various materials including mud, twig bundles, or soft fiber. Some use ground burrows, tree cavities, or rock clefts. Ovenbirds are insect eaters, and some are foliage gleaners. They have mostly muted, brownish to reddish-brown plumage, and some have wing bars, pale breasts and bellies, or other subtle markings.

Figure 12.1 The Streak-chested Antpitta is a small, short-tailed bird that spends most of its time on the ground in the understory (author's collection).

Other families in this large clade include spinetails, canasteros, and thornbirds, small birds with long tails that are mostly found in scrub habitat. Miners, earthcreepers, and cinclodes are ground dwelling birds, but the latter group is often found near water. Horneros are thrush-sized reddish-brown birds that forage on the ground, but some use dense scrub, mudflats, or mixed habitats. Foliage-gleaners, woodcreepers, leaftossers, and treehunters [Figure 12.2] are forest species, and birds within these families are uniformly brown and have a very similar appearance.

Woodcreepers were formally a separate family, but were always recognized as being closely related to ovenbirds and are now part of the *Furnariidae* family. They are not related to the brown creeper of North America, which has similar behavior and represents a case of convergent evolution (see Chapter 3).

Like other ovenbirds, the woodcreepers range from central Mexico to the northern part of Argentina. With more than 50 species of woodcreepers, it can be difficult to tell one from another. They are all brown birds, many usually streaked, with bills ranging from short and straight to long and curved. The Black-billed Scythebill [Figure 12.3], for example, has a bill that is strongly curved and one-third the length of its body.

The bills of woodcreepers are narrow in width and not adapted to chiseling wood like a woodpecker, so they nest in natural tree cavities. They remain in an upright posture and move up and down tree trunks and branches with jerking movements while they search for insects by probing, and rarely remain in one place for

Figure 12.2 The Streak-breasted Treehunter is a large foliage-gleaner that frequently forages for insects in epiphytes in tropical forests (author's collection).

more than a few seconds while searching for prey. They have stiff tails to brace themselves on tree trunks. They range in size from 6 to 15 inches in length (15 to 38 cm).

There are a number of small woodcreepers with straight bills, and they will feed on the ground and join birds following ant swarms. The plain-brown woodcreeper will feed on ants. The scythebills, with their long bills, are able to find soft-bodied prey like spiders and insects in crevasses that other species cannot reach.

Figure 12.3 The Black-billed Scythebill uses its long curved bill to capture insects by probing into crevices in the same manner as other woodcreepers (Nigel Voaden).

Eleven species of gnateaters (*Conopophagidae*) are predominately small ground birds of tropical understory. These small, plump birds have long legs and short tails. They are solitary, staying in small territories throughout the year and rarely venturing more than a few feet off the ground. They are unusual for the bright white feather tufts behind the eye, which can be erected to attract females.

Tapaculos (*Rhinocryptidae*) range in size from a small wrenlike bird to about the size of a small quail. Named by Charles Darwin, they received the Spanish name tapaculo—which means "cover your rear"—because some of them carry their tails upright. They have short wings, large feet, and a movable flap covering their nostrils. Some inhabit open scrub habitats, but most are found on the ground, in forest thickets, and are very secretive and difficult to see. There are 65 species, and they are considered closely related to the gnateaters.

Tyrannida: Tyrant Flycatchers, Cotingas and Manakins, and Others

This clade of suboscines includes five families: manakins, cotingas, tyrant flycatchers, tityras, and a smaller family that includes the royal flycatchers and their allies (*Oxyruncidae*). Although these families appear to be very different, genetic analysis has shown that cotingas and manakins are closely related to tityras and flycatchers (Ericson *et al.* 2006).

Tyrant flycatchers (*Tyrannidae*) are the largest family of passerine birds. They will be discussed in Chapter 13.

Cotingas (*Cotingidae*) and manakins (*Pipridae*) are fruit-eating birds known for their bright colors. They feed like a flycatcher, plucking fruit by sallying—leaping out and rushing back. Manakins carry out elaborate mating displays, usually at a lek site, while cotingas have a variety of mating practices, sometimes using a lek gathering to attract a mate, where the males perform displays to attract females.

Manakins are small, most have short tails, and they are dimorphic: males are very colorful, and females are muted olive. Species of manakins have elaborate stereotyped courting displays. In these displays, the males select a small sapling and display there to a single female. Their performances vary by species: for example, two to six males will carry out an elaborate ritual in which they dance and fly in short loops, making a sharp, snapping crack with their wings and a variety of continuous chattering sounds. Males will follow each other, performing the same looping routine. In this process, the more experienced male mates with the female, and the apprentice males hone their routine for future dominance of the lek. These performances should be seen to be appreciated, and have been recorded on various internet sites.

Cotingas range widely in size, from small to moderately large. They have variation in their breeding and nesting practices. Most are brightly colored, very beautiful, and some have very unusual plumages; some are dimorphic. Cotingas are predominately arboreal and are also frugivores and insectivores. The cotingas include the capuchinbird [Figure 12.4a], bare-necked umbrellabird, cock-of-the-rock, fruit-eaters, bellbirds, cotingas [Figure 12.4b], pihas, fruitcrows, plantcutters, and other

Figure 12.4a The Capuchinbird has a song like a chainsaw and raises its cowl feathers in display (Kester Clark).

species; they are so diverse that ornithologists believe they might represent more than a single family.

The breeding displays of the cotingas are not as elaborate as those of the manakins. Some are lek breeders, gathering in one place and using flight and feather display routines to attract mates; they also use calls and songs much more than do manakins.

One of the smallest suboscines is also a cotinga: the kinglet calyptura is just three inches long, and its weight has

Figure 12.4b Lovely Cotinga, a midsized fruit-eating canopy bird, has spectacular blue plumage (Bruce Hallett).

never been measured. This very rare bird was first discovered in 1818 and was only seen once in the twentieth century, so our knowledge of it is very limited.

Oscines Passerines

The word "oscine" comes from the Latin *oscen*, meaning songbird, and these birds are often collectively referred to as the songbirds. They represent the majority of our familiar backyard birds, many of which are seasonal migrants. The oscines include 123 families worldwide—so many families and species that it is difficult to make any general statements about them. The order is so large that some ornithologists group the families into superfamilies, and these are further grouped into six infraorders; the taxonomy is constantly updated to reflect new information discovered through molecular and behavioral studies. Just a few of the families of oscines include larks, swallows, corvids (crows, ravens, jays, and others), birds-of-paradise, orioles, bulbuls, thrushes, Old World flycatchers, pipits, waxwings, drongos, wrens, wood-swallows, shrikes, starlings, flowerpeckers, vireos, American wood-warblers, Old World warblers, tanagers, Hawaiian honeycreepers, finches and many more.

Nineteen of the largest families have close to or more than 100 species. Most of these large families are distributed around the world; we are familiar with starlings, corvids, wrens, swallows, larks, thrushes, finches, and others. Tanagers, thrashers, New World warblers, sparrows and flycatchers are restricted to North and South America. Bulbuls, sunbirds, laughing thrush, and weavers are found in Africa and southern Asia. Cisticolas, white-eyes, Old World flycatchers, and waxbills are distributed over Africa, Asia and Australia. Among the passerine families, the flycatchers of both the oscines and suboscines are the largest families, and the next chapter will focus on them.

A total of 34 of these families consist of just one (monotypic family) or two species. Families with very few species have, relatively speaking, uncommon names. Some familiar monotypic families include olive warbler and yellow-breasted chat in North America; wallcreeper and bearded reedling in Eurasia; in Southeast Asia, crested shrikejay; in South America, black-capped donacobius and rosy thrush-tanager; wrenthrush in Costa Rica and Panama; two oxpeckers in Africa; in northern China, Przevalski's pinktail; and in Australia, crested shrike-tit. The two warblers in Cuba, the oriente and yellow-headed, are in their own family. The waxwing family contains only three species: cedar, Bohemian and Japanese, and they have a range across all of North America, northern Europe and Asia.

Thirty-one oscine families have between 30 and 90 species, and include such well-known groups as shrikes, chickadees, vireos, wrens, wagtails and pipits, as well as birds-of-paradise.

One of the mysteries about the oscine songbirds is how they became so diversified. Fish are the most diversified vertebrates on earth with about 34,000 species described as of 2018. Birds are the second most diversified vertebrate, and half of all birds are the songbirds. Adaptation to habitats is the major factor in the development of species, but why so many species of songbirds compared to all other orders of birds?

Recent genetic work has led to speculation that their variety might have a genetic origin (Wong 2019). All songbirds have an unusual extra chromosome that does not appear to exist in other birds (Torgasheva *et al.* 2019). This chromosome appears in males and females and is transferred by the female. The exact role of this gene is not understood, but it is considered a rarity among birds and may play a role in speciation.

The oscines range in length from 39 inches (1 m) to just 3 inches (8 cm). The superb lyrebird of Australia is the largest at 39 inches long (1 m) from bill to tail and weighs 2.2 pounds (1 kg). Common raven is even heavier at 3.3 pounds (1.5 kg). At the opposite end of the spectrum, small oscines are about 4 to 5 inches (10–13 cm) long, with the weebill of Australia as the smallest at 3.1 to 3.5 inches in length (8–9 cm), and weighing 0.2 ounces (6 grams). This is not the smallest passerine, however: that honor goes to the short-tailed pygmy-tyrant of South America, a suboscine only 2.6 inches long (6.5 cm).

Oscines' diversity carries over into their diet. Most feed on insects, fruits, nectar, and/or seeds, while at least two families—shrikes and butcherbirds—are carnivores. Many are omnivores, like the corvids, and a few of the larger species will prey on bird eggs, nestlings, and small vertebrates. Oscines are similar in their breeding behavior: they generally build their nests out of twigs or grass, and place them on the ground or in trees or bushes. A few are gregarious and some use cooperative breeding, wherein the community supports newly hatched birds. However, in most cases, the solitary male and female raise the chicks.

Many oscine species are migratory, returning to the tropics in the non-breeding season. This is true for both northern and southern hemisphere birds. Ridgely and Tudor (1994) list 25 austral species that migrate north to the tropics, and 41 species of northern hemisphere Neotropical species that migrate from the United States to the South American tropics. Yet most oscines migrate shorter distances to simply warmer climates. Many northern hemisphere oscines only go as far as the Caribbean, Mexico and Central America. Others have adapted to the winter conditions in the northern and southern hemispheres.

The Most Colorful Passerines

Many passerine families are made up of very colorful birds. The most spectacular are the birds-of-paradise, sunbirds, honeyeaters, fairywrens, bowerbirds, Old and New World orioles, monarch flycatchers, leafbirds, thrushes, waxbills, cardinals, grosbeaks, New World warblers, and tanagers.

Tanagers

The tanagers (*Thraupidae*) are the second largest family among the passerines with 381 species, and they are very beautiful birds. The tanager family is so large and diverse that it is split into subfamilies, all of which are restricted to Mexico and Central and South America. The tanager family is mostly omnivorous; the birds' diet can be made up of fruit, seeds, nectar, or insects, although fruit and insects are their primary foods. Most tanagers are small or medium-sized birds, but the saltators of Central and South America are larger, averaging about 8 inches long (20 cm) compared to about 5.5 inches (14 cm) for most tanagers. South American honeycreepers, flowerpiercers, dacnises, and conebills are small groups of very colorful members of the *Thraupidae* that feed on nectar as well as insects. Also included in the *Thraupidae* are a variety of less colorful, small finch-type birds: seed-finches, seedeaters, grassquits, grass-finches, ground-finches, yellow-finches, and the famous Darwin's finches of the Galápagos Islands.

Some species with the common names of tanager are actually in separate families. The tanagers of North America (*Piranga*), for example, were split from the *Thraupidae* and are now considered part of the cardinal family.

Warblers

Seven families in Eurasia and Africa with almost 500 species are small insectivores similar in behavior to our North American warblers: African warblers, bush warblers, leaf warblers, reed warblers, grassbirds, cisticolas, and Sylviid (*Sylviidae*) warblers, but they are not necessarily closely related. These seven families are made up of 493 species. Those of Eurasia, Africa and Australia have more plain and cryptic plumages, and many are very difficult to identify. In South America the antwrens also share warbler-like behaviours and have cryptic plumages. Another family, the *Acanthizidae*, are referred to as the Australian warblers. They include thornbills, gerygones, scrubwrens, field wrens, and other names, including warbler.

The colorful New World warblers of the western hemisphere include 111 species. These brilliantly hued birds, primarily forest species, are popular with birdwatchers throughout the Americas as migrants and residents. A number of these species will live within the same area, but co-exist by stratifying and segregating their home range.

Finches

While some finch-like species are classified among the tanagers, other families that have finch-like characteristics include the finches, longspurs, Old World buntings, and New and Old World sparrows.

The finches (*Fringillidae*) are one of a number of passerine families that are called the "nine-primary oscines," because they all have nine visible primary feathers on their wings and almost all of them have a tenth primary feather, diminished in size and hidden under the ninth primary.

The nine-primary oscines also include Old World buntings, New World sparrows and warblers, cardinals, tanagers, blackbirds, and Hawaiian honeycreepers. Genetic analysis tells us that these families are closely related.

The *Fringillidae* also include a number of subfamilies that are all quite different. In the past, this family has gone through many changes in taxonomy as scientists learn more about their genetic makeup. The finches, for example, tend to have conical bills, primarily adapted to feeding on seeds. Many of them feed on a particular type of seed, and their bill, palate bones, and head and limb muscles are equipped for the task of extracting and managing these specific seeds (Austin, Jr 1961) from grasses, sedges, shrubs and trees. Some North American sparrows, *Passerellidae*, are seed eaters, as are some members of the tanager family, and there has been a significant degree of convergent evolution among the families of seed-eating birds. These will be discussed in Chapter 18.

Other familiar *Fringillidae* include canaries, seedeaters, grosbeaks, crossbills, euphonias, rosefinches, and Hawaiian honeycreepers.

A large percentage of passerines as well as other orders of birds are dependent on forests habitats. Passerines have different lifestyles within this environment. They can be solitary with either individual territories or shared territories, or live communally. One very interesting behavior is the mixed flock.

Mixed Flocks—Birds in Forested Landscapes

Mixed-species flocking of birds is a well-recognized phenomenon studied throughout the world. It occurs in many different kinds of ecosystems, including pelagic, wetlands, grasslands, forested habitats, and in many different groups of birds including waterfowl, wading birds, granivores, and passerines (Goodale *et al.* 2015). Most research has focused on feeding flocks in forested habitats, but feeding flocks in grasslands and on oceans have also been documented.

Birds use forested habitats in different ways, as evidenced by their behavior in tropical versus temperate forests. Over the past forty years, ornithologists have worked to understand the patterns of life in these habitats. Steve Hilty (1994) gives many examples of the research on the lives of forest birds. He describes his own work and that of other researchers including John Terborgh, Ted Parker and many others, and their graduate students.

When comparing the density of birds in tropical rainforest to that in mature deciduous forests of the northeast, Terborgh (1992, 1990) found that the density of breeding birds was about equal in the two types of habitats. However, in the tropical forest, the number of different species is greater, due to the diversity of plants and greater total biomass of tropical forests. Scientists counted from 100 to 300 species of trees in 2.5 acres (1 hectare) of rainforest. Within the tropical forest, plants do not become dormant in fall and provide food throughout the year, so the forest supplies more diverse food sources and more microhabitats. When compared over an area of about 240 acres (100 hectares), this diversity of plants supports at least five times the diversity of bird species over temperate forests. Forest structure also exhibits remarkable diversity when viewed on a smaller scale, with areas of tall trees, shorter trees, forest openings that create edges, swampy areas, riverine floodplains, grassy areas, and pools.

If you are birdwatching in tropical forests, you might think that birds occur randomly wherever you happen to be standing, but that is not the case. Ornithologists have found that there are complex patterns in how birds are distributed. Many types of birds congregate in mixed flocks, and coexist by exploiting microhabitats—small areas that differ in character from the larger area around them. Studies have classified the mixed flocks into two different types: those that occupy the lower elevation and feed primarily on insects, and those that feed mostly on fruits nearer to the canopy. In both cases, the feeding flocks move rapidly through the forest, making it difficult for bird watchers to see many of the different species.

The lower elevation feeding flocks are primarily birds with muted coloration. Flocks usually have a "nuclear species" that is central to the formation and movement of the flock. The nuclear species tends to initiate and lead foraging flocks that then attract members of other bird species. Some work suggests that gregarious species—those that live communally—are more likely to be a nuclear species (Goodale and Beauchamp 2010).

Mixed flocks also have "sentinel species," usually flycatchers or other species that search for predators and give warning calls. Research has shown that one objective of mixed feeding flocks is predator avoidance (Thiollay 1999), especially for those species that are gleaners—birds that feed on seeds that other birds have left behind. Foraging birds that are most vulnerable to predators are most likely to join these feeding flocks. This makes foraging more efficient for the entire group, making a mixed flock desirable for many birds. Membership in a mixed flock need not be permanent: some species will join the flock when it is within their territory and abandon it when the flock moves on. Various species of antbirds, especially antshrikes, will serve as nuclear species. Nuclear species are territorial and will try to exclude conspecifics—that is, others of the same species.

As many as 25 to 100 birds of many different species will associate with a feeding flock. The feeding flock stratifies in the lower canopy. This stratification is a concentration in space and time impacted by interactions such as competition, prey avoidance, food sources, and how different species are beneficial to each other (mutualism) (Goodale *et al.* 2015). In the flock, a hierarchy exists within the various species. Many species will feed on insects, but specialize their search area: some will feed on insects in curled dead leaves, since dead leaves are always present; while other species find insects on leaves, vine tangles, tree bark and branches, shrubs, and buds, and within air plants (epiphytes).

Feeding flocks will also join antshrikes that are exclusively (obligate) army ant followers. In these cases, the antshrike will hold the central area above the ant swarm. Smaller antbirds are stationed on the periphery, and other members such as antbirds, antvireos, antwrens, foliage-gleaners, flycatchers, and woodcreepers will make up the flock (Rice and Hutson 2003). The make-up of these feeding flocks can be complex, varying considerably among species and geographical locations. These birds will follow army ant swarms and feed on insects flushed by the ants, but the obligate antshrikes occupy the most favorable position. The ant column has a finite life of about three weeks, then regenerates over another three-week period (Hilty 1994), during which the obligate antshrikes look for other active columns within their territory. Figure 12.5 shows the Black-crested Antshrike.

At higher elevations, mixed feeding flocks focus mostly on fruits. In South

Figure 12.5 The male Black-crested Antshrike has a conspicuous crest and is easily recognized by its bold pattern. Note the hook on its bill (author's collection).

America, for example, about fifty percent of the birds in these flocks are tanagers. They appear and move through the forest independently of the understory flocks, covering larger territories than the understory flocks as they move from one group of fruiting trees to another. They do not have sentinel species for predator detection. Apparently, they rely on their individual ability to find predators. The fruit-eating birds are more colorful than the insect eaters of the lower canopy.

In Mexico, some canopy flocks are made up of insectivores. In winter, Neotropical migrants join both canopy and ground feeding flocks, depending on their preferred habitats. And some species are found in scrub and edge habitats, and are not part of feeding flocks. Analysis has shown that a higher proportion of Neotropical species were edge-avoiders compared with temperate species, and a higher proportion of temperate species were edge-exploiters compared with Neotropical species (Lindell *et al.* 2007).

Asian feeding flocks are notable for the high numbers of individuals per flock, because nuclear species tend to be those that are gregarious (Goodale *et al.* 2015). There is a large difference between northern temperate flocks and tropical ones in seasonality, and the species richness is correlated with latitude. In southern China, the fulvetta species, small songbirds of the Old World babblers, play a nuclear role, and as many as 30 will be found within the flock. In northern China, flocks are dominated by the tits and bushtits. In tropical Asian forests, drongos and some flycatchers serve as sentinel species.

In the northern forests, temperate forest birds also form feeding flocks. The big difference between them and those of the tropics is that tropical flocks feed communally throughout the year, while those of the temperate forests are cohesive only outside of the breeding season. During breeding, temperate species establish and defend territories. In spring and fall, migrant flocks of warblers and flocks of other mixed species are common, to the delight of birdwatchers. Blackbirds form large mixed flocks in fall and also on their wintering grounds. During winter, other small birds that remain in the northern latitudes in winter form feeding flocks.

In the tropical forest, there are many species that do not associate with feeding flocks. Mixed flocks are most likely found when the food source is dispersed. Hummingbirds represent an example of a family that uses a centralized food source of nectar producing plants. Their flight style allows them to avoid predators. Sit-and-wait ambush predators such as a phoebes, some flycatchers, and puffbirds tend to be solitary, and their feeding strategy is based on optimum use of energy. Ground birds such as doves, tinamous, and ground-cuckoos also do not engage in mixed flocks.

Conservation of forests has been the goal of responsible forest management. Aboriginal people have always depended on forests for their livelihood, and did so without endangering the health of this environment. Over-logging, clear-cutting, mining, and slash-and-burn to clear forests for agriculture have led to losses and degradation of forested lands. This has had a significant impact on bird populations, as mixed feeding flocks are very dependent on large tracts of undisturbed forest. Some species, like obligate army ant followers, remain in the dark understory and never leave this habitat. They will not cross boundaries like rivers and roads, so fragmenting forested habitats is permanently destructive for these species. It is only through government policies that we will ever be able to maintain the diversity of forest life. Environmental scientists estimate that at the current rate of tropical forest loss, all the world's rainforest will be gone a century from now.

13

Flycatchers

Insects are a class of invertebrates in the arthropod phylum. It is believed that they first appeared on earth about 480 MYBP in the Ordovician period of the Paleozoic era, though the earliest fossils of terrestrial insects are from about 400 MYBP, in the Devonian period. Winged insects followed about 80 million years later. Their class underwent numerous expansions and contractions in numbers of species, surviving repeated climate changes and the greatest extinction event in history, the Permian–Triassic boundary, about 250 MYBP.

Today there are about one million insect species that have been described by science, but the earth is estimated to have another five million insects that have not been described. Since insects are ubiquitous, it is not surprising that many bird species seek them as a primary food source.

Today the largest number of birds on earth feed on invertebrates, including insects, spiders, aquatic larvae, worms, ants, beetles, flies, butterflies and many others. Martin Nyffeler (2018) and co-workers analyzed 106 technical papers on birds that eat invertebrates and estimated that terrestrial insectivorous birds consume at least 450 million tons (400 metric tons) of insects per year. Seventy percent of this total is consumed by forest birds.

Over time, birds developed various techniques to capture insects. Flycatchers are, perhaps, the most well know insectivore among birds. "Flycatcher" is a general name most frequently applied to two large groups of birds that primarily feed on insects.

To be clear, flycatchers are far from the only family of birds that eat insects; nor are they the only ones that catch them on the wing. Warblers, vireos, swallows, wood-swallows, swifts, drongos, trogons, and nighthawks use aerial capture of insects, each with their own method for doing so. Nighthawks and swallows use a completely different feeding technique compared to flycatchers. They hold open their large mouths in flight as they make many passes while gathering insects. Other types of birds, such as small falcons and some species of kites, routinely feed on large flying insects, and hummingbirds are extremely efficient at hawking small insects, their major source of protein. Gulls will feed on insects, and it is comical to watch their seemingly uncoordinated attempts at aerial capture.

The wagtails (*Montacillidae*), a large genus of small birds more closely related to pipits, act like flycatchers by using their long tails to flush insects. The todies of the Caribbean are very small, extremely colorful birds that feed primarily on insects, employing the same method used by some flycatchers, but they are related to kingfishers and bee-eaters. Bee-eaters capture medium to large insects, their long beaks protecting them from stinging insects.

The term "flycatcher," however, normally applies to the new and old world families of flycatchers. The suboscine New World flycatchers (*Tyrannidae*) are the largest family of birds on earth, with about 425 species in 97 genera. They occupy every habitat in South America and range from Alaska to Tierra del Fuego. The Old World Flycatchers (*Muscicapidae*) of Europe, Asia, Africa, and Australia are oscines and number about 317 species in 50 genera, and they are not related to the *Tyrannidae*. These two families of birds represent examples of adaptation leading to extensive species diversity.

In many ways, these two families are alike: they have similar behavior, share a common lifestyle, and they have similar feeding habits. Their ancestries, however, are very different. The New World flycatchers are closely related to cotingas and manikins, while the Old World flycatchers are related to the thrush family.

Generally, flycatchers feed by hawking from an exposed perch, by gleaning, by hovering, or by flush-and-capture (Robinson and Holms, 1982). When hunting from an exposed perch, they spot a flying insect, sally out for aerial capture, and then return to a perch. They move from perch to perch, but research indicates that many small flycatchers move just far enough on average to take them into areas they have not previously searched visually. In gleaning, they search for prey among the foliage. They can snatch prey from an exposed perch by hovering, or hover to search for prey. Flush-and-capture is the least common method used by flycatchers; this is more prevalent among warblers.

Some flycatchers are known to flick their wings and tails, and to bob their tails—all of which improves foraging performance. While this behavior has been researched specifically with flycatchers, a study of hooded warblers has shown that the loss of white tail spots significantly reduces foraging success at capturing aerial prey, and that tail flicking behavior develops in juveniles and is important for foraging success (Mumme 2014). This same technique of tail flicking is used by phoebes and other flycatchers.

Many flycatchers have flattened bills that range from broad to narrow, and most have bristles around the bill—but some, such as the elaenia of Mexico, the Caribbean and Central and South America, do not have them. We mentioned in Chapter 4 that they are sensitive to touch. The role of the bristles is not known for certain, but studies show that they do not aid in insect capture. Other studies suggest that they might be important in protecting the eyes of flycatchers from thrashing legs and wings of insects (Conover and Miller 1980). Many species of birds have these specialized feathers, so they might serve different purposes in different species (Delaunay *et al.* 2020).

Suboscine Flycatchers

The suboscine flycatchers of the Americas are varied in physical size, plumage, and behavior. Most are in drab colors of gray, green, olive, brown, and one, the white monjita, is almost all white. They have few distinguishing features, but some have distinctive facial markings; a few have long, showy tails. Within specific genera, many are so similar in appearance that they can be difficult to identify, even for ornithologists. Some can be identified by their song and sometimes by their breeding range.

In many, the central crown feathers are colored white, yellow, or red, or they may be the same color as the head. In some species, these feathers are elongated and form a crest, and the birds raise the crown when they are excited. In the case of the Royal Flycatcher [Figure 13.1], the crest is orange-red and very decorative, but is rarely displayed. A few flycatchers are very colorful, such as the Vermillion Flycatcher [Figure 13.2], the only one with extensive red coloration; the kiskadees with their distinctive facial markings; and the many-colored rush-tyrant.

So diverse is the flycatcher family that it is hard to sort them into broad categories. There is still much controversy over the *Tyrannidae* family itself; questions abound about whether it is a single family, or if it should be split into multiple families. Within the family, there are a few larger groups of species. For example, about 40 species—many of these are distant relatives of each other—are referred to by the common name of tyrannulets. These are among the smallest flycatchers, almost all of which have muted and indistinct plumages. Most have off-white to yellowish

Figure 13.1 The Royal Flycatcher has the most elaborate crest of any flycatcher; note the bristles around the beak (Dr. J. Waud).

Figure 13.2 One of the most colorful suboscine flycatchers is the male Vermillion Flycatcher of the Americas, which makes use of a sit-and-wait strategy to capture flying insects (author's collection).

breasts, olive backs, and wing bars. In contrast, a group of species called the elaenia are similar in plumage to each other, larger in size than the tyrannulets, and number about 25 species. Elaenia are also muted in coloration with brown to grayish backs and lemony to plain breasts and wing bars.

Another group within the *Tyrannidae* family, the Empidonax, number about 15 species and are mostly confined to North America in their breeding range, though they migrate farther south in winter. They are among the most difficult flycatchers to identify because of their uniform appearance. Like other groups, their overall plumage has a limited theme: most are olive to grayish-olive backed, with pale breasts, eye-rings, and wing bars.

The *Myiarchus* are larger flycatchers with pale to bright yellow breasts, olive backs, and wing bars. Birds in this group have reddish undertails, and they lack eye-rings, which are frequently found in flycatchers. There are 25 species of *Myiarchus*, six of which can be seen in the United States.

Most flycatchers are birds of the forest, but one group of *Tyrannidae*, the kingbirds (*Tyrannus*), however, prefer more open country. *Tyrannus* are bigger flycatchers with larger bills, and most have brighter plumages with bright yellow breasts and grayish backs; they usually lack wing bars. These birds hunt from open perches, wires, and posts. They total 13 species; seven of these breed in the United States. Three others have been recorded but are very rare.

Among the suboscine flycatchers, the Great Kiskadee (shown in Figure 13.3) is perhaps one of the most unusual. Its varied diet and aggressive foraging techniques set it apart (Skutch 1997). Besides capturing flying insects, it eats fruits, especially bananas and palm tree fruits. It will come to feeders and readily consume some peppers, cooked rice, and bread along with its fruit diet. It will dive into water to capture small fish and swallow them whole, and also wade into shallow water and completely immerse its head to capture tadpoles. From an exposed perch on a wire, it will pounce on grasshoppers, lizards, frogs, small snakes, or even a mouse, which it consumes after beating it against a branch—and it also will pick up and eat broken snails. Kiskadees will capture and consume nestlings of small birds, and will follow a plow to take advantage of the flushed insects and critters, a behavior similar to foraging gulls. It is clearly an omnivore, and has the skills of a corvid (crow, etc.), with the additional advantages of a flycatcher. It is very close in size to the largest flycatcher in the Americas, the Boat-billed Flycatcher. It has a large range, extending from Texas through northern Argentina, and covers almost all of South

Figure 13.3 The Great Kiskadee is one of the largest suboscine flycatchers and exploits a remarkably wide range of food sources (Bruce Hallett).

America, and it is a non-migratory species, thus able to tolerate moderately lower temperatures.

Both oscine and suboscine flycatchers use a variety of feeding strategies. Although they are insectivores, most will also take fruit and berries. As they are fly-catchers, we imagine that they primarily use aerial sallying to capture prey—and some do, but some use many other techniques, according to Skutch (1997) and Tray-lor and Fitzpatrick (1980), who have described the various ways some of the subos-cine flycatchers prey on invertebrates. For example, the Sulphur-rumped Flycatcher (and probably other flycatchers) joins feeding flocks of other species that follow army ants, and uses aerial capture to feed on flushed insects.

Among flycatchers, the most common method of feeding is to glean insects from the underside of leaves where arthropods are more common. The flycatchers will sit in the forest and inspect the underside of leaves above them for prey. One species, the appropriately named Southern Bentbill, has a bent bill especially suited for this purpose (see Figure 13.4). Some flycatchers will feed like warblers and vireos by hopping and flitting through the foliage, and then glean from the leaves, branches and flowers. Capture by foliage gleaning is about equally widespread among species of temperate and tropical forests (Terborgh 1992), while sallying capture is much more common among tropical species.

Figure 13.4 The Southern Bentbill has a bill shape tailored to snatch insects from the under-side of leaves (author's collection).

Another feeding technique used by flycatchers is to sit on low perches in open areas and capture insects on or near the ground, returning to the same perch. This is a favorite approach of black phoebes, a technique referred to as ground-grazing. Kingbirds will follow a similar pattern of sitting on wires, dropping to the ground to capture insects below them. They also use aerial capture. Ground tyrants (*Muscisaxicola*) and a number of other flycatchers will hunt from the ground, and several species can run down insects before they can fly. Torrent flycatchers of South America will wade into shallow water and capture prey from the water's surface.

Several suboscine flycatchers have long tails: strange-tailed tyrant, streamer-tailed tyrant, and scissor-tailed and fork-tailed flycatchers. In these species, the two long tail feathers are the outer feathers. In the long-tailed tyrant, the innermost two tail feathers are the longest. The scissor-tailed flycatcher will spread its long tail feathers frequently and use them in aerial displays. The suboscine long-tailed flycatchers prefer open habitats, while the paradise flycatchers, also long-tailed, are more common in forested areas.

Oscine Flycatchers

While the suboscine flycatchers tend to be dull in plumage, the oscine flycatchers of the Old World (*Muscicapidae*) are made up of some very colorful and beautiful species, as well as a few with less distinctive plumage. They are a very diverse family with dozens of common names. These birds never invaded the western hemisphere except for a few species that breed in Alaska, and their taxonomy is not totally understood and is confusing at best. Some of what we traditionally considered flycatchers—formally part of the Old World flycatchers, such as the monarchs and the fantails—are now thought to be related to crows and ravens (*Corvidae*). Recently Sangster *et al.* (2010), using genetic analysis, has proposed dividing the remaining Old World flycatchers into four subfamilies that each encompass a multiplicity of species and a diversity of characteristics: scrub robins and old world flycatchers; flycatchers and the blue flycatchers of Southeast Asia; African forest robins; and chats. Figure 13.5 shows an example of one of the blue flycatchers.

The taxonomy of the 317 species of Old World oscine flycatchers is, no doubt, complex. Unlike the suboscines, there are only a few large groups of a single genus. The oscine flycatchers take on a wide variety of names: robins, scrub-robins, robin-chats, magpie-robins, niltavas, akalats, flycatchers, blue-flycatchers, jungle-flycatchers, forktails, shortwings, chats, stonechats, bushchats, wheatears, whistling thrush, rock thrush and nightingale. Some of these birds are medium sized like the rock-thrush, whistling thrush, and the shamas. The largest genus, the *Ficedula* of Asia, Africa and Europe, has a number of colorful flycatchers and a total of 34 species. The blue flycatchers of Southeast Asia (*Cyornis*) total 27 species and are very small with deep blue color. The *Muscicapa* flycatchers are also small in size, and are grayish overall.

Many in the flycatcher family are called robins and chats; these are also small with small bills, and are colorful with pleasant songs. The European robin was the

Figure 13.5 The Chinese Blue Flycatcher is a colorful oscine flycatcher of the *Cyornis* genus in the *Muscicapidae* family, which has over 300 species that are found in Europe, Asia, and Africa (Chris Chafer).

inspiration for the name of the American robin; they have some similarities in appearance, but while the American robin is a thrush, the European robin is not. Small, colorful flycatchers in this group range across Europe, Asia, Africa, and Indonesia. Three of the oscine flycatchers—the Bluethroat (shown in Figure 13.6), Northern Wheatear, and Siberian Rubythroat—breed in the northern parts of North America.

Although the monarch flycatchers were separated from the *Muscicapidae*, the monarchs are literally flycatchers. Within this group, the males of the paradise flycatchers (*Terpsiphone*) are exceptionally elegant in their stunning plumages. Fifteen species of these long-tailed flycatchers are widely distributed across Southern Asia, India, sub-Saharan Africa, the Philippines, Taiwan, Japan and Madagascar. The northernmost species are migratory, but the majority are permanent residents of warmer climates. They are dimorphic, and the males grow long central (inner two) tail feathers. The tails do not reach their full length until the third winter molt, and the feathers can get to be 20 inches long (51 cm). Females select mates based on the tail length.

Suboscine flycatchers have one more unusual feature: the peculiar shape of their primary feathers. A large variation in the shape of the outer primaries is unique

Figure 13.6 The colorful Bluethroat finds its home among the dwarf willows across all of Europe and Asia, extending into northwestern Alaska (Bruce Hallett).

to males and specific to certain species or subspecies. The variations are of several basic types: either shortened inner or outer primaries or a series of notches, or thinning on the inner edges of multiple primaries. Since they only occur in males, it is believed that these primary feathers are used to make sounds with wing displays for courtship or territorial defense (Traylor and Fitzpatrick 1980).

Flycatcher Nests

Perhaps one difference between suboscine and oscine flycatchers is their nests. Most of the oscine flycatchers build cup nests in branches of trees or bushes. Cup nests vary in complexity, size and neatness; some are very ragged, while others are quite tidy. Some oscines nest in tree cavities or under cliffs (Mead and Lundberg 2003), weaving their nests with flexible material using shuttle-like movements to insert and tie the material together.

The suboscine flycatchers build the greatest diversity of nest types of any family of birds (Skutch 1997). In almost all cases, suboscine flycatchers do not weave their nests as do oscine songbirds. The female does all the work in nest building, beginning the hanging nests as a mass of randomly tangled fibers. When the mass grows large enough, the builder pushes in from the side and makes a pocket, occupying the nest and pushing with her feet and breast to stretch out the interior of the nest. As the interior is expanded, the flycatcher adds more material inside to strengthen the walls and builds up the exterior. The female also winds strands until the structure reaches its final shape.

The many-colored rush-tyrant creates an unusual nest as well, weaving an elaborately constructed deep cup nest affixed to a reed. Black and eastern phoebes attach their nest to a vertical surface with plant material and bits of mud, which are also used in forming the structure. Some species, such as yellow-bellied flycatcher, place their nests on the ground. Many construct completely covered nests with side entrances that give protection from the rain, sun and predators. These covered nests come in a wide variety of shapes and sizes. A common spherical nest with a side entrance stays in place using a tree fork or reeds.

Hanging nests also come in a variety of shapes: some are conical with a round bottom and pointed top; some are spherical, and several are elongated. The elongated nest of the royal flycatcher can be two to six feet in length with a side opening to a small nest chamber. These hanging, covered nests are attached to dangling roots, leafy twigs, epiphytes or small branches. Nest interiors are lined with soft material. The exterior of the nest can be decorated with mosses, likens, or vegetable fiber, and flycatchers often use spider silk and cocoons to help hold the structure together. Some species build covered cup nests in woodpecker holes. One member of the family, the piratic flycatcher, does not build a nest but finds a suitable victim, evicts them, and takes over its nest.

Flycatchers collectively represent a very large number of the perching birds—about 740 species, more than ten percent of the passerine order. In the next chapter, we will discuss another rather unique group of passerines, the Hawaiian honeycreepers. You will meet passerines again in this book because they occupy so many habitats and feed not only on insects, but on other food sources as well.

14

Hawaiian Honeycreepers

Adaptive Radiation

Charles Darwin set out on the voyage of the *Beagle* in 1831. Throughout this journey in the South Pacific and while in the Galápagos Islands, he collected specimens of animals and plants. On the voyage home in 1836, he began to look carefully at the specimen, and that's when he realized that 14 species of finches from the Galápagos were different from each other, but had some shared characteristics. It was not until 1845, after much more effort, that Darwin first suggested that these finches had a common origin. These are referred to as the Galápagos finches, and they have been studied for many years by numerous researchers.

The ancestral species of the Galápagos finches is believed to have arrived there from South America at least 3 MYBP (Grant and Grant 2002). Today these finches are considered an example of *adaptive radiation*, a process in which organisms diversify rapidly from an ancestral species into a variety of new species, particularly when a change in the environment makes new resources available or opens new environmental niches. Adaptive radiation is the result of divergent natural selection between varying environments. There are many examples of this process in fish, lizards, marsupials, and plants, but in addition to the Galápagos finches, the most well-recognized example in the world of birds would be the Hawaiian honeycreepers.

When the Polynesians arrived on Hawai'i they found islands teeming with birds. The only mammal useful to them was the Hawaiian seal, but they found both practical and social uses for birds in their culture, particularly as a food source. They also found uses for their bones, and used their feathers for ceremonial purposes. When Captain James Cook arrived in Hawai'i in 1778, the naturalist in his party took specimen samples of the birds that they found for scientific purposes; over the next 150 years, other biologists also collected specimens. Analysis of the Hawaiian birds in the twentieth century from specimens collected over time and from fossil evidence yielded a detailed picture of the family of Hawaiian honeycreepers.

Relatively few families of passerine birds are found on the Hawaiian Islands besides the honeycreepers. These included the elepaio flycatchers (3 species related to monarch flycatchers), the thrush family (6 species), the nectar-feeding Hawaiian honeyeaters (4 Hawai'i 'ō'ōs and kioea), a millerbird, and a single crow. The Hawaiian honeyeaters are not related to the Australian honeyeaters, but to waxwings and to Neotropical silky flycatchers (Fleischer *et al.* 2008). Given the size and diversity of the islands, it is remarkable that so few families of passerines established permanent residence there.

Analysis of the Hawaiian birds in the twentieth century from specimens collected over time and from fossil evidence yielded a detailed picture of the family of Hawaiian honeycreepers. Today, 57 species of honeycreepers are recognized; of these, only 20 remain in existence. The remainder are extinct due to a wide variety of causes, all induced by human occupation. It is difficult to determine which causes had the greatest effect on the bird population, because so many changes took place at the same time. All the lowland rainforests were eventually cleared as a source of lumber and for agricultural purposes. Marshes were drained to be used to grow taro. Pigs were introduced, became feral, and eradicated many of the plants important as food sources for honeycreepers. Birds were captured for their feathers, especially for red, yellow and black plumes.

As if all of these factors were not catastrophic enough for the honeycreeper population, when the islands became a commercial seaport for whalers and the export of lumber, the traffic brought avian malaria to the islands, where it remains to this day. Still later, the mongoose was introduced to control rats; unfortunately, the rats were nocturnal and the mongoose was diurnal—so the mongoose preyed upon ground nesting birds. And finally, birds were exploited as caged pets because of their bright colors, and also as game. It is hardly surprising that so many of the honeycreepers went extinct so rapidly, but there were no conservation measures in Hawai'i until at least the later part of the twentieth century.

Today the honeycreepers are included in the subfamily of cardueline finches (crossbills, goldfinches, etc.), of the *Fringillidae*. The species of this large family tend to be nomadic. Research based on DNA has identified the rosefinches (*Carpodacus*) of Europe and Asia as the closest living relatives of the honeycreepers, and today it is estimated that they arrived on the Hawaiian Islands around 5.7 MYBP, the time when Kauai and Niihau had formed. Most of the distinctive birds diverged after the island of Oahu emerged (Lerner *et al.* 2011), and this occurred before the emergence of the more easterly islands. The Big Island of Hawai'i is only 700,000 years old.

Adaptive radiation allowed birds to diversify rapidly, creating new taxa of honeycreepers for 5.3 million years until one of the latest, the Hawai'i amakihi, appeared. Lerner *et al.* (2011) estimated the time for the formation of a new species on Hawai'i to be as long as 1 million years, but some new taxa appear in as little as 200,000 years. As a comparison, estimates suggest that the 14 to 15 species of Galápagos finches have emerged in the space of two million years (Uppsala University 2015).

The classification of honeycreepers is a complex subject with a long history. Douglas Pratt provides a historical summary of their classification, and a current consensus of their taxonomy (Pratt 2014). In the early nineteenth century, many researchers began to study the approximately 39 extant species of the honeycreepers of that time period. Pratt (2005) provides a comprehensive life history of this entire subfamily, *Drepanidinae*, of the finch family, and he summarizes the earlier work. *Drepanidinae* are made up of colorful birds ranging in size from very small (4 inches, 10 cm.), to medium size (9 inches, 23 cm.). From a simplistic view, they can be sorted into five general groups—nectarivores, seed and fruit eaters, foragers, bark-pickers, and generalists (Pratt 2005). The diversity of their bill shapes gave rise to the adaptation to specific food types on a set of islands with relatively restricted food sources. One of the most unusual bills is that of the akiapola'au, shown in Figure 14.1.

There were about fifteen species of nectar-feeding honeyeaters, and six of them survive to the present. The two mamos, all colored black and now extinct, were the largest nectarivores. The 'I'iwi [Figure 14.2] is one of the largest surviving nectarivores. Nectar feeders rely on a large variety of flowering plants; important among these are the red flowers of the ohi'a lehua tree (*Metrosideros polymorpha*), the flowers of the mints endemic to Hawai'i (*Stenogyne*, 22 species), and the lobelia (*lobelioids*, 125 species). These plants also attract insects, so they attract insectivores as well. Many of the mints and lobelias have been driven to extinction by some of the causes already mentioned: feral pigs and clearing forests.

The seed-eating birds are among the most specialized, with bills adapted to the specialized food sources found in Hawai'i. The palila, for example, feeds on the seeds of a variety of plants including the mamane tree (*Sophora chrysophylla*). The birds use their feet and bill to extract the seed from this tree, as the pod is difficult to open. The seed is toxic to other birds, but the palila apparently has enzymes that counteract the toxin. Another seed-eater, the Nihoa finch, fed on birds' eggs, arthropods, flowers and seeds. There were at least six yellow and greenish seed-eating birds, some with very large bills—like the now-extinct Kona grosbeak, an olive-colored

Figure 14.1 The Akiapola'au, a Hawaiian honeycreeper, has an elongated lower mandible to probe tree bark for insects (Eric VanderWerf).

Figure 14.2 The scarlet 'I'iwi relies on nectar from the ohi'a lehua tree (*Metosideros polymorphis*), and from flowers of the Hawaiian lobelioids (Aaron Maizlish).

grosbeak that would consume the hard seeds within the fruit of a large shrub (Pratt 2005). Two species of seedeaters had crossed mandibles similar to crossbills. Despite their sophisticated adaptations to their food sources, the seed-eating birds were the most specialized and among the earliest to disappear (Austin and Singer 1961).

A large number of small to medium-sized insectivores have thin, pointed bills, while others have curved bills. Many of the insect eaters also feed on nectar, but are mostly gleaners with their warbler-like bills. The endangered 'akepa has crossed mandibles to pry open leaf buds.

The bark-pickers have a large diversity of bill types. Most unusual among them are the long curved-billed nukupu'u, a critically endangered species, and akialoa, which is extinct. The Kauai Akialoa [Figure 14.3] had a 2.5 inch (6.2 cm)-long curved bill, and the bird plus its bill measured only 7.5 inches long (19 cm) (Munro 1960). It used its long curved bill to probe deep into the cracks of bark and decaying wood for insects. The endangered and highly localized Maui parrotbill feeds mostly on wood-boring beetles on the windward side of Haleakala volcano, captured by peeling away bark and splitting twigs to probe for beetles. It also feeds by surface gleaning. Among the bark-pickers were several short-billed species such as the Hawai'i creeper, which runs up-and-down trees like a nuthatch, and picks from cracks and crevices in the bark of limbs and tree trunks (Pratt 2005).

Figure 14.3 The bill length was more than half the length of the body plus tail of the Kauai Akialoa (author's collection).

An unusual species, the Po'o-uli (the black-faced honeycreeper mentioned in Chapter 3), was a generalist that fed on snails, insects, and fruit. Other generalist honeycreepers included the three species of 'amakihi, small yellowish birds with short, curved bills.

These honeycreepers vary in their feeding styles, bill types, size, and colorations, but other anatomical characteristics, such as their tongues, show they all belong to the same family. For example, the honeycreeper tongue structures are non-tubular, resemble those of cardueline finches (such as crossbills and redpolls), have fringes, and are somewhat unique at the base where the tongue meets the back of the throat (Pratt 2005). This helps ornithologists determine that they are all of the same family, even though their outward appearance suggests otherwise.

The Hawaiian honeycreepers are an exceptional example of how birds can evolve to specialize on different food sources over a relatively short geological time period. Because of their strong attachment to particular food sources, they were vulnerable to a rapid loss of that resource—which is exactly what happened when people arrived on the Hawaiian Islands and harvested these important plants for human use, or removed them altogether.

15

Grassland Birds

The way we define grassland is deceptively simple: an area with a virtually continuous covering of grasses, but generally lacking in trees and shrubs. Widespread throughout the world, grasslands provide habitat for many different kinds of birds, from species that live and nest on the ground to perching birds that feast on the seed heads many grasses produce. These lands provide benefits to people as well. Combined with rolling hills, blue sky, and puffy white clouds, grasslands offer their unique scenic beauty and a feeling of serenity. They can be particularly beautiful in spring when clumps of wildflowers peak out of the new grass.

Like oceans and forests, grasslands represent a significant area of the earth, and many different species of birds and animals have adapted a lifestyle to this habitat. Lands dominated by grasses are referred to as prairies in North America, steppes in Eurasia, savannas and veldts in Africa, savannas and rangeland in Australia, and pampas, llanos and cerrados in South America. Plains, meadows, and fields are more informal names for them.

Perhaps because they are wide open, flat, and easier to bend to our will than forests or oceans, humans tend to see grasslands as little more than vacant lots waiting to be developed. This is our own loss. To preserve more of this habitat, we need to better understand its value to the planet's ecosystem.

The Nature of Grasslands

Grasslands are the result of natural processes in which there is sufficient moisture to sustain the plants for part of the year and drought for the remainder. Fire and rainfall are factors controlling the extent and health of grasslands. Some deserts have sufficient rainfall to produce grasslands, and they are referred to as semi-desert. In areas where there is more abundant rainfall, natural events—particularly fire and animal grazing—maintain the grasslands so that they don't become forested. Although tundra is not normally considered grassland, it does contain grasses, sedges, and mosses, and can be a wet habitat.

Grassland ecosystems have developed in part as a response to natural grazing animals, such as bison and prairie dog in North America, wildebeest and elephant in Africa, and other mammals.

Based on their biology, we can describe many different types of grasslands.

The tallgrass prairies of North and South America and Nepal, for example, have moderate to high rainfall. Temperate grasslands are found in the mid-latitudes with weather that cycles between periods of very warm and very cold temperatures. Savannahs contain about ten percent woody plants, making them distinct among grassland types. Shortgrass prairies of the western United States and Eurasia are found in semi-arid regions, but they are different from desert grasslands, which occur in even more arid regions with shorter, less dense grasses interspersed with desert shrubs; they are highly susceptible to overgrazing by domestic cattle.

Grasslands vary considerably from tall to short grass and from dense to sparse, and all are mixed with a wide variety of other wild plants. Perhaps one of the most unusual grassland types is the tallgrass prairie of North America. It originally covered 170 million acres (266,000 sq miles), stretching from northern Canada to Texas. The grasses averaged about six feet tall (2 m), and reached as high as nine feet (3 m). Only a few small tracts of this prairie remain. One of the important attributes of this habitat is that it supports a high level of biodiversity.

While agriculture and commercial/residential development have been the most formidable threats to grasslands, they now face an even greater threat: climate change has made these lands highly susceptible to fire. Lightning causes fires naturally, but some fires are set on purpose, using controlled burns to control the habitat—a complex subject with no simple rules. The effects of grassland fires have received much attention, particularly in Australia (Low 2014). An unprecedented 46 million acres (18 million hectares) burned in Australia during the 2019–2020 dry season, caused by a number of brushfires in the southeast. It has been estimated that over one billion animals were killed in a single season. Fires in the United States have rivaled the Australian burn, with increasing numbers of wildfires in the past decade as the drought that began in 2000 in western North America has altered the landscape. From 2011 to 2015, fires naturally caused by lightning accounted for 17 percent of forest or woodland fires, but only 4 percent of overall grass, brush, and forest fires in the United States. All other fires have deliberate or accidental human causes such as fireworks, trash burning, and so on—and 19 percent were set intentionally (Ahrens 2018).

Birds and Grasslands

A wide variety of bird species make use of grasslands. While other environments are the domains of large families of birds—pelagic birds on the oceans, for example—no large families of birds dominate the grasslands. Instead, many families of birds have individual species that adapted to grasslands either exclusively or partially. Birds that use grasslands for breeding migrate to the same types of habitat for winter.

Small birds are the most numerous occupants of grasslands. They include a large number of songbird species from different families of the Passerine order (*Passeriformes*), but sparrows and finches are the most diversified grassland species. Most grassland passerines nest on the ground, but a few build nests above the ground on

small shrubs. The process they use to travel to and from their nest in the homoge-
nous landscape is something of a mystery, but research suggest it is very likely that
they use migratory navigational skills. It is even difficult for field ornithologists to
find their nests (Winter *et al.* 2003). Many sparrows will walk on the ground to their
nest rather than approach by flight, thus drawing less attention to its location.

Grassland birds typically have brownish plumages with streaking, making
them difficult to detect [Figure 15.1]. They will sing atop grass stems and on woody
plants or in flight, but become more secretive and retiring after breeding commences. They feed on insects or seeds depending on the species, and more on insects
when breeding. Foraging for food usually involves short flights.

Asia and Africa have a wide array of grassland passerine birds of the family
Locustellidae, which includes the grassbirds and warblers, mostly long-tailed brownish birds. Grasshopper-warblers and bush-warblers are in this family. The passerine
grassland birds include larks, pipits, various warblers, cisticolas and apalises, prin-
ias, and many more.

Australia lists 71 species of grassland birds (Olsen *et al.* 2006), of which 55 are
seed-eating birds of the tropical savannah. The majority are passerines such as
bushlark, cisticolas, fifteen species of finches, and pipits. Non-passerines including

Figure 15.1 The Henslow's Sparrow, an uncommon grasslands specialist of the United States,
is secretive in its habitat and has the brownish plumage with streaking typical of grassland
sparrows. It also has an unusual greenish tint to its facial features (author's collection).

parrots and pigeons also make use of grasslands. Unique grassland passerines of Australia, the grasswrens, emu-wrens, and fairy-wrens, *Maluridae*, are all small birds with cocked tails.

On the North American Great Plains, twelve species of birds are endemic to grasslands, and 25 species evolved to be dependent on grasslands. Eighteen of these species are passerines including sparrows, longspurs, larks, meadowlarks, dickcissel, bobolink, and others.

Throughout the world, many other bird families beside passerines use grasslands. A few shorebird species prefer this habitat: in North America, the long-billed curlew, mountain plover, marbled godwit, and upland sandpiper are examples. In Australia, the masked lapwing, banded plover, and inland dotterel fit this category. Pratincoles of Eurasia and Australia occur in grasslands near water.

Quails, pheasants, francolins, grouse, sage grouse, and doves also may be grassland birds. In Australia, the plains-wanderer, a monotypic family, is a quail-like bird restricted to short grass habitats. A large bird like the emu is an obligate grassland species; cranes utilize grasslands for feeding.

In Africa, Zimmerman *et al.* (1996) identified eight different types of grassland habitats. They range from short grass plains and moist and mountain grasslands to heaths, semi-desert, and a variety of grasslands with sparse trees and shrubs. Each of these habitats hosts a different group of birds. The common ostrich, bustards, sandgrouse, francolins, southern ground-hornbill, secretarybird, and tawny eagle are more specialized grassland residents.

Among predators, some hawks and owls specialize in grasslands. In North America, ferruginous hawk, Mississippi kite, Swainson's hawk, and northern harrier utilize this type of habitat, as do harriers throughout the world. Many species of hawk and eagle search for prey in open country. A few examples include steppe eagle and tawny eagle of Asia and Africa, savanna

Figure 15.2 **Short-eared Owls prefer to hunt over open country and grassy fields (author's collection).**

hawk of South America, secretarybird of Africa, and the common buzzard of Eurasia, Africa and Australia. The tyto owls (barn owl, grass owl, and others) and short-eared owls [Figure 15.2] rely on grasslands.

Worldwide, between six and ten percent of all bird species are endemic to grasslands or evolved to be dependent on grasslands. In addition to these obligate grassland species, there are many species of birds that utilize grasslands but are not solely restricted to this habitat. Grassy marshes are homes to bitterns, rails, wrens, and blackbirds. Shrubby grasslands are home to some species of sparrows, blackbirds, and shrikes.

Problems Facing Grassland Birds

Data from the U.S. Breeding Bird Survey, which began in the 1960s, shows that grassland birds have significantly decreased in numbers over the past 50 years. McCracken (2005) shows that 86 percent of North American grassland birds are declining. Surveys done in Australia in the 1980s show that 37 percent of grassland birds were already decreasing then (Olsen *et al.* 2006). The problem is worldwide and also has been documented in Great Britain and Western Europe, largely due to the intensification of agricultural operations (Peterjohn 2003).

Habitat Loss

The largest problem facing grassland birds is the loss of adequate habitat. In the very distant past, driven by the changing environment, grassland coverage expanded from tropical forests. Historically, grasslands covered a significant portion of the habitable land mass of the earth. The decrease in bird populations in these habitats has coincided with changes in land use and the spread of agriculture.

Habitable land on the earth is just 43 percent of its total land mass, covering 24.6 million square miles (64 m sq km); deserts and mountainous areas make up the remaining land area (Pianka). As of 2000, analysis of satellite imaging estimated that the world's croplands cover about 5.8 million square miles (15 m sq km), and another 10.8 million sq mi (28 m sq km) is pastureland (Ramankutty *et al.* 2008). These account for 67 percent of all habitable land. The amount of natural undisturbed grasslands that have survived to the present are a very small percentage of the total crop and pasture lands. In the United States, just 1.6 percent of the native grasslands are protected, while Mexico protects 1.8 percent of its natural grassland, and Canada does a better job with 15 percent of its grasslands protected (McCraken 2005). South Africa protects 1.8 percent of its grassland and has started a program to increase this to 5 percent (SANBI 2010).

How did we lose so much grassland? The northeastern United States was almost completely forested prior to the 1700s, before the arrival of Europeans and their agricultural needs. Deforestation peaked around the late 1800s and has increased since then. Even in areas where agriculture has decreased and cropland has been reclaimed for natural use, this land is more likely to become successional forest than open grassland. Norment (2002) cites the abandonment of farmland and its subsequent reforestation, decline of hayfield areas, and increased hay cropping during the nesting season as principal causes for the loss of grassland birds. A significant portion of fallow fields have been converted to agriculture, and today hay is harvested twice annually—and the need for the most nutritious feedstock leads farms to time the mowing of these hayfields with the peak season for birds that nest on the ground in these fields. The result is devastating for bird populations, often wiping out entire generations of nestlings of bobolinks, meadowlarks, and grassland sparrows. Organizations like the Bobolink Project in New England are working with farmers to implement possible solutions, including mowing some hay fields later, or leaving portions of hay fields until the young have fledged (Minetor 2022).

Grasslands are one of the most threatened habitats on earth. More than eighty percent of the western North American grassland-obligate species winter in the Chihuahuan grassland desert of northern Mexico, and in the Sonoran desert of the southern U.S. and Mexico. These grasslands are under developmental pressure, especially in Mexico, where these areas are converted to center-pivot irrigated agriculture (Macias-Duarte *et al.* 2011).

Land Usage

Agricultural practices, cattle grazing, and development for extracting natural resources all constitute threats to grassland species.

Intensive industrial agriculture leaves no open space or habitat for wildlife. In addition, this form of agriculture is based on the use of toxic, synthetic pesticides and fertilizers. Pesticides are only one class of biocides used to treat soils and plants; these chemical treatments also include fungicides and herbicides.

Pesticides are frequently mentioned as a problem for grassland birds. Restrictions on the use of toxic substances are often delayed until serious losses of wildlife are uncovered, such as the effects of DDT on nesting birds. Carbofuran is a neurotoxin that kills any vertebrate, even when used in extremely small amounts. It was eventually banned in the United States, but is still manufactured in other countries and used illegally in the U.S. (Ebersole 2020). When another class of toxic substances, organophosphates, were used to spray sunflower seeds, Swainson's hawks in Argentina died from the seeds and from ingesting grasshoppers contaminated with this pesticide (Goldstein *et al.* 1996). Another example is a class of pesticides called neonicotinoids, which are used to spray seeds and are toxic to birds (Gibbons *et al.* 2015). These are only four examples of pesticides toxic to wildlife—there are many others that have been banned in the U.S. over the last several decades, but that may still be in use beyond North America.

Some pesticides cause poisoning of wildlife with short term exposure; this acute poisoning has been seen in birds and fish. Chronic poisoning is caused by exposure over an extended time period—resulting from, for example, chlorinated aromatics in dieldrin, endrin, and chlordane. Chronic poisoning is sometimes considered a secondary effect. Secondary poisoning, on the other hand, occurs when an animal consumes prey species that contain pesticide residues—a frequent cause of mortalities of birds, fish, and other aquatic animals.

While testing continues to safeguard humans and animals from these chemicals, based on these well-known examples, it does not seem likely that the toxicity levels and risk assessment to wildlife are adequate to fully understand the short-term and long-tern impacts. In 2020 the Environmental Protection Agency reduced testing of pesticides on birds, citing that its objective is to ensure the protection of public (human) health (EPA 2020). However, states can enact further restrictions beyond federal guidelines.

Industrial development within grasslands does sometimes leave open space, but industrial practices in these open spaces can be destructive. Practices in oil and gas extraction with its associated roads and spills, open oil pits, habitat conversion and fragmentation, weed encroachment, spread of exotic species, and fire suppression all degrade the environment. Mining practices for coal and minerals are also very destructive to the environment. Incompatible recreational use, such as off-road vehicles, further destroys habitat already damaged from natural resource extraction. There is no doubt that these development and recreational needs are valid. The issue is how we preserve some balance between the needs of humans and of nature. For the most part, this depends on each state's willingness to make rules that may restrict industry or recreation while benefiting wildlife (Wyoming Game and Fish Dept. 2017).

The quality of grazing and crop lands are an important factor for grassland species. Croplands used for monocultures such as cotton, corn, wheat, and soybeans are of little or no value for breeding birds, as these birds do not feed on these introduced plants or use them for nesting. In the west, alfalfa fields are often flooded before plants begin to grow, providing feeding areas for herons, ibis, and curlews; however, the resulting alfalfa crops have little value for breeding birds.

Conversely, open spaces created by hayfields, abandoned fields, and pastures provide the types of habitats beneficial to breeding birds, though these fallow fields often foster woody vegetation that needs to be controlled before it becomes successional forest. Harvesting hayfields needs to be delayed until after fledging; unfortunately, hay farmers object to this, because hay left in the field loses its protein and nutrient content quickly, making it less useful as a feedstock for cattle and horses. A paradigm shift needs to take place to make hay farming more compatible with breeding birds.

Cattle grazing does not support the same ecology as do the indigenous grazers. Domestic cattle tend to destroy endemic grasses and plants, and these are replaced with foreign grasses that change the ecosystem, such as buffel-grass (*Cenchrus ciliaris*) which is a problem in the Sonoran Desert. In drier climates, cattle and sheep change the plant communities and leave woody plants. Studies have shown that cattle can be beneficial to grassland plant communities if stocking rates of pastures are low enough to maintain grass cover. In spite of this, overgrazing has been the rule throughout most of the world.

Habitat fragmentation is a threat to breeding birds, because size and quality of the habitat are important to territorial species. Some species, like hawks and other large birds, require large open fields, 250 acres (100 ha) and larger for optimal hunting. Research has shown that even small birds like sedge wren, bobolink, clay-colored sparrow, grasshopper sparrow, Baird's sparrow and LeConte's sparrow strongly prefer larger grassland patches (Johnson and Igl 2001). Other species including eastern kingbird, common yellowthroat, savannah sparrow, and western meadowlark are less sensitive to the size of the field. A single, optimum size of a grassland cannot be determined because of the variable area sensitivity among species, but patch sizes of less than 25 acres (10 ha), or those that are long and narrow, are of little to no benefit to grassland birds (McCraken 2005). In the northeast, Vickery *et al.* (1994) estimated the minimum area and the area that has a fifty percent chance of supporting a specific grassland species. They found, for instance, that grasshopper sparrows were present in 50 percent of the fields of about 250 acres (100 ha). Upland sandpiper, which has the largest field requirement, is rarely found in a field as small as 125 acres (50 ha), but are most likely to occur in fields of about 500 acres (200 ha). Grassland fields of this size that are managed for avian needs are rare; however, Vickery suggested that grasslands need to be at least 250 acres (100 ha) in size to support a diversity of grassland bird species.

In the United States and in some other countries, conservation measures have been developed to benefit birds. The Farm Bill Conservation Programs provided funding for conservation to improve habitat, access, and soil and water quality on private lands. The U.S. Department of Agriculture's Natural Resources Conservation Services (NRCS) have worked with private land owners to manage range land to support the needs of wildlife. The survival of grassland bird species is totally dependent on how private landowners and governments manage range and croplands—and this is especially true in the eastern United States, which has little federal land. This subject is explored further in Chapter 24.

No one denies the fact that we need agriculture and industrial development. Our challenge is to find a balance between what is needed for the economy and what can be managed for conservation. History shows us that private land owners are not usually willing to adhere to conservation measures for any extended time. Because of the need to produce income to cover taxes, expenses, and profits, there are always incentives to ignore wildlife needs. Ultimately, we rely on state and federal governments to set aside land for conservation measures.

16

Nectar-Feeding Birds

Flowering plants provide nectar for many animals including insects, moths, bats, and birds. These plants are distributed throughout the world, but they are most diverse and concentrated in the tropical and subtropical regions. Nectar provides energy through its sugar content, but it is deficient in minerals and vitamins, so all nectar-feeding birds must supplement their diet with protein food sources, particularly insects. The nectars contain about 20 percent carbohydrates in the form of sucrose, and also small amounts of amino acids (Nicolson and Fleming 2014). Since the quantities are so small (in the micro liter range), birds have to expend a lot of energy visiting sources.

Anyone with hummingbird feeders has seen many different bird species beyond hummingbirds routinely visit to drink the nectar. Orioles, woodpeckers, finches, tanagers, warblers, mockingbirds, verdin, and titmice take nectar directly from the flower or feeder, or from trees that drip sap. Sapsuckers chisel shallow holes into tree bark to cause sap to run, feeding on the sweet sap and the insects that are trapped in the sap.

Nectar-feeding birds play an extremely important role in pollinating flowering plants. Some plants have evolved to be selective of whom they attract to influence the spreading of pollen, forming symbiotic relationships with specific bird, butterfly, and bee species to ensure that their pollen gets distributed to other flowers of their own species. The flowers have developed tubular bases, matching the narrow bills of their selected pollinators.

Nectarivores are found on all continents except Europe and Antarctica, though recent studies have shown that some European bird species that feed primarily on insects and seeds will visit flowers and spread pollen (de Silva *et al.* 2016). Hummingbirds, sunbirds, sugarbirds, honeyeaters, honeycreepers, lorikeets and flowerpiercers are the most well know nectarivores. Hummingbirds and lorikeets are not passerines, but sunbirds, flowerpiercers, honeyeaters, and most of the other birds attracted to nectar are. In tests, one-fifth of individual birds carried pollen from different plants, demonstrating that they were taking advantage of flower nectar.

While European insectivores and granivores can dabble in nectar when it suits them, nectarivores must seek that resource daily to maintain their energy levels. Nectarivores found in tropical regions have highly energetic lifestyles with high metabolic rates. Generally, most of them are very colorful; they have specially adapted tongues and bills for extracting fluid from tubular blossoms. The lorikeet's tongue is stiff and tipped with bristles, making it easy to gather quantities of nectar, while honeyeaters have a bush-tipped tongue fringed with bristles, allowing them to draw lots of nectar from large surfaces with one lick. Sunbirds and hummingbirds have tubular shaped tongues, most lined with fringes, and some bifurcated. In the case of hummingbirds, high speed photography shows that after inserting the long tongue into a flower, it flattens the tongue; when it reaches the nectar, it closes the tongue to capture the fluid.

Nicolson and Fleming (2003) describe how different families of nectar-feeding birds deal with problems associated with subsisting on and processing the different types of nectars, and why some species resort to a torpor state. Around ten percent of all bird species may use nectar as a resource at some time. Since nectar feeding is so common among birds and present in so many different families, it must have evolved independently multiple times.

Because of their diet of carbohydrates and proteins, nectarivores use a variety of habitat types that provide nectar and insects. They will visit fragmented landscapes (natural habitats converted into isolated patches) for their needs. Some research has shown that urbanization has reduced the diversity of nectar-feeding birds (Pauw and Louw 2012). In the United States, hummingbird numbers are declining, like those of other pollinators, because of habitat loss, changes in the distribution and abundance of nectar plants, the spread of invasive plants, and pesticide use (USFS 2017).

Hummingbirds

Hummingbirds are well known for their spectacular colors and remarkable flight skills. They are among the smallest of all birds—in fact, the bee hummingbird of Cuba is the world's smallest bird; it is only 2.2 inches long (5.6 cm) and weighs 0.065 ounces (oz) (1.8 gm). By comparison, the weight of a United States dime is 0.077 oz (2.2 gm), considerably more than the bee hummingbird. The average hummingbird, such as a ruby-throated, weighs about 0.11 oz (3.1 gm). Larger hummingbirds like the mountain gems, Rivoli's, and the larger hermits, weigh about 0.27 oz (7.6 gm). These larger hummingbirds are about the same weight as the smallest flycatchers (the pygmy-tyrants) and just slightly heavier than a U.S. quarter-dollar by about one gram. The giant hummingbird of the Andes is the largest, weighing about 0.7 oz (21 gm) and is 9 inches long (23 cm), about the same as a sparrow. Because of their small size and the speed with which they use up the energy they consume every day, hummingbirds go into a torpor, a state of reduced metabolic activity, to conserve heat during the night or in cold conditions at high altitudes.

A remarkable 350 species of hummingbirds live only in the western hemisphere, where they are widely distributed throughout the Americas and the Caribbean islands. They are in their own family, *Trochilidae*, and are placed in the same order as swifts and nightjars. Although the majority are found in tropical regions, some hummingbirds occupy very cold habitats. A few breed in the high Andes, for example, where temperatures are below freezing after sunset: the Andean hillstar is well adapted to the cold conditions of the Andean high plains, thriving at an altitude of 15,000 feet (4,560 m).

Most hummingbirds are non-migratory except, for instance, those species that breed as far north as Alaska and as far south as the southern tip of South America. These species carry out long annual migrations back to the tropical regions to winter. In spring, for example, the rufous hummingbird migrates 2,000 miles (3200 km) to Alaska from its wintering range in Central Mexico. In the Southern Hemisphere, the green-backed firecrown breeds as far south as Tierra del Fuego and migrates to central Argentina for the winter, a distance of about 1,800 miles (2900 km). The ruby-throated hummingbird crosses the Gulf of Mexico in spring migration, a non-stop flight of 600 miles (960 km), on its way to breeding grounds in the eastern half of the United States and southwestern Canada.

Hummers are nature's most accomplished fliers; they are the only bird that can fly backward. They have the greatest ability to maneuver of any bird family: they can hover with their body motionless, and fly upside down and sideways. They accelerate and decelerate to and from full speed very rapidly. They position themselves with needlelike precision when feeding, and are physically strong enough to feed while hovering.

These flight skills are the result of their unique shoulder, arm and fingers. The shoulder joint is extremely mobile and controls the wing rotation—we rotate our hand with our forearm, but hummingbirds rotate their wing with their shoulder while the wing itself remains straight. The arm bones and the hand (humerus, ulna, and metacarpus) are all very short compared to other birds, and they have fewer secondary feathers on the arm and hand compared to passerines. All hummingbirds' primary feathers are attached to the long finger.

Crawford Greenewalt (1960) was one of the first to study the flight characteristics of hummingbirds. More recently, researchers have investigated more detailed aspects of their flight dynamics. These birds generate power on both the up-stroke and down-stroke of the wing, as opposed to other birds that generate power on the down-stroke. Hummingbirds' unique shoulder joint gives them the ability to control flight direction by rotating the wing in different planes. For forward flight, the wings rotate towards the back of the bird; while hovering, the wings rotate in a figure eight on the side of the bird. The ability to fly backward is accomplished by rotating the wing above the head. These changes in wing motion with forward, backward, and hovering flight are accompanied by slight changes in body posture, wing beat rate, and wing angle (Sapir and Dudley 2012). Hummingbirds are the only nectar-feeding birds that routinely feed while hovering.

Female hummingbird plumage tends to be more cryptic and nondescript compared to males. In males, the colors and the adornments of their crests, decorative tails, and occasionally ear tufts are part of their spectacular beauty. The shimmering metallic colors of the head, gorget (throat), and green back are all the result of iridescence, described previously in Chapter 4.

Hummingbird crests and tails come in a wide variety of shapes and sizes, depending on the species. An example of an unusual tail adornment is seen on the Peruvian Racket-tail [Figure 16.1]. Male hummingbirds use their brilliant colors and adornments to attract a partner, as well as for territorial defense (Tilford 2014). Some hummers have plain plumage, such as the brown-violetear, sicklebills, and the hermits. The hummers with more cryptic plumage tend to occupy the dense under-growth, and the bright hummers with startling adornments are most often found in sunny exposures. Almost all hummers have dark brown to black wings.

Helmuth Wagner (1946), in his studies of the foraging behavior of humming-birds, showed that all the hummingbirds he studied relied more on insects than on nectar. Insects are particularly important nourishment for immature birds. We rarely see hummingbirds capture insects, since most are observed at feeders or flowers, but Connor (2010) summarized a number of field studies that show that hummingbirds spend far more time capturing insects—both by hawking and glean-ing—than they do feeding on nectar. For hummingbirds, the process of capturing insects is unique. High-speed motion videos show that their lower mandible flexes downward and their gape widens when they open their bill, which then snaps shut in less than a hundredth of a second (Yanega and Rubega 2004).

Figure 16.1 Male Peruvian Racket-tail hummingbirds will display with their tail and orange leg feather during competition with other males and to attack females (author's collection).

Other Nectarivores

Eucalyptus trees and shrubs include approximately 900 species in three genera, and most of these are found in Australia. These flowering plants of the myrtle family produce nectar, and serve as an important source of nourishment for the honeyeaters. Australia has the largest number and most diversified nectar-feeding birds: 75 species of honeyeaters, and also many parrots and other birds that are attracted to nectar, such as metallic starling and woodswallows.

The honeyeaters (*Meliphagidae*) are a diversified family of 186 species in 42 genera, with a range that includes Papua New Guinea, Australia, New Zealand, and the Pacific Islands. They are a very large, diversified family of passerine nectarivores, and few generalities apply to them. Many honeyeaters are colorful, and they have a wide range of sizes: the smallest, the weebill, is about 3.5 inches long (8 cm) and weighs 0.21 ounces (6 gm). Amazingly, the largest are the wattlebirds and friarbirds; the yellow wattlebird reaches 18 inches long (45 cm) and weighs 7.0 ounces (200 gm).

Some honeyeater species, like chats, are entirely insectivores even though they have a brush-tipped tongue suited for gathering nectar from plants (Clark and Wooller 2003). Those with short bills feed more on insects; those with longer bills tend to feed more on nectar, while some also feed on fruits and other sugar sources. The colorful Eastern Spinebill [Figure 16.2] uses its long, curved bill to probe into

Figure 16.2 The Eastern Spinebill, an Australian honeyeater, feeds on nectar and will store fat when nectar is abundant for periods of low nectar availability (Celeste Morien).

flowers for nectar. The larger honeyeaters have large bills. Some species are predators and will take chicks, lizards, and frogs.

As a group, the honeyeaters can be extremely aggressive. Research has shown that when nectar is extremely abundant, different species will feed in common in large eucalyptus trees covered with thousands of flowers. The various species divide the tree into small territories, returning to the same tree as long as the blossoms last. Like nectar-feeding birds elsewhere, they will move as flowers mature in other trees. When nectar is scarce, one species will dominate a tree and drive off all smaller species (Armstrong 1992); in fact, some species of minors will defend their territory throughout the year with total dominance.

Lories and lorikeets are small to medium sized parrots of the family of Old World Parrots (*Psittaculidae*) that specialize in nectar feeding. Their diet also includes pollen, flowers, and some fruit. They are among the most colorful parrots, and range widely from Southeast Asia to Australia. Large parrots in Australia consume nectar as well, and some eat whole flowers.

Manna, one of two popular food sources for Australian honeyeaters, is crystallized plant sap rich in sugar. Several types of Australian eucalyptus trees produce this sap, which is eaten by a variety of birds, some honeyeaters, and mammals—including Aboriginal people, who collected it for their own use for thousands of years. Early settlers to Australia also discovered manna's high nutritional content. Another food source, lerp, is a flake-like solid produced by larvae of an insect in the *Psyllidae* family, and contains starch and sugar (Low 2014).

Sunbirds and spiderhunters of Africa, India and Southeast Asia make up the *Nectariniidae* family of the passerine order. Spiderhunters are small, long-billed birds with muted plumage. Sunbirds are similar in many respects to hummingbirds, though not related to each other; the two families represent another example of convergent evolution. Sunbirds are larger than hummingbirds and do not have the same degree of flight control; however, they have strong feet, giving them the ability to feed while perched, something hummingbirds struggle to do. They harvest nectar from flowers by perching on the flower stem and plunging their long bill and brush-tipped, tubular tongue into the blossom to reach the nectar; if they are not able to reach the nectar, some sunbirds will pierce the flower at the base. Insects, especially spiders, are important in their diet as well.

Like hummingbirds, the males sunbirds are brightly colored, and many have iridescent heads and backs; the forest species tend to be more muted than those of open country. They also go into a torpor to conserve body heat in cold conditions, just as hummers do. The Crimson Sunbird [Figure 16.3] is among the smallest and is about the size of a larger hummingbird.

In southern Africa, the two species of the sugarbirds family (*Promeropidae*) look like large, long-tailed sunbirds. These nectar-feeding birds are related to the Australian honeyeaters, however, and not to sunbirds.

Flowerpiercers are members of the tanager family (*Thraupidae*) in the passerine order. Most of their 18 species are found in shrubby areas and forest borders of the Andean highlands. These small birds, which feed on nectar and small insects, are generally plain: some have gray or cinnamon bellies, but most are uniformly

Figure 16.3 The brightly colored Crimson Sunbird is one of the smallest sunbirds of tropical southeastern Asia, and the national bird of the Republic of Singapore (Bruce Hallett).

Figure 16.4 White-sided Flowerpiercers of the mountainous areas of northern South America cut the base of flowers to get at the nectar (Diane Henderson).

dark in color. They have an upturned bill with a hook at the end of their upper mandible, allowing them to hold the flower and take nectar while piercing its base with their lower mandible, as shown in Figure 16.4. They do not help to pollinate the plant (Clark and Wooller 2003).

In South and Central America, four species of honeycreepers, the *Cyanerpes* genus, are members of the tanager family and are nectarivores: red-legged, purple, shining, and short-billed honeycreepers. Like most nectar-feeding birds, they are small; males are brightly colored, and they have curved bills. The golden-collared and green honeycreepers are closely related to tanagers. They are larger than the *Cyanerpes*, less dependent on nectar, and feed on fruits and berries. In addition to the honeycreepers, the tanagers of these regions are omnivores, but they routinely consume nectar and flower petals.

The Hawaiian honeycreepers are well diversified and are part of the finch (*Fringillidae*) family, discussed in Chapter 14. Of the 57 identified species of honeycreepers, 15 were believed to be nectarivores, and only seven survived to the present time. These brightly colored birds with curved bills feed on mint and lobelia flowers and the ohi'a lehua (*Metrosideros polymorpha*) tree. This large tree with bright red flowers attracts many species of birds. The extinct o'os, mamos, and Hawai'ian kioea—the largest nectarivore in Hawai'i at 13 inches (33 cm) long—are believed to have been nectarivores as well.

17

Wetland Birds

When we visit a wetland, we can only begin to conceive of the many life forms within this habitat. Mammals, reptiles and amphibians, fish, aquatic insects, dragonflies, butterflies and many species of birds choose this form of habitat, and most of these are not obvious to the human eye. Wetlands are like tropical forests in that they support a very high diversity of life. It takes some effort to see, understand and appreciate the complexity of this smaller world.

Wetlands are areas in which the soil is saturated by water for all or part of the year. Today, wetlands cover only six percent of the world's landmass, but they are important because they protect populated areas from flooding, and they have natural water management cycles that purify water, recycle nutrients, and sequester carbon. They also include a wide variety of habitats important for birds and wildlife.

Despite the considerable value that wetlands provide to human life, a vast majority of the world's wetlands have been drained off or filled in for industrial and residential development, as well as for agriculture. Environmental organizations estimate that 50 percent of all the world's wetlands have been destroyed in the past 100 years. In 1991, Dahl and Johnson determined that in the United States, 22 states had destroyed more than 50 percent of their wetlands (Dahl and Johnson 1991). More recent analysis covered a five-year period between 2004 and 2009, and discovered that wetland area in the United States declined by an estimated 62,300 acres (25,200 ha) (Dahl 2011)—an area nearly the size of the state of Washington.

Indiana, Illinois, Missouri, Kentucky, and Ohio have lost more than 80 percent of their original wetlands, and California and Iowa have lost more than 90 percent. Marshes along the Great Lakes have been significantly degraded or completely lost as a result of back filling, pollution, invasive species, and water management policies. The Clean Water Act and the Emergency Wetlands Resources Act in the United States provided guidelines to try and stem the loss of wetlands, and results show that the rate of loss has decreased (Dahl 2011). Nevertheless, these human-caused changes have led to a precipitous decline in the native species that are adapted to these habitats. The Natural Resources Conservation Service (NRCS) of the U.S. Department of Agriculture has two programs directed at the restoration of wetlands, and these are discussed in Chapter 24.

Many kinds of ecosystems qualify as wetlands:

- **Freshwater marshes**: Wetlands dominated by plants without woody stems.
- **Hemi-marshes**: Wetland with mixed vegetation found in deeper water, especially on the edges of lakes, ponds and rivers.
- **Saltwater tidal marshes**: Marshes near oceans, representing the extreme in terms of salinity.
- **Marine estuaries** with tidal and brackish marshes: mixed freshwater and saltwater, forming a habitat for specific bird, plant, and animal species.
- **Bogs**: peatlands—areas that remain so waterlogged that plant matter does not decompose completely in it—usually found in formerly glaciated areas.
- **Swamps**: Wetlands dominated by trees.
- **Wet meadows**: usually low-lying farmland with poorly draining soil, which may be dry most of the year but acts like a marsh during wet seasons
- **Floodplains**: Flat areas near rivers, lakes, and streams, that flood during the wettest seasons
- **Prairie potholes**: Depressions in glaciated landscapes, especially in the upper midwestern United States, where temporary ponds fill with snowmelt and rain in some seasons
- **Tundra pools**: Wet spots that emerge in summer in areas with frigid temperatures for most of the year, stunting tree and shrub growth
- **Vernal pools**: Pools that emerge only in spring, so that they support small amphibians, some insects, and birds, but no fish.
- **Playas**: Shallow temporary lakes with no outlet, found primarily in deserts.

Worldwide, more than 900 bird species are highly dependent on wetlands. Birds including rails, herons, ibises, flamingos, spoonbills, ducks, shorebirds, grebes, gulls and terns, cormorants, and some passerines use wetlands as a place to breed, for feeding, and as a migratory stop. In addition, many species of waterbirds use wetlands, but are more associated with other habitats. The osprey, for example, has a worldwide distribution and is often found around lakes and oceans, but it breeds near wetlands.

Gruiformes—*Rails, Coots and Others*

The birds that are well known for their association with wetlands are in the order *Gruiformes*, including a variety of species: rails, wood-rails, crakes, flufftails, swamphens, moorhens, gallinules, coots, finfoots, sunbitterns, limpkins, trumpeters, and cranes. The *Gruiformes* are a loosely connected order that lack distinctive shared characteristics, but are united based on internal anatomy.

The *Rallidae* (Rallids) are a subset of the *Gruiformes* that can be divided into three loosely related groups: rails and crakes, gallinules (moorhens) and swamphens, and coots. Rallids tend to have muted colors, favoring cryptic patterns. They are all associated with wetlands—most are marsh birds, and they are widely distributed throughout the world except in the polar regions. They vary from the size of a sparrow to that of a turkey. The Gray-necked Wood-Rail [Figure 17.1] is one of the largest and can be found in forests as well as swampy areas. Birds in this family are

Figure 17.1 The Gray-necked Wood-Rail ranges over Central and South America. Wood-Rails are the largest members of the rail family and are frequently found in open habitats (author's collection).

precocial, raise large broods, and have variable mating practices within the order, but most are monogamous.

Most rails have long bills and compressed bodies in order to move efficiently through reeds. Crakes have short, conical bills, as do others including sora, yellow rail, and black rail. Rails and crakes are shy birds that rarely leave the marsh. They are small with short, rounded wings, long legs, and long webless toes that allow them to walk on broad leaf marsh vegetation and to grip reeds. Many rails are sensitive to the depth of the water in marshes. For instance, yellow rail requires a maximum water depth of 6 in (15 cm) and vegetation dominated by grasses and sedges (Leston and Bookout 2015). They are particularly vulnerable in deep water because they cannot flee.

Within the family, the Rallids are very vocal and have loud distinctive songs, with a variety of calls that sound like clicks, barks, or grunts. They tend to be most active at dawn and dusk (crepuscular) and feed on a variety of vegetable matter and aquatic invertebrates. Rails also probe the mud for invertebrates with their long bills, and will search the underside of vegetation for small animals. Many states allow rail hunting, including Alabama, Texas, and others.

Corn crake has, perhaps, the largest distribution among the Rallids. Unfortunately, its population fell precipitously in the twentieth century. It breeds across Europe and most of Asia and winters in Africa, and it is one of the few members of the *Rallidae* that adapted to life in coarse grasslands. Flufftails—very small crake-like birds of Africa with short pointed bills—look like small quail. The nine flufftail species have restricted ranges and are very secretive and difficult to see.

Rails colonized many oceanic islands, adapted to island life, and some became flightless. The introduction of rats and feral animals and the conversion to agriculture doomed most of these flightless birds to extinction. To date, 31 species of rails and gallinules have gone extinct since 1600. Guam rail, for example, was common on Guam in the 1960s with as many as 80,000 individuals, but the unintended introduction of the Australian brown tree snake doomed this species. It survives today only in small numbers, thanks to human conservation measures.

The largest rail is the South Island takahē of New Zealand. The last specimen was taken in 1855 and it was thought to be extinct, but it turned up again in 1948 in a small, isolated valley at 2,000 feet (600 m) in elevation on South Island (Austin and Singer 1961). This bird looks like a large, dark swamphen with a heavy, conical bill; it feeds exclusively on vegetation. The males weigh six pounds (2.7 kg) and have a length of 25 in. (63 cm).

There are many small rails, including the black rail of the Americas and the dot-winged crake of Argentina. The smallest is probably the yellow-breasted crake of South America, which weighs about 0.9 oz. (25 g) and is 5 in. (13 cm) long. The smallest flightless surviving bird on earth is the Inaccessible Island rail, found on the island of that name in the Tristan Archipelago, in the mid–Atlantic Ocean. This rail averages about 1.5 oz. (42 g).

Swamphens, wood-rails, moorhens, and gallinules are larger marsh birds. Most of these have charcoal gray plumage with brightly colored beaks that extend over the head, forming a shield. Some, like the swamphens and purple gallinule, have strikingly bright iridescent shades of blue, blue-green and purple. The purple swamphen complex consists of six species ranging over Europe, southern Asia, the Philippines, and Australia. Gallinules, moorhens and swamphens are frequently found at the marsh edges. They are less secretive than rails, but their lifestyle is similar.

Some Rallids are forest birds, and among these are the eight specie of wood-rails of South and Central America and Mexico. Wood-rails are larger in size than the gallinules, and build nests in shrubs or trees.

All the world's coots share dark gray plumage. Coots are more aquatic than other Rallids, favoring open water, but they will scurry to take refuge in reeds when disturbed. They are entirely vegetarian and dive for food. The lobed membrane on their toes and foot allows them to swim as well as walk on land. During molt, they are flightless, and after molt they can withstand reasonably cold temperatures; many will remain in the Great Lakes region during winter in the absence of ice and will migrate south only if necessary. They are gregarious during non-breeding.

South America has the highest diversity of coots, with six species. Horned Coot of the Andes breeds above 13,000 feet (3,000 m). This species builds a conical mound of stones and covers it with vegetation that results in a nest 13 feet (4 m) across, weighing as much as 1.6 tons (1.45 metric tons).

Ardeidae—*Herons and Others*

About 76 species of wading birds—herons, ibis and spoonbills—are adapted to feeding in wetlands. The herons, of the family *Ardeidae*, are widely distributed on earth, and some, like great egret, black-crowned night-heron, and cattle egret, have extensive distributions. Herons are known by a number of titles: bittern, heron, tiger-heron, reef-heron, night-heron, pond-heron, and egret. These names imply something about the habits or appearances of these long-legged waders. Most are colonial breeders, except the bitterns, and the vast majority nest in or near marshes. Herons range in size from the small bitterns, about 12 in long (31 cm), to the goliath heron of Africa, which is 53 inches long (135 cm) and when standing erect is 5 feet (1.5 m) tall.

A few species of herons are sub-tropical forest dwellers, such as tiger-herons, the Agami heron of South America, and great-billed heron of Australia. They feed in forested rivers and streams.

In most herons, the males and females have the same plumage. They develop long filoplumes in breeding season for display purposes, and they undergo a change in the color of their bare facial skin. The color becomes intense; it then reverts to muted tones later during the year.

Many herons also have powder down feathers, a special type of feather that frays at the end. Oliver Austin, Jr., (Austin and Singer 1961) provides this story about the use of powder down.

> Herons use their powder down for removing oil, grease, and slime from their feathers. Some herons apply it with their beaks. The bitterns, who are great eel eaters, rub their heads in the breast powder down patches after a meal until their head feathers take on a powdery appearance. They leave the powder on for a while to soak up the slime and dirt, and then comb it out by vigorous scratching, mainly with the serrated nail of the middle toe. When the feathers are clean again they add waterproofing with their own oil from the preen gland.

Bitterns are solitary birds that are most adapted to life in the marsh. They have camouflage plumages that help to blend in with their surroundings, they are more secretive than other herons, and they rarely leave the marsh. They have shorter legs and generally have a different profile from other herons. The four medium-sized bitterns (*Botaurus*) are distributed around the world. They are similar to one another in appearance and behavior: American bittern in North America, pinnated bittern in Mexico and South America, great bittern in Eurasia and Africa, and Australasian bittern in Australia. All have a striped breast and belly that resemble the marsh grass when they stand erect. They remain erect when threatened and sway with the moving reeds to avoid detection.

These bitterns have similar territorial songs as well as other calls: usually a single phrase with one to three notes, depending on the species, repeated between two and six times. These low frequency booming sounds carry a long distance. The American Bittern [Figure 17.2] can be heard from 400 yards away (365 m) (Bent 1926).

The bitterns also include ten species of the *Ixobrychus* family, the smallest herons. Five of them are similar in appearance and average about 12 inches (30 cm) long,

Figure 17.2 The American Bittern with its erect posture and striped breast resembling reeds; here it is shown with its wing coverts and neck feathers extended in display (author's collection).

and have more noticeable sexual dimorphism: little bittern of Eurasia, black-backed bittern of Australia, stripe-backed bittern of South America, least bittern of the Americas, and dwarf bittern of Africa. It is surprising that even though the two basic groups of bitterns are so widely distributed, they still have the same lifestyle and similar plumages, and they fit into two basic size groups. It is likely that they evolved from single ancestors that adapted to the various continents, and diverged over time.

Anseriformes—*Ducks, Geese, and Swans*

The waterbirds, *Anseriformes*, are made up of one large family, the *Anatidae*, and two other ancient families, the magpie goose (*Anseranatidae*) and three species of screamers (*Anhimidae*). Ducks, swans and geese make up the *Anatidae* (Anatids), which include about 160 species. The *Anseriformes* order has a long history with ancestors dating back to the Mesozoic era, prior to the extinction of dinosaurs at the K–Pg boundary.

The Anatids are diversified by size and global regions, and use some form of wetlands to breed or for feeding. They share many traits: long necks, webbed toes that allow for control in water and enable diving ducks to maneuver underwater, and short legs placed to support walking and landing and taking off from land. They all have "duck" bills with comb-like structures on the bill edge (lamellae), which allow them to strain small animals and plants from the water. They are densely feathered with sufficient insulation for body warmth, their feathers naturally oiled to prevent water penetration to the skin. Most are gregarious; they migrate and feed in flocks and gather in groups during non-breeding.

These waterfowl have evolved a life in water and on land, though some are more aquatic than others. Many ducks winter in cold areas and spend all their time on water. Ducks and geese feed on animal matter including fish, insects, small amphibians, and small mollusks, and many are vegetarian. A number of duck species have developed special mechanisms for gathering and consuming food: some have a nail on the upper mandible, for example, that they use to remove seeds from plants and manipulate vegetation. Diving ducks use this to pry shells out of the mud of lake bottoms. Geese and some diving ducks have straight bills with tough lamellae to grasp underwater vegetation. Mergansers have tooth-like serrations on the bill for grasping fish. Overall, there are over 100 species of ducks within the Anatids, divided into nine subfamilies, and 15 species are not assigned to subfamilies. Taxonomic changes are always possible between these groupings. Those genera that include more than one species are summarized in Table 17.1. There are a number of species named "goose" that are actually ducks: Egyptian goose and pygmy-geese are examples of this.

Within the Anatids, 29 species are monotypic genera and include Egyptian goose, muscovy, musk duck, and torrent duck. Hooded merganser and smew are also monotypic, but they are certainly closely related to the mergansers. The whistling ducks are tropical birds and distinct from other ducks in a number of anatomical characteristics. Magpie goose of Australia is the most archaic form of the order of waterbirds: it is the last living member of its family (*Anseranatidae*) and is related to screamers. Once common in Australia, it now survives in much smaller numbers because so much of its habitat has been lost.

Also within the Anatids are many adaptations to different habitats and food sources. The dabbling ducks of fresh water are surface feeding ducks, and are by far the most diverse of all the ducks. In contrast, the diving ducks have their feet further back, enabling better diving ability. Because of this, they sacrifice the ability of rapid takeoff, which is a quality of dabbling ducks.

Table 17.1 Genus and Subfamilies of the Major Groups of Ducks, Geese and Swans

Type	Subfamily	Genus	No. of Species	Species (examples)
Geese & Swans				
Whistling Ducks	Dendrocyginae	Dendrocygna	8	Fulvous, West-Indian
True Geese	Anserinae	Anser	11	Graylag, Snow, White-fronted
True Geese	"	Branta	6	Brant, Canada, Cackling, Barnacle
Swans	"	Cygnus	7	Trumpeter
Geese, (S. Ame.)	"	Chloephaga	4	Upland, Kelp, Ashy-headed
Goose, Perching	Plectropterinae	Plectropterus	1	Spur-winged Goose
Ducks				
Shelduck	Tadorninae	Tadorma	6	Australian Shelduck
Steamer-Duck	"	Tachyeres	4	Flying, Flightless, White-headed
Pygmy-Goose		Nettapus	3	Green, Cotton, African
Dabbling Ducks	Anatinae	Aix	2	Wood, Mandarin Duck
Dabbling Ducks	"	Spathula	10	Blue-winged, Cinnamon Teal
Dabbling Ducks	"	Mareca	4	Widgon, Gadwall
Dabbling Ducks	"	Anas	29	Mallard, N. Pintail, various teal
Diving Ducks	Aythyinae	Netta	3	Red-crested Pochard
Diving Ducks	"	Aythya	12	Canvasback, scaup, Common Pochard
Sea Duck	Merginae	Somateria	4	Eider—Common, Steller's
Sea Duck	"	Melanitta	4	Scoters—White-winged
Sea Duck	"	Bucephaia	4	Goldeneye, Bufflehead
Mergansers	"	Mergus	5	Common, Red-breasted
Stiff-Tailed Ducks	Oxyurinae	Oxyura	6	Ruddy, White-headed

Sea ducks spend most of their life in salt water, and some will winter in fresh water habitats. The four species of steamer ducks are restricted to southern South America, and three are flightless.

Ducks are dimorphic, the males of many species are extremely beautiful with distinctive plumages, and the females are all more cryptic. They are semi-aquatic; nesting, feeding on land, but also requiring open water for feeding, for their needs of immersion and cleaning, for resting, and for their tendency to gather together. They are strong flyers, flying very fast for their size, and some of the more northern species migrate long distances in winter. Most noteworthy is the fact that the female ducks take all the responsibility for bringing in the next generation.

Ducks have been domesticated for thousands of years, so their life history is well understood. There have been efforts to reverse the decline of ducks, so today their populations are generally stable. Nevertheless, the International Union for Conservation of Nature (IUCN) lists six species of ducks as critically endangered, including two that are possibly extinct: the Laysan duck of Hawai'i and the Brazilian merganser.

A few species of ducks have learned to exploit the urban environment we have created. A number of wetland species are now considered pests because their populations have increased beyond historical levels. Canada geese, mute swans, muscovy, and mallards have adapted to live in close proximity to human development, and they have few predators.

Canada geese decreased in the nineteenth century through hunting and loss of habitat, and were re-introduced into most of the United States in the 1950s and again in the 1970s (Mowbray *et al.* 2002), as well as into Europe and New Zealand. Prior to 1900 in North America, Canada geese were seasonal migrants, heading north in spring to their breeding range in Minnesota, throughout Canada, and as far north as Alaska. Geese have site fidelity to their breeding range, and the immatures learn the migration route from the adults. With the reintroduction, however, this natural pattern changed: introduced geese ceased to migrate north, and are now non-migratory except for short movements to areas of open water and fields. They have proliferated to the point where they are now considered a nuisance, feeding on grassy lawns with few threats in our urban environment. As a result, municipalities and states have taken measures to reduce local populations. They are a protected species in the United States, but certain legal measures have been used to decrease their numbers, including harassment, habitat modification, and egg addling by coating the eggs with corn oil.

Snow geese have become a problem similar to Canada geese. They have dramatically increased in numbers over the past fifty years because, during winter, they have adapted to feeding in agricultural fields. Historically, they wintered at coastal locations where their food supplies were limited (Mowbray *et al.* 2020). Today experts estimate that the total number of snow geese exceeds five million birds, far more than any other time in history. They are destroying their arctic breeding habitat through the large quantity of organic matter they consume. Governmental agencies have tried to reduce their population through hunting and other measures, but their numbers continue to increase.

Mute swans were introduced into the United States from their native Eurasia before 1900 for their aesthetic value. Now, however, they have increased to such a degree that they are considered an invasive species. Because they are not native to North America, these swans are not protected by national statutes, so they fall under the jurisdiction of state game laws. The same control measures used to limit Canada geese can be employed for the swans. These non-migratory birds tend to dominate wetlands, threaten native wetland species, and consume excessive amounts of native vegetation.

Table 17.1 lists the types of waterfowl in the *Anatidae* family. This is an overview of the genera within the various subfamilies, and it shows the diversity of waterfowl. This table is an example of the use of the taxonomic classification of subfamilies, which provides another level of the evolutionary relationship within the family. The genera have adapted to different diets, types of wetlands and bodies of water, and environmental conditions.

More than fifteen other species of ducks are unresolved at the subfamily level. Most of the unassigned are monotypic genera. The spur-winged goose is monotypic and has been included in Table 17.1 because it is so unique among waterfowl, and because it was mentioned in Chapter 10.

Ducks and their larger close relatives breed in a wide variety of habitats, most of them associated with water. Most ducks breed on dry ground along the edges of marshes or ponds, and conceal their nest in tall grass or under bushes or spruce trees. They also occasionally breed far from water on the prairies. Some breed in marshes (gadwall, wigeon, American black duck, redhead). Others breed in prairie marshes and potholes (mallard, blue-winged teal, northern shoveler, canvasback, ruddy duck). Some are found in wooded ponds (green-winged teal, bufflehead). Other breeding habitats include coastal sedge tundra, upland tundra, grassy tundra, coastal deltas, inland lakes, small ponds, swamps and alkaline sloughs (Johnsgard 1975). And a number of ducks nest in tree cavities (goldeneye, wood duck, hooded merganser). Common merganser breeds both on the ground and in tree cavities. In order to protect ducks, we must set aside the habitats where they breed as well as the ponds, lakes and streams where they feed.

Other Wetland Species

The rail, heron and duck families are the most adapted to wetlands. Shorebirds also use wetlands for feeding. Besides these, other species rely on this type of habitat.

The 22 species of grebes, *Podicipediformes*, are distributed throughout the world. Three species are found across the northern hemisphere: eared, horned, and red-necked. Several others have wide distributions, such as little and great grebe, while many others have very localized ranges. Both genetic and morphological analysis suggests that grebes are closely related to flamingos (Mayr 2004).

Grebes are strictly waterbirds. Their feet are placed back on their bodies so they are unable to walk on land, and they have webbing on their toes, similar to coots, that allows them proficiency and maneuverability in underwater pursuits. They have very short tails that are unsuitable for flight, and use their feet and toes to steer in flight (Austin and Singer 1961).

Grebes breed in freshwater wetlands, mostly in marshes. They gather plant material and create a floating platform and anchor it to reeds. Larger grebes will build their nests in the open. In the far north, they nest in hemi-marshes dominated by lakeshore bulrushes *(Schoenoplectus lacustris)*, a reed that can grow to 11 feet (3.5 m) (Bent 1919). They will carry their young on their back and have been recorded diving with young in place. Measurement of the underwater speed of great crested grebe has been recorded at 4.2 ft/sec (1.3 m/sec) (Johansson and Norberg 2001); that is about the speed of a person walking.

Grebes spend most of their breeding season in shallow water, but in fall and winter, they occupy much deeper waters. On Lake Ontario, for example, they are commonly found in water ranging from 10 to 60 feet deep (3 to 20 m) and on the ocean in deeper water. On the surface, grebes ride high in the water, but when threatened, they can increase their density by eliminating air from their feathered insulation and possibly from their air sacs, and remain submerged with just the head out of water.

Like grebes, loons have adapted to a life on water. They breed on the shores of secluded freshwater lakes and large ponds; they use emergent vegetation for nests and reed beds as well as other natural plants to help conceal their nest. Although they cannot walk, common loons can amble on land by supporting themselves in an upright stance and advancing with wing assistance. Red-throated loon, the smallest species, can walk upright.

Loons spend their winter on the ocean. They are large birds and have difficulty taking off from water, but once airborne, they are strong fliers in spite of the fact that they have a very high wing loading—the ratio between their total mass and the size of their wings. Loons and grebes use similar underwater skills to capture fish, reaching underwater speeds of 5 ft/sec (1.5 m/sec), and they are adept at making sharp turns. Loons can vary their swimming speeds by synchronized foot paddling, turning by increasing the speed of the outboard foot and banking their bodies (Clifton and Biewener 2018). Once they capture a fish, they swallow their prey while submerged.

Many other species use wetlands in addition to those already mentioned. Among the hawks, harriers routinely hunt wetlands, marshes, and grasslands. African marsh-harrier, Eurasian marsh-harrier and northern harrier are all common in

Figure 17.3 The Saltmarsh Sparrow, found in tidal salt marshes of the Atlantic seaboard, is susceptible to nest failures at excessively high tides (author's collection).

these habitats. In Africa, the marsh owl has adapted to hunting in wetlands as well. Among terns, four species breed in marshes: black, white-winged, black-fronted, and whiskered terns. Franklin's gull breeds in colonies in prairie potholes of the United States and Canada.

Many species of passerines breed in marshes as well—some exclusively so, while others, such as red-winged blackbird and common yellowthroat, breed in marshes and in other scrub habitats. In the United States, 4.5 percent of passerines breed in wetland habitats; the most common are some species of sparrows, marsh wren, flycatchers, and others. In Australia the passerines are dominated by the honeyeaters, and only 2.9 percent of the passerines are found in marshes; several fairy-wrens, emuwrens, chats, grassbirds and reed warbler, among others.

East Africa has almost 800 passerine species, and approximately 4.4 percent are marsh species. Among these are the swamp and reed warblers, the wetland streaked-backed cisticolas, at least five species of weavers, several species of bishops and a variety of species that breed in papyrus reeds. The average percentage of passerines in wetlands for these three world regions is about four percent. That would imply about 240 passerine species worldwide breed in marshes, reed beds and wetland habitats.

Since passerines using marshes are small birds, they build cup nests anchored to reeds. In South America, passerines including pied water-tyrant build a fully covered nest. Many-colored rush-tyrant builds a long, conical-shaped nest attached to a single reed. Salt water marshes that are subject to flooding from high tides and rising ocean depth present risks to nesting due to inundation, so passerines are careful about where they build their nests and their timing of nest initiation in cordgrass salt-meadows (Gjerdum *et al.* 2005). Nevertheless, nest flooding does occur. Chicks and eggs can tolerate inundation for a short time period (Gjerdum *et al.* 2008), but if they cool too much, nests will fail. Tidal inundations is a major cause of breeding failure for passerines as well as other tidal salt marsh species, such as the Saltmarsh Sparrow shown in Figure 17.3.

18

Seed- and Fruit-Eating Birds

When we examine the world's birds collectively, their food groups fall into a few basic categories that we all share. Some birds feed mostly on vertebrates, small to medium-sized mammals, amphibians, reptiles and carrion. Seabirds, waterbirds and some terrestrial species (like kingfisher) rely on fish and crustaceans. Some prefer vegetable matter: plants, buds, and flowers consumed by waterbirds and other species, including parrots. Many bird species feed on nectar.

And finally, many birds feed on seeds, nuts and fruits. Seeds and nuts present a unique challenge to birds, because birds have no ability to chew, and the dry quality of seeds makes it necessary for seed-eating birds to drink more water for digestion. In some species, finding water is a major portion of their feeding routine. Fruits provide birds with energy, but they also can present nutritional and toxicity challenges for birds.

Birds are easy to attract to our backyards if we are willing to provide foods they prefer, so backyard bird feeders are very popular in many countries throughout the world. Many different types of bird food are offered at feeders, but seeds are the most common. In the United States, estimates suggest that 53 million people feed birds (Reynolds *et al.* 2017). In Britain, a very high percentage of households feed birds at least some time during the year: in 2006, the British Trust for Ornithology estimated that Britons spent $110 million on feeding birds, and the number of different species attracted to feeders rose from 18 in 1987 to 130 by 2006 (Knapton 2017). Bird feeders bring birds to backyards, where they are closer to people's windows, patios, and decks, making them easy for the general public to watch and appreciate.

Granivorous Birds

Many bird species from different orders and families eat seeds. Seeds are a plant's unit of reproduction, and they are one of the most important food sources for mammals as well as birds. They offer good nutrition because they contain carbohydrates, fats, fiber, and small amounts of some minerals, including phosphorus and calcium. However, seeds are low in proteins, and they lack water, making them difficult to digest. As a result, many birds eat nuts, which are higher in protein and minerals than seeds. Moreover, even birds that primarily eat seeds or nuts have to augment their diet with other foods, such as vegetable matter and insects.

Granivores—birds that eat seeds—have a well-developed gizzard for digesting hard material, especially birds that ingest the seed's entire shell. The gizzard is the muscular part of the digestive system in the stomach, lined with a series of hard leathery ridges. Birds with well-developed gizzards eat grit—sand, gravel, bits of shells, and minerals—and the heavy, muscular wall of the gizzard uses the grit to grind the seeds, mixing them with digestive enzymes (Van Tyne and Berger 1976). Some of the grit is expelled, while strong stomach acids and enzymes make it digestible. For many birds, the grit is an important source of calcium and other minerals.

Larger birds like chickens ingest larger grit, including small stones. The stones become abraded with use, but they can remain in the digestive system for days, or even weeks. This form of digestive system harkens back to some dinosaurs, which appeared to use stones to grind plant matter in the same way; these small stones, called gastroliths, have been found in the abdomens of fossils (Ji *et al.* 1998).

Ducks and shorebirds ingest grit as part of their feeding routine. Diving ducks that eat clams and other mollusks swallow them whole, and stomach acids break down the shell material. In this way, calcium is not only useful to the bird, but it is also recycled.

People with feeders will notice that certain species prefer certain types of seeds, and this depends on their bill size. Seed-eating passerine birds have cone-shaped beaks. Some with larger bills are especially adapted for breaking open the hard shells of seeds, allowing them to ingest the nut. The northern cardinal and all parrots, for example, can remove the shells of large seeds by using their bill and manipulating the seed with their tongue. For the *Fringillidae* and *Cardinalidae* finches, a groove runs the length of the bill on the edge of the upper mandible. When they grasp a seed in their bill, they maneuver the seed lengthwise along the edge of the bill, so the bill edge holds the seed while they shear it, moving the bill from side-to-side to break the husk. They finish the process by extracting the nut with their tongue (Tudge 2008).

Other seedeaters also have bill adaptations for opening seeds. Smaller birds like tits will take the seed to a surface (a tree branch, for example) where they can hold the seed under their toes and crack open the husk by hammering it with their beak. Most small seed-eating birds ingest small seeds of grasses and weeds, dispersing undigested seeds through their excrement or by accidently dropping seeds.

Some birds, including accentors, chickadees and tits (*Paridae*), as well as longspurs and snow buntings (*Calcaridae*), feed on insects during the summer and switch to seeds during the winter (Perrins 2003). This seasonal variation is common among birds, as they have a greater need for protein when breeding. Also, many types of seeds become available in fall, thus allowing birds to cache seeds for winter. Some birds, like American goldfinch, delay breeding until the thistle seeds they prefer have ripened in late summer.

While many of us enjoy watching birds feed on seeds, farmers who grow cereal crops consider seed-eating birds to be pests. Blackbirds, crows, and grackles in the United States and the European starlings in U.S. and Europe are considered agricultural pests. Some other birds have become targets of agricultural interests as well: the dickcissel, for example, is not usually considered a problem in the United States because it breeds on the prairies, but in their wintering grounds, dickcissels gather in large flocks of over one million birds, primarily on the Llanos of Venezuela and Colombia. Here they are considered an agricultural pest. The problem arose as industrial farms converted natural grasslands to grain farming, and the birds switched to feeding on cereal crops. Consequently, commercial farmers poison them in their nocturnal roosts to avoid agricultural losses (Basili and Temple 1999). Because of this, some work has been done to try to assist in dickcissel conservation. Research has shown that dickcissels do not cause as much crop damage as some commonly perceive; however, changing current agricultural practices is difficult at best. Poisoning has taken its toll: The bird's population dropped by 40 percent over a ten-year period.

In Sub-Saharan Africa, the red-billed quelea is a small weaver that lives in large colonial nests. Their population in the past was estimated at 1.5 billion breeding pairs. They gather in flocks of over one million and feed on cereal crops. Because they are nomadic and so prolific, population control measures have not been successful—but that has not prevented farms from taking action, fire-bombing and poisoning the colonies to reduce their numbers.

In addition to grain-eating birds, fruit-eating birds are also treated as agricultural pests in some areas and with some crops. The Eurasian bullfinch in Europe is considered an orchard pest because it takes flower buds before they produce fruit, reducing the orchards' yield and directly affecting profits. Its population has been greatly reduced through control measures. The house finch, American robin, waxwings, and some blackbirds are considered by farmers in North America to be fruit pests (Brittingham and Falker 2010), as are some species of grain-eating doves and pigeon. Not only has the population of many bird species decreased significantly because of human actions, but some have become extinct, such as the Carolina parakeet of the United States, which was considered a fruit and grain pest.

Many of the passerine bird families that rely on seeds and nuts all year long or at least for part of the year are listed in Table 18.1. Within these families, there are over 1,000 species of passerines that rely primarily on seeds. In addition to finch-like species, there are many other passerines and birds in other orders that rely on seeds or have varied diets that include seeds, such as woodpeckers.

Table 18.1 Families of Seed-Eating Passerines

Family	Number	Ranges	Example Species
Fringillidae	228	Worldwide	True finches: canaries, buntings, goldfinches, grosbeaks, Hawfinch, Hawaiian Honeycreepers, African seedeaters, serins, rosefinches, House Finch, Chaffinch, chlorophonias, euphonias
Emberizidae	313	Old World buntings	buntings, crossbills, grosbeaks, Yellowhammer
Estrildidae	140	Africa, Asia & Australia	twinspots, waxbills, whydahs, munia, avadavats, firetails, firefinch, quailfinch, parrotfinch
Ploceidae	120	Africa, India, Malaysia	weavers, quelea, bishops, widowbirds
Passeridae	42	Europe, Asia, Africa	House Sparrow, sparrows, snow-finches
Passerellidae	130	New World sparrows	sparrows, towhees, juncos, ground-sparrows, bullfinches, chlorospingus

Family	*Number*	*Ranges*	*Example Species*
Cardinalidae	43	North & South America	Grosbeaks, cardinals, buntings, Dickcissel, Piranga tanagers, ant-tanagers chats
Calcaridae	6	Northern Hemisphere	Snow Buntings, longspurs
Paridae	51	Worldwide	Tits, Chickadee, titmice
Prunellidae	13	Europe & Asia	accentors, Dunnock, Siberian Accentor
Thraupidae	240	Mexico, Central & South America	True tanagers; Sierra-finches, Inca-finches, warbling-finch, yellow-finches, grass-finches, hemispingus

Some of the most common names for seed-eating passerines are finches, sparrows, buntings, and grosbeaks. Examples of very unfamiliar names applied to granivores throughout the world include the citrils, nigritas, bluebills, pytillas, malimbes and petronias, to name a few.

The *Fringillidae* are classical finches and the family we most often associate with a seed diet. They are very colorful birds, but the females are usually cryptically colored. The males are adorned with red or yellow, contrasting with darker tones or black. Euphonias and clorophonias, tropical frugivores, are also included in this family—very colorful birds with bright blues and greens as part of their plumages.

Many of the northern species of *Fringillidae* are gregarious in winter, gathering in small, nomadic flocks to search for sources of seeds. They have the classical conical beaks, with variations in beak sizes tailored to particular seeds. Goldfinches and siskins feed on the small thistle seeds; in contrast, the large hawfinch has a powerful bill and exerts enough force to crack cherry stones.

Crossbills represent one of the most interesting examples of a bill adapted for a specific purpose. The tips of the crossbills' upper and lower mandibles don't meet; they cross over each other. This allows them to pry open the scales of a pinecone and extract the seed with their tongue, a feat they can accomplish this in a single motion. Among the crossbills, the bill sizes and crossing angles vary depending on the type of pinecone they prefer.

The Estrildids are small colorful finches of Australia, Africa and Europe, diversified into 31 genera. One of these, the Gouldian finch of Australia [Figure 18.1], provides a fine example of the vivid colors found in some finches. The Estrilidids are extremely beautiful, making many species targets for collection by the caged bird trade. Unfortunately, more and more colorful birds, especially finches, parrots, and parakeets, are becoming rare in their native habitats—in part due to habitat loss, but also from the major problem caused by the illegal pet trade (Bird Life International). Some species have been reduced to threatened status because of these practices.

Within the *Estrildidae* family are the seedcrackers (*Pyrenestes*) which have an unusual characteristic: some adults have large bills, while others of the same species have small bills, depending on the type of seeds they eat. This polymorphism in beak size is the result of gene expression affecting the development of the bills. (Clabaut *et al.* 2009).

Granivores of the Americas include a number of families. The Cardinalids

Figure 18.1 The Gouldian Finch is one of the colorful Australian finches that is threatened due to illegal pet trade and habitat loss (author's collection).

(*Cardinalidae*) are restricted to the Americas and include several cardinals, grosbeaks, seedeaters, small colorful buntings, and chats. In recent years the *Piranga* tanagers and ant-tanagers of North and Central America were added to this family, based on genetic evidence.

The Hawaiian honeycreepers, discussed in Chapter 14, are in the Fringillid family, and a number of these finches are seed eaters, while the remainder rely on nectar or insects. Like the Darwin finches of the Galápagos Islands, the seed eaters have developed a range of bill sizes and adapted to specific plants. Perhaps the largest of the Darwin finches is the Española ground-finch, shown in Figure 18.2.

Although the taxonomy of the 13 species of Galápagos finches has been uncertain for some time, ornithologists place them in the *Thraupidae* family and believe their closest relative is a seedeater. They are very similar in plumage, but vary in size, bill size and shape, and diet. Some of these finches are seed eaters. Peter and Rosemary Grant (2003) have been studying the seed eaters for 40 years, and they have determined how they evolved when the food supply was changed due to drought. The finch's beak and body size changed in response to the availability of seeds and to competition. It has always been thought that evolution is a very gradual process, but the Grants established that the evolutionary process among the finches is variable. They have demonstrated that evolution can happen rapidly with beak and

Figure 18.2 The Española Ground-Finch of the Galápagos Islands is the largest of Darwin's Finches. It has a varied diet but gets most of its food from the prickly-pear cactus (*Opuntia helleri*) (Lucretia Grosshans).

population sizes. Finally, they showed that the nature of selection can change over time. (Zimmer and Emlen 2013).

Game birds, the *Galliformes* family, are a large order of about 290 species distributed throughout the world, except for some islands. Within this family, perhaps the most well-known is the chicken, which was domesticated at least 5,000 years ago in Southeast Asia. Most of the species in this order feed on a diverse diet of vegetable matter and insects. Those that inhabit the temperate regions also consume seeds, grains and nuts. Game birds are not migratory except for a few species of small quail, so their diets change throughout the year depending on the seasonal food availability.

Pigeons and doves are members of the *Columbidae* family and can be divided into two sub-groups. The larger group are the *Columbinae* that number about 175 species; they are ground feeders that rely primarily on seeds as their food source. The remainder of the 300 species feed on fruit and mast—the fruit of certain trees, such as acorns and certain kinds of nuts. Mast can be hard or soft. The seed-eating doves and pigeons and the ones that eat mast rely on their gizzard and grit, but they need to consume water for digestion.

Many other bird species eat seeds besides the passerines listed in Table 18.1. In addition to the pigeon, tits, and game bird families mentioned above, the following families rely heavily on seeds as part of their diet: tinamous, guineafowl, ducks, buttonquail, sandgrouse, ferry wrens, nuthatches, larks, and blackbirds. Other families of birds supplement their diet with seeds, even though they do not rely on them as a critical food source. Species such as cranes, jacanas, mockingbirds, kinglets, and pipits are just a few that fall into this category.

Frugivorous Birds

There are many tropical birds that feed primarily on fruits (as well as mammals), making them frugivorous birds. A brief list of the South American birds that rely on fruit includes oilbirds, bellbirds, manakins, cotingas, trogons, tanagers, fruit-crows, tityras, saltators, toucans, parrots, macaws, fruit-pigeons, fruit-doves, and even some flycatchers. In Asia, barbets, green pigeons, parakeets, koel, hornbills and birds-of-paradise are just a few that consume fruit. Cedar waxwings and phainope-plas are two species of the temperate latitudes that survive on fruit throughout the year. Oilbirds and bellbirds are obligate frugivores. Many species have a diet high in fruit, and others rely on additional food sources. Since fruits are lacking in protein, nestlings that receive only fruit are slow growing.

Very few bird species are exclusively frugivorous, because of the potentially toxic compounds found in fruits. For example, many species of passerines and woodpeckers eat poison ivy berries (*Toxicodendron* species), which are toxic to humans. The birds must have enzymes that deal with the urushiol toxin (the oily toxin in poison ivy), because they have no way to avoid it—the entire berry is coated with the oil. A number of other toxic substances also have no deleterious effect on birds, in part because of adaptation. For some substances that produce strong organic acids, for example, birds can produce a bicarbonate to neutralize the acid (Levey and Martínez del Rio 2001). In other cases, mildly toxic substances in the pulp can prevent consumption of too many fruits at one time, and birds can regulate seed retention time by expelling the pulp or seed. This short retention time is benefi-cial for seed dispersal (Barnea *et al.* 1993).

Just as in the case of insectivores, the frugivores have adapted to the challenges of harvesting this food source (Hilty 1994). They capture fruit either by sallying or by plucking from a perch. Birds that capture fruit by sallying have less developed legs and proportionately wider and flatter bills than those that pluck fruit from a perch. Trogons and oilbirds, for example, feed by sallying, while tanagers have strong, short legs and feed by perching.

Frugivores' digestive system is adapted to certain types of fruits. Small birds such as tyrannulets, mistletoebirds, chlorophonias, euphonias, and even west-ern bluebirds will routinely feed on mistletoe berries, which are difficult to capture and digest (Hilty 1994). Fruit- and insect-eating birds have a much smaller gizzard that is different in shape from those of the granivores, making it easy to digest the fleshy part of the fruit and disgorge the seeds. This seed distribution is beneficial to the plants, because frugivores disperse the seeds away from the parent plant. Some species will disgorge the seed in flight. Researchers have measured seed dispersion many miles away from the host plant by this mechanism. Seed dispersal by fruit-eating birds can also link disconnected patches in fragmented forests (Mueller *et al.* 2014), creating larger areas of continuous habitat for birds and mammals.

The Thraupids (*Thraupidae*) of the Americas are a large family of the true tan-agers, and also include a wide variety of finches. The tanagers consume mostly insects and fruit, but they also take seeds. The finches fall into a number of gen-era that include grass-finches, Sierra-finches, seedeaters, mountain-finches, and seed-finches. The only finch from this family found in the United States is Morelet's seedeater (formerly called white-collared seedeater).

Many bird species throughout the world will feed on fruit as a supplement to their diets. However, research has shown that the fruit consumed by these species reduces the efficiency of the birds' nitrogen metabolism, compromising their ability to convert food into protein and make new cells—thus leading to an inability to subsist on fruit exclusively (Izhaki and Safriel 1989).

More than 90 percent of tropical trees rely on fruit-eating animals, especially birds, for the dispersal of their seeds. Muñoz *et al.* (2016) found that many of the important species that disperse seeds in tropical forests are disappearing through habitat loss and hunting. Their work also shows that large birds are more important than small and medium-sized birds for plant regeneration. In Thailand, 17 families of tropical forest birds are made up of species that rely on fruit for a large part of their diet. Tropical forest cover in Thailand was 53 percent in 1960, but has been reduced to about 20 percent today. Although the Thai government banned commercial logging in 1989, illegal logging continues and agricultural land is expanding. Some land has recovered with degraded forest, but frugivorous species are absent in these habitats (Blakesley and Elliott 2000).

Frugivorous birds are a conservation priority, because their natural habitats have been transformed into agricultural ecosystems worldwide. In ecosystems that are devoted to growing fruit, birds have a negative impact on this industry, so commercial fruit growers in the United States work to deter birds from eating the fruit—most commonly by netting, lethal shooting, and auditory scare devices, but they also use chemical repellants, visual scare devices, and predators' nest boxes (Anderson *et al.* 3013).

Planting wild fruit-bearing trees and shrubs in small plots or hedgerows within orchard areas—a technique recommended in olive groves in the Mediterranean region (Rey 2011)—has become an important approach to conservation for frugivorous birds. This concept is also in use in plantations that grow shade-grown coffee, which incorporates indigenous trees within an agricultural environment.

19

Trogons

Trogons are one of the many non-passerine orders of tropical and subtropical birds, and they present a prime example of the lifestyle of larger tropical birds. Their lives contrast with the suboscine and oscine passerines previously mentioned. Like many tropical species, they rely on a forested habitat.

Trogons are popular among bird enthusiasts because of their colorful and splendid beauty and their unusual features. They are members of an ancient family of tropical birds dating back as far as 50 MYBP, and have persisted to the present. These moderately sized birds have bright breast colors; in the Americas they have iridescent green mantle colors, but those in Asia do not show this iridescence.

Mostly forest birds, trogons seldom fly into bright sunlight, but when they do, their back appears to be very bright, like a bright light. The eared quetzal of northern Mexico, a member of the trogon family, likes to inhabit steep canyon walls and will occasionally sally into bright sunlight, providing one of the most spectacular sights in the world of birds as its emerald green back and azure blue tail glisten in the sun. Creating this iridescent color requires a high level of structural complexity in the melanin substrate and layered structure on the surface of the feathers. The iridescent green can change to violet-blue, bronze-yellow, or even reddish, depending on the angle and diffusivity of light (Forshaw and Gilbert 1994).

Living in tropical zones, trogons are normally not difficult to observe, but they rarely draw attention to themselves except if they are calling. Males and females are dimorphic—specifically, the females are mostly brownish with gray heads and yellow bellies, while males usually have green or brown backs, bright red, orange or yellow bellies, and dark heads of blue, green, indigo, or red. The upper surface of the tail is usually the same color as the back, and the underside is either white with black barring, or black with or without large white spots. Most species have wide eye-rings of bright yellow, cyan, or gray, which are helpful in identification. Many species show grayish wing coverts with vermiculation (wavy lines) or black stripes, although in some the coverts are plain.

In the current taxonomy, there are 43 trogons divided into six genera (Clements *et al.* 2021). Trogons are in their own order, *Trogoniformes*, and a single family, *Trogonidae*. Within the trogon family, 37 species are referred as trogons, and these are similar in size, structure and lifestyles. Six species are known as quetzals, which are restricted to Mexico, Central, and South America. Unlike trogons, quetzal wings are iridescent in color. They are larger than trogons, about 14 inches (35 cm) in length, excluding their tail and decorative covert feathers.

Trogons are found in North, South and Central America, and in Asia and Africa. The majority of trogons are distributed between 15 degrees north and south latitudes, mostly in humid or wet forests. A few species venture as far as 34 degrees north and south, and these are adapted to more diverse habitats (Forshaw and Gilbert 2009).

The earliest fossil related to trogons was found in the Fur Formation in northwestern Denmark, and was dated from the late Paleocene to early Eocene epoch (Kristoffersen 2002), about 50 MYBP. The climate in Denmark in the early Eocene was very warm, cooling in the mid–Eocene (Lindow and Dyke 2006). Another fossil specimen found in France from the time of the Oligocene, about 33 MYBP, has been determined to be related to trogons as well (Mayr 1999).

These discoveries suggest that trogons have not evolved significantly from their early history. They may have followed the movement of the tropical climatic regression toward the equator through the Cenozoic era, rather than adapting to cooling conditions at the end of the Oligocene epoch (Kristoffersen 2002). Trogons no longer breed in Europe but are spread around the globe in three tropical and subtropical regions: central to northern South America, India-Asia-Indonesia, and Africa. Just one species, the elegant trogon, reaches the United States as an annual breeding bird. The northern part of South America and Central America have the highest diversity of trogons, but their origin is believed to be either South America or Africa.

The relationship of trogons to all other avian species has been uncertain because of their unusual characteristics. Recent whole-genome analysis places their closest relatives as the kingfishers (*Coraciiformes*), woodpeckers (*Piciformes*) and hornbills (*Bucerotiformes*) (Jarvis *et al.* 2014).

Trogons are non-migratory, perhaps an indication of why they reside in patchy groups across the globe. Some high altitude species will carry out short distance altitudinal migrations in winter for better feeding opportunities. The Elegant Trogon [Figure 19.1] in Arizona is an exception because it retreats to Mexico for the winter, although a few will remain in Arizona and seek more productive winter habitats. This is one of the few trogon species that carries out any noteworthy migration. While the short wings of trogons allow them to do tight, quick aerial maneuvers and achieve rapid acceleration, they seldom engage in long flights.

Trogons have very unusual feet. Most birds have three toes pointing forward and one backward, but trogons have two toes pointing forward and two backward—with the third and fourth toes that pointing forward, a foot arrangement called *heterodactyl*. They are the only avian family on earth with this arrangement of toes. As a result, they have limited use of their feet. They do not walk easily and when perched on horizontal limbs, they move sideways. Their feet are strong enough to support their weight when they grasp onto a tree or other object; however, when resting, they spend most of their time sitting erect with the body's center of gravity over their feet. Woodpeckers and parrots also have two toes pointing forward and two backwards, but in these groups, the second and third toe point forward and the

Figure 19.1 The Elegant Trogon male is brightly colored with an iridescent head and back (author's collection).

fourth toe is rotated backwards, an arrangement called *zygodactyl*. Because of this, these birds have stronger feet and are capable of more versatile movements.

Trogons feed on insects and fruit, which they catch or pluck while sallying. They have the strength to hover for several seconds, giving them the ability to take fruit and insects from the outermost branches of trees and shrubs, an advantage over other competitive species (Taylor 1994). They will sit for long periods of time observing and catching insects, beetles and small lizards, occasionally going to the ground for a beetle. They have flat, wide bills to capture insects in flight, with bristles at the base like flycatchers. Also unusual, trogons' upper mandible has serrated edges that assist in grabbing insects, lizards and fruit. In Arizona, the Elegant Trogon likes to feed on the berries of the pyracantha, an evergreen shrub with red berries, and madrone (*Arbutus*) berries in winter. Analysis of the stomach contents of an Elegant Trogon in Mexico in October showed that it contained 68 percent insects and 32 percent fruit (Bent 1940).

The body feathers of trogons are only weakly attached to their thin skin, unlike all other birds. Trogons are occasionally displayed in zoo aviaries but usually lose feathers and appear disheveled. Resplendent quetzals will expire in captivity. The brilliant coloration of male trogons and quetzals is lost after death, so specimens do not retain their colors (Forshaw and Gilbert 2009). Generally, they do not survive well in captivity.

Trogons are usually not thought of as aggressive with each other, but they will compete for nesting sites and control of a territory. In Arizona, territories average about one-half mile along the length of riparian canyons. It is not unusual to have several nesting pairs in the same canyon (Taylor 1994). In Central America, as many as five different species of trogons will occupy the same regional area.

Many trogons are monogamous; the males seek out a breeding territory and attract females by calls. All trogons are cavity nesters and frequently use old woodpecker holes. The Elegant Trogon makes its nest in existing tree cavities in forested canyons. Males will wander up and down the canyon, inspecting the various options before choosing a nest site.

Alexander Skutch (1999) has studied the nesting habits of Central and South American trogons. Most trogons carve their nest by digging with their bills in soft, partially rotted wood, in wasp or termite nests, or in epiphytes. Slaty-tailed trogon will excavate a termite nest. The termites will tolerate the trogon and wall off the trogon portion of the termite mound, then reclaim it after the trogons depart. The gartered trogon of Central America will readily excavate a cavity in an active wasp's nest. The wasps will harass the trogons; the trogons pick off and eat the wasps until they eventually give up and abandon their paper nest. The trogons also consume the wasp larvae and pupae. Skutch (1999) has also observed this same trogon species successfully excavating and creating a nest in an active ant nest.

Both males and females take turns in excavating a nest. They must find soft material to dig into because they have limited ability to dig with their bill. If the decaying tree is too hard to dig as they proceed, they will abandon that site and search for a more suitable alternative.

Taylor (1994) gives examples of Elegant Trogons defending themselves against accipiters (short-winged hawks) and snakes. In the latter case, he describes an incident when a pair of Elegant Trogons defended their nest from a five-foot-long Sonoran gopher snake. The snake was climbing the sycamore tree towards the young trogons, which were giving food-begging calls. The male and female trogon hovered around the snake, battering it with stiff-wing blows. The snake struck at the female as she was diving at it, but missed her and fell down the trunk, ending the event.

Quetzals

Quetzals are strikingly beautiful because of their size, coloration and decorative covert feathers that adorn five of the six species. The most well-known is the Resplendent Quetzal [Figure 19.2] of southern Mexico and Central America. They were revered by the Mayans; they are the national bird of Guatemala, and Guatemalan currency is named for them.

Figure 19.2 The Resplendent Quetzal of Central America is a stunning member of the trogon order (Diane Henderson).

The male Resplendent Quetzal has long wing coverts that drape over its body, and two long tail coverts that extend over its long, white tail. The bird is 14 to 16 inches (36–40 cm) long, and its tail coverts are over two feet long, giving the bird an overall length of about 30 inches (76 cm) for a few months each mating season. In the forest, the long tail coverts appear cyan in color in the shade and brilliant green in sunlight. In flight the tail coverts stream behind the bird and undulate with sinusoidal motion, like movement of waves on water.

This species feeds on fruit, insects and small animals, but its primary food is fruit from many different types of trees. Resplendent Quetzals prefer the wild avocado (*Persea americana*), which is about two inches long (5 cm). They swallow the fruit whole and disgorge the pit.

In Central America, the Resplendent Quetzals gather in groups of males and females and exchange a variety of calls, some soft and others harsh, while the females search for a nest site. Like trogons, resplendent quetzal also use old tree cavities for nesting.

The Eared Quetzal is restricted to the Sierra Madre Occidental region in western Mexico, but is occasionally found in southern Arizona, and individuals may have bred in Arizona's Chiricahua Mountains. This quetzal lacks the long covert feathers, making it sufficiently different to merit its own monotypic genus. It breeds in pine-oak forests at upper and mid-elevations, usually on steep canyon walls along stream drainages. Unlike trogons, eared quetzals are strong fliers, and their territory can exceed four miles (6.5 km) (Taylor 1994).

The focus of previous chapters of Part IV has been on the passerines, the largest order of birds. Trogons differ significantly from the passerine order, and they evolved to take best advantage of tropical forest habitats. Trogons are non-passerines, as are barbets, toucans, rollers, jacamars, and many other species, all of which share these forest habitats with passerines. Most of these families inhabit one environment year-round and find ample nourishment without relocation. That being said, we are losing tropical forests rapidly to agricultural and other commercial interests. Tropical forests contain the largest proportion of imperiled bird species (Sodhi *et al.* 2011), and they are a conservation priority.

20

Oceanic Birds

While humans live almost exclusively on land, the oceans contain an incredible universe. Oceans cover 71 percent of the earth's surface, and are home to hundreds of millions of sea birds.

As discussed in Chapter 1, oceanic birds first appeared about 90 MYBP in the Cretaceous period. The 90-million-year-old *Ichthyornis*, the pigeon-sized, tern-like bird with its bill adapted to catching fish, and the 70-million-year-old *Hesperornis* are important indications of the exploitation of the sea by birds before the extinction of dinosaurs. *Hesperornis* was flightless, but *Ichtyronis* had a full sternum, so it had the ability to fly.

Over time, many different families of birds adapted to life on the sea. The total number of species dependent on the sea today is estimated to be only about five percent of all species of birds—but this only accounts for truly pelagic species, which are birds that spend most of their lives on the open sea. Many more species of birds are dependent on the ocean in coastal areas within the continental shelf.

Pelagic birds live most of their lives beyond the continental shelf in water called the pelagic or ocean zone. These birds have adapted to a marine environment, and they often have a similar morphology of dark plumage on top and light plumage on the underside—or they are all dark. Some species have dark and light morphs. Their plumage is denser than land birds, and waterproofed to provide better insulation.

These birds have long life spans, small clutches, and invest a great deal of time and effort in raising one chick on their breeding sites, which are usually on islands or along remote seacoasts with little to no human activity. They locate food in the vast ocean using their keen sense of smell. Their wing shapes, flight styles, and webbed feet are adapted to their ocean habitat.

As most of the oceans are devoid of life, many pelagic species cover vast distances to find optimum feeding locations. Repositories of food occur in areas where ocean currents transport large amounts of heat and energy around the world, interacting with the air through surface contact and creating downdrafts and upwellings near continental shelf boundaries. Sea life concentrates in these regions of upwelling, and fish and nutrients come to the surface. Plankton, a collection of microorganisms particularly nutritious for seabirds, is seasonal and is most concentrated in the colder seas around the poles, as well as in the currents that bring the nutrients from the polar regions.

Threats to Seabirds

Adam Nicolson (2018), in his book *The Seabird's Cry*, provides insights and stories about the lives of various seabirds, as well as an understanding of exactly how difficult it is to successfully live off the sea. Not only do ocean birds face difficulties with the limitations imposed by nature, but they must also contend with all the barriers created by humankind.

Because of the problems facing seabirds, their populations are significantly declining. Since 1950, overall seabird populations have decreased by 70 percent, based on a survey of the total number of seabirds in an analysis of 3,213 breeding populations (Paleczny *et al.* 2015). The mortality rate of seabirds is dependent on a number of important factors: ocean storms, starvation, human causes, and predation.

Ocean Storms

Ocean storms are a major cause of seabird mortalities. The warm waters of the El Niño-Southern Oscillation produce severe storms that cause high mortality rates of shearwaters. The effects of El Niño are variable, but it causes prey species to move into deeper water, making them less available for surface feeding. The loss of available food resources caused by these storms results in high rates of mortalities in penguins and albatross (Tavares *et al.* 2020). It also affects many other seabirds, such as cormorants and petrels in the tropical Atlantic. There are no survey results that separate the number of fatalities caused by starvation versus those from the severity of the storms, but the two phenomena have an inextricable link in reducing populations of pelagic birds.

Storms that strike nesting areas cause considerable and measurable fatalities. A storm struck the Falkland Islands in 2010, where two of the islands are the largest breeding sites for black-browed albatross, and are major sites for breeding rockhopper penguins. The storm destroyed between 84,000 and 146,000 nests of these two species. Thirty percent of the nests in elevated areas were lost, but in certain low-lying areas of these islands, the storm wiped out 100 percent of the nests (Wolfaardt *et al.* 2012).

Starvation

Seabirds only survive because they are naturally long lived, have a long maturity, are numerous, and nest in colonies. Long life allows them more seasons to successfully fledge progeny. Like most bird species, however, they suffer high mortality rates during their first year of life. Seabirds undergo a long process of learning where to find food at sea—and failure to master this can lead to starvation of immature birds.

Other circumstances contribute to the known causes of seabird starvation. If adults cannot find enough food for themselves, their fitness deteriorates. If they are breeding, they may not be able to find enough food for themselves and their chicks.

This has been documented in one case from South Africa, where the condition of adult cape gannets deteriorated and chick growth rate declined when commercial overfishing did not leave enough prey species to sustain the birds (Grémillet *et al.* 2016).

Another cause of loss of prey resulting in starvation occurred in the spring of 2019, when thousands of alcids and tufted puffins died from starvation in the Bering Sea, following periods of elevated sea temperatures (Jones *et al.* 2019).

Storms, changes in ocean temperatures and currents, and competition with commercial fisheries—these causes of the loss of prey species can all have consequences for both immature and adult seabirds.

Human Hazards

The most malevolent factor in seabirds' decline, however, is a suite of human activities: introduced predators at their breeding sites, the loss of food sources due to overfishing, death by entrapment in nets, losses due to incidental bycatch (unwanted species) from longline fishing, climate change affecting food sources, oceans becoming more acidic due to increased adsorption of carbon dioxide, toxic pollution including plastic debris, oil spills, pesticides in coastal waters; and harvesting of birds and eggs as a food source.

Seabirds' breeding sites have become infested with introduced mammalian predators that arrived with human activity: house cats, rats, foxes, mice, and raccoons. The introduction of grazing animals like cattle and sheep has also cause a threat to indigenous species (Micol and Jouventin 1995). As far back as the nineteenth century, dogs and cats were primary predators; more recent research has shown that islands that have rats (*Rattus norvegicus*) are almost completely devoid of storm-petrels (de León *et al.* 2006). Other predators are also well documented. In more recent times, other introduced mammalian predators include shrews, mice, hedgehogs, opossum, pigs, fox, mongoose, weasels, and even squirrels (Towns *et al.* 2011). Some predators prey on chicks and eggs, causing the adults to abandon the nest; while larger alien species kill the adults, leaving the eggs or nestlings unprotected.

Pollution from the development and production of oil and other manmade toxins have had a well-documented effect on seabird populations. Oil products coat the feathers of birds and destroy their waterproofing properties, which leads to hypothermia. When birds preen feathers contaminated by petroleum oil, they ingest the oil, which is toxic. Research has shown that even small amounts of oil on feathers impacts the ability of birds to fly (Maggini *et al.* 2017). The Deepwater Horizon oil spill in the Gulf of Mexico in 2010 is estimated to have killed between 600,000 and 800,000 seabirds (Haney *et al.* 2014). Thousands of oil spills occur annually in coastal waterways, but most of these are small. In the last two decades, the amount of oil leaked by tankers and major oil spills (>770 tons) has generally declined, but there were a number of major spills in 2010 and 2018 (ITOPF 2020). Furthermore, increased pollution by hydrocarbons from oil spills is a major hazard to seabirds, especially those that occupy waters within the continental shelf, where most of the spills occur. All of this pollution can be prevented.

A research study published in 2018 found that seabirds and commercial fishermen compete for the same prey species—and research has shown that the problem is broadly based. Data averaged over two decades and spanning a 40-year period, 1970–2010, showed that seabird prey consumption decreased by 19 percent at the same time that commercial harvests increased by 10 percent. Over this four-decade period, seabird populations showed a continuous decline. Higher competition with seabirds was seen in the Southern Ocean, Asian shelves, Mediterranean Sea, Norwegian Sea, and on the California coast (Grémillet *et al.* 2018).

Human-caused climate change has decreased the annual ice coverage of the Bering Sea, allowing warmer ocean temperatures in that region; 2018 was first year ever with no winter ice coverage. The decreasing ice has markedly changed the abundance of prey species that support seabirds, causing a significant die-off of seabirds from starvation, and not from disease in each year from 2016 to 2018 (Gramling 2019).

Similar concerns are being raised about the Antarctic. Up to 90 percent of the warming caused by human carbon emissions is absorbed by the world's oceans, and they are warming more rapidly than originally predicted (Harvey 2019). The colony sizes of chinstrap penguins from 1980 to the present time showed that they have declined by 45 percent, while 18 percent of colonies have increased in areas where comparisons were monitored (Strycker *et al.* 2020). Changing climate around Antarctic has caused shifts in the populations of Adélie penguins, with decreases in the West Antarctic Peninsula but with stable or increasing populations in the other parts of the continent (Cimino *et al.* 2016). Projections of future warming of Antarctic seas are expected to adversely affect penguins.

Worse, hungry birds do not understand that the increasing quantity of plastics they find in ocean waters are not food—and, in fact, are affecting the health of seabirds (Lavers *et al.* 2019).

Literature published between 1962 and 2012 reported that 80 to 135 (59 percent) species of seabird species had ingested bits of plastic. On average, 29 percent of individuals had plastic in their gut; estimate predict that by 2050, 90 percent of seabirds will have ingested plastic (Wilcox *et al.* 2015). Eating these materials can kill birds—by choking and asphyxiation, or from ingestion of toxic substances (Lopes *et al.* 2022).

Commercial fishing has an impact on pelagic bird species, especially several species of albatross. On a global basis, longline fishing clearly has an impact on seabird populations, as the 60-mile-long fishing lines, with their thousands of baited hooks set at intervals, attract birds as readily as they do fish. Seabirds dive to grab the bait, become impaled on the hooks, and get dragged underwater, where they cannot free themselves. The birds drown in the very environment in which they are normally most comfortable.

The American Bird Conservancy (2015) reports that in Alaskan waters alone, more than 20,000 seabirds die every year from longline fishing encounters. As many as 210 million hooks are set by longline operations annually in Alaska and Hawai'i, with many millions more used around the world.

Fairly simple measure have been identified that can reduce these losses (NOAA

2021). Additional "tori" lines—lines designed to scare birds away from the long-lines, using colorful streamers that flap in the wind—have been used in Japan for some time and are very effective in deterring birds from diving for the fish bait. In addition, properly weighted longlines disappear below the ocean surface, where surface-feeding birds will not follow them. The American Bird Conservancy notes that these two solutions "virtually eliminate" bird mortality from longline fishing. This positive result has not led all commercial fishing operations to adopt these measures, however, so birds continue to die, particularly in Alaskan waters.

Natural Predation

Skuas and gulls are natural predators of smaller seabirds. When their usual food sources of pelagic fish become scarce, they switch to preying on seabirds. Great skua will take fledgling dovekie and petrels (Bearhop *et al.* 2001), and brown skua will predate blue petrels and slender-billed prions (Mougeot and Bretagnolle 2000). Many published accounts tell of gulls preying on seabirds: herring and lesser blacked-gulls prey on Leach's storm-petrels, western gulls prey on ashy storm-petrels, and yellow-footed gulls prey on black storm-petrels.

Sharks are also natural predators. When albatrosses make their maiden flight to sea, like those at Midway Island, the process begins by trial and error as they flounder in the sea until they are able to stay airborne. Tiger sharks (*Galeocerdo cuvier*) visit this site annually to take thrashing immatures—the same way tiger sharks prey on trans-gulf passerines in North America. These are not the only sharks that target birds: great white sharks (*Carcharodon carcharias*) prey on seabirds off the South African coast (Johnson *et al.* 2006).

There are no estimates of the annual loss of seabirds due to this long list of problems—and indeed, many of the fatalities directly related to these causes cannot even be estimated. But the sizes of breeding colonies can be accurately measured; our knowledge of the losses of seabird populations is based on how these colonies have diminished over time.

Acknowledging these losses, the United States government has been active in the protection of oceans that lie within its continental boundary. The National Marine Fisheries Service (NMFS) is responsible for ensuring that fisheries do not overexploit the ocean's fish resources. Government regulation of oil spills through the Oil Pollution Act of 1990 has helped reduce the number and size of oil spills. The Marine Protection, Research and Sanctuaries Act of 1988 (MPRSA) is intended to safeguard the ocean by preventing or limiting the dumping of any material that would adversely affect human health and the marine environment. This act created 13 national marine sanctuaries and one national monument, covering a total of 150,000 square miles of marine waters. The Environmental Protection Agency has been managing ocean dump sites to minimize the risk of disposed wastes having a negative impact on the environment. Passage of the Break Free from Plastic Pollution Act, introduced in Congress in 2021, would trigger a wide range of plastic reduction strategies. The United States Coral Reef Task Force (USCRTF) works towards preserving and protecting coral reefs. These are not all the U.S. government programs working to protect the oceans and all marine life. Besides the governmental efforts, many activities and private organizations are focused on conservation of oceans as well.

Many other countries have also worked towards protection of their ocean boundaries. However, two-thirds of the world's oceans lie beyond national boundaries. The Convention on the Law of the Sea (UNCLOS) was initiated by the United Nations, signed in 1982, and ratified by 133 nations, and it covers international law pertaining to the treatment of the open oceans (Churchill *et al.* 2022). The UN law provides a regulatory framework, but it has no mechanism for the conservation of marine diversity. Obtaining a consensus on regulations that go beyond national boundaries is conceptually difficult because of existing inequalities among the parties involved (Vadrot *et al.* 2021). Enforcing any statute on a worldwide basis is also a very difficult challenge.

Avian Species Utilizing Oceanic Habitat

Many different families of birds feed in the pelagic and coastal waters. Each has adapted different lifestyles in terms of how it uses these habitats.

Seabird Order—Procellariiformes

Albatrosses, shearwaters, fulmars, petrels, storm-petrels, and diving-petrels make up the *Procellariiformes* order. Recent phylogenetic analysis confirms that *Procellariiformes* and *Gaviiformes* (loons) are sister groups, and share a common ancestor with *Sphenisciformes* (penguins) (Pacheco *et al.* 2011). They have a long breeding cycle, and spend a great deal of effort to raise their single chick. After breeding, the adults disperse, many to distant locations.

All seabirds must deal with the problem of ingesting salt water. Species of *Procellariiformes* (tubenoses) remove salt with a gland and discharge it from their nostrils as a saline solution. In the albatross, the nostrils are at the side of the base of the bill, whereas in all other tubenoses, the nostril forms a tube running along the top of the bill. Their bill is formed from a series of distinct plates, and has a hook at the end that they use to grasp prey species.

Most species of albatross are dispersed over the ocean, but a few species tend to remain in coastal waters. The great albatrosses (wandering and royal) are the largest pelagic seabirds, with a wingspans of up to 12 feet (3.6 m); the smallest albatross has a wingspan of 6 feet (2m). These birds capture fish, squid, and crustaceans on or near the water's surface, and feeding can occur during day or night. Albatrosses are mostly southern hemisphere birds, but a few species have colonies in the northern hemisphere.

Species in this family use a variety of techniques for breeding, raising their young, and encouraging them to become independent. Albatrosses and fulmars breed on land in the open, some on simple scrapes. Laysan albatrosses breeding in the Hawaiian islands travel 1500 miles (2400 km) north to the Aleutian trench to find food, making two-week synchronized trips just to feed their chick once a week. The full breeding cycle for Laysan albatross is 270 days from egg laying to fledging. In contrast, the wandering albatross spends 356 days from hatching to fledging one chick, and can only breed once every other year. After breeding, albatrosses and shearwaters abandon their chick and wander the ocean until the next breeding season. The chick must fledge on its own to find food and a life on the ocean.

Fulmars, shearwaters, petrels, gadfly petrels, prions, and diving-petrels represent progressively smaller species of the *Procellariidae* family, with lifestyles characteristic of sea birds, but with different strategies for survival. There are two species of giant petrels and 31 species of shearwaters. Species with the name "petrel" fall into two groups: the largest group, the *Petrodroma*, are made up of 34 species, referred to as gadfly petrels. The remaining petrels are placed in eleven genera—about 18 species of named petrels, as well as the fulmars, to which some of these petrels are closely related. The diving-petrels and prions make up the smallest members of this large family.

Most of the *Procellariidae*—except the albatross, giant petrels and fulmars—nest in burrows or rocky crevices and feed on fish, squid, offal, krill, and plankton. All species of this family render food into an oily substance to regurgitate for their juveniles. Fulmars breed on cliffs, and giant petrels breed on bare ground.

The two species of giant petrels are the largest petrels, resembling the albatross. Scavengers and predators, they regularly feed on carrion as well as squid, and they prey on penguins. Shearwaters and fulmars feed by diving and by swimming beneath the surface, sometime to considerable depth, to pursue their prey. The gadfly petrels and larger petrels feed from the ocean surface.

The number of species and the relationships among the petrels have been revised in the past by ornithologists, and they continue to be subjects of study. Determining exactly how many actual species of petrel there are is difficult, because many are very similar, some have subspecies, and many have allopatric ranges.

The storm-petrels are sparrow-sized sea birds, and are divided into their own families: 18 species in the northern hemisphere (*Hydrobatidae*), and nine species in the southern hemisphere (*Oceanitidae*). Storm-petrels dig their burrows in soft soil to depths of one to three feet (0.3 -1.0 m). This means their nests are easily predated by mammals, as well as by gulls and skuas. The highest predation occurs during the breeding months. To minimize this threat, the adults avoid daylight exposure and forage much less on moonlit nights. Storm-petrels fly with a fluttering flight just above the water, feed by hovering, and pick morsels of food and oil droplets from the surface; occasionally they immerse their entire head. The small prions are filter feeders (Klages and Cooper 1992).

Many of the *Procellariidae* are dark mantled above and white on the body and underside, possibly for countershading. Nevertheless, a few are all dark plumage, and some have light and dark morphs. Their flight styles vary from the larger species with slower flapping and gliding to the smaller species with quick wingbeats. Steve Howell (2012), in his book *Petrels, Albatross & Storm-Petrels of North America*, gives details of the flight style and identification of the members of this family, as well as the albatross.

Fulmars are mainly surface-feeding birds. They will dive to shallow depths, but have not been observed doing so very often (Wahl 1984). Unlike other seabirds that live colonially, they are active during the daylight. During the breeding season, a pair will alternate trips to find food.

Edwards *et al.* (2013) placed a geolocator on an adult male northern fulmar, and tracked him as he left his nest site on Eynhallow, an island in Orkney off the Scottish coast. After the winds came up and he had fed, he traveled nonstop for two and a half days westward to the Mid–Atlantic Ridge, a linear distance of about 1400 miles (2200 km), where he fed in the Charles-Gibbs Fracture Zone, a location known for its fertile waters. He did not follow a straight path, but traveled 3,850 miles (6200 Km) on the round trip in 14.9 days. This shows the importance of these rich waters, which are feeding grounds for other North Atlantic pelagic birds. It also illustrates the difficulty large seabirds face in obtaining food.

Shearwaters

Shearwaters are about as large as a medium-sized gull, and they are the great wanderers of the sea, similar to the albatross. Shaffer *et al.* (2006) tracked sooty shearwaters in their annual ocean wandering from their breeding area off New Zealand during their non-breeding season. Over a 300-day period, they traveled an average of 40,000 miles (64,000 km), crossing the Pacific Ocean in a figure eight pattern, and stopping in either one of three regions: off Japan, Alaska, or California. These are important feeding location for them in their non-breeding season.

During the breeding season, sooty and short-tailed shearwaters forage locally off southern Australia and New Zealand to take advantage of the closest food sources. These local sources do not furnish all the required nutrients, however, so they supplement their diet by making periodic trips of several thousand miles one-way, going south to the Polar Front zone, an area where the cold Antarctic air meets tropical air at about 60 degrees south latitude. Upwellings along this front bring lantern fish (*Myctophidae*) and other species to the surface, which they consume and render into a stomach oil to feed their chicks. They feed more frequently at sunrise and sunset, when these small fish are most likely to come to the surface (Raymond *et al.* 2020).

In another study, Brooke (2010) examined banding ring recoveries that show that Manx shearwaters migrate around the Atlantic Ocean on circular routes, taking advantage of the prevailing trade winds. They begin by heading south as fledglings, and eventually reach Brazil; large numbers have been recorded off Argentina. Later in the year, they head north along the coast of North America, and follow the westerly winds back to their breeding grounds in Europe.

Shearwaters [Figure 20.1] inhabit both hemispheres, but most breed in the southern hemisphere. A few, including Audubon's, black-vented and tropical shearwater, breed in the tropical areas of the northern hemisphere. Manx shearwater is the most northern species, breeding as far north as the Faroe Islands, south of Iceland. Sooty shearwater is one of the most common species; it breeds on small islands off Australia, New Zealand and southern Chile, and spends its non-breeding season

Figure 20.1 Cory's Shearwater flies with bowed wings; it breeds on islands off the west coast of Europe and Africa, and winters in the southern hemisphere of the Atlantic Ocean (Gary Chapin).

in the Northern Hemisphere, feeding in areas where ocean upwellings concentrate small fish.

Shearwaters feed below the surface by diving and swimming underwater, and some species can reach depths of 230 feet (70 m) (Weimerskirch and Cherel 1998). However, most shearwaters dive to shallower depths depending on food availability, and process the fish in their stomach into an oil, which they feed their chicks by regurgitation.

Petrels

Gadfly petrels, *Pterodroma*, are the largest genus of petrels—small to medium in size with relatively long, narrow wings, rapid flight and a stout bill. The *Pterodroma* are generally smaller than shearwaters with very slightly shorter wings, and a flight style with rapid wingbeats, weaving, and high arching glides. Some, such as Hawaiian, herald, and black-capped petrels, breed in the northern hemisphere, but most of them breed in the southern hemisphere. A number of species are endangered because they breed on heavily populated islands. Like the shearwaters, petrels have very poor ability to walk on land.

Storm-petrels are like the sparrows of the sea; they are the smallest sea birds, weighing as little as 0.74 oz (21 gm). Most are about the weight of a towhee—a large sparrow—but unlike a towhee, they have long, pointed wings. Ornithologist Brian Sullivan (2006) notes that it is hard to believe that these small, delicate birds could spend most of their lives on the open ocean, one of the most inhospitable environments when earth. They feed on small prey items and oil droplets at the surface of the sea, and will peck at the surface of the water to retrieve morsels of drifting organic matter; they also feed by hovering, and like to follow groups of whales. The Wilson's Storm-Petrel [Figure 20.2] is one of the most abundant seabirds on earth.

Storm-petrels nest in burrows, leaving their nests before dawn, and returning after sunset to avoid possible predators. During daylight, crows, hawks, and gulls will take them. Fox and mink attack their burrows, but the greatest land predators in the nineteenth century were pet dogs and cats (Bent 1922), and that problem has remained up to the present time. These pets have had a devastating effect on breeding colonies on some small islands.

Figure 20.2 The Wilson's Storm-Petrel is widespread over the world's oceans. Like other storm-petrels, it flies low over the water in searching for food (Brad Carlson).

The *Procellariidae* family also includes six species of prions, small petrels found in the southern oceans. They are blue-gray in color with a distinctive "M" mark on their backs, and their bills have serrated edges. Among the southern petrels is the unique snow petrel, which breeds in Antarctica. This dove-like bird is all white with long wings, a black bill and black eyes.

Finally, the four species of diving-petrels are small birds of the southern hemisphere. Although they are chunky in shape and only seven to ten inches long (18–25 cm), they use their short, rounded wings to fly underwater while diving for prey. They can dive below the surface from 30 feet (10 m) to as deep as 200 feet (60 m). The diving-petrels look and behave much like the smallest alcid, the dovekie of the northern hemisphere.

Life on the Sea

During their long annual journeys, tubenoses are faced with the challenge of finding food on the vast oceans. Food sources along this journey are variable in time and location. These birds cannot rely on randomly locating prey, because the oceans are so vast and prey congregate in specific areas. The birds must rely on the sense of smell and on the prevailing winds to locate important breeding and feeding areas. This is a learning process, and when they locate an important site, they retain the location and will return to these preferred sites annually.

Odors are also important to these birds in locating their own nesting burrow and bonding with their chick. Shearwaters in particular have poor vision, and rely on odors when they return to their nest site after dark. Nevitt and Bonadonna (2005) showed that shearwaters are sensitive to very small concentrations of dimethyl sulfide emitted by phytoplankton, and pyrazines emitted by krill (Nevitt *et al.* 2004). These scents can persist for days or weeks, providing navigational clues to the *Procellariiformes* seabirds.

Early observations of the life of seabirds during the nineteenth and early twentieth century are summarized by Bent (1922). The larger pelagic birds all have a similar flight style at sea. Albatrosses, shearwaters, and fulmars fly gracefully with much buoyancy, skimming along the surface of the waves. In the air they are swift, strong, and graceful, gliding over the surface and using dynamic soaring to take advantage of the updrafts deflected by the waves. Once in a higher position, they convert their height gain to momentum, gliding down into the wave trough only to rise again on the next wave. This makes them very comfortable and active in stormy weather—they maintain perfect control while flying headlong into gale winds. The constant tradeoff between altitude and velocity results in a zigzag flight pattern, taking advantage of wing gradients and wave motion while minimizing energy expenditure. In light winds or when molting, they have difficulty becoming airborne, and will sometimes use their feet to smack the water as if running in order to get some lift.

While pelagic species prefer moderately high winds and are exceptionally adapted to withstand storms, those on the periphery of a hurricane will move outward to avoid severe winds, and sometimes will be blown inshore. Fatalities occur, especially for birds that get near the center of the storm (Weimerskirch and Prudor, 2019). Most fatalities from tropical cyclones (hurricanes) in the Indian Ocean are juvenile petrels (Nicoll *et al.* 2017), based on breeding bird surveys. Shearwaters are known to be strong fliers and adult normally circumvent the strong winds of hurricanes. In contrast, however, researchers in the Sea of Japan found that when the cyclones get near land, adult streaked shearwaters fly directly into the storm towards the eye to avoid being trapped on land (Lempidakis *et al.* 2022).

In the breeding season, most shearwaters leave their burrows and go to sea before dawn, returning at dusk. They will rest on the sea in rafts, quietly floating on the surface during the night, and are sometimes found in rafts during daytime. In heavy winds, the birds will remain aloft and sleep on the wing; they are at home in the air as much as on the sea.

Storm-petrels have a different flight style from the larger shearwaters. Some have been described as having a bat-like fluttering flight; others are swallow-like, but they are all comfortable in stormy weather. Some of the longer winged storm-petrels, such as Leach's, fly like a small shorebird or erratically, like a nighthawk. Storm-petrels bound over the water with their feet dangling as if on springs; their weak legs prevent them from perching or standing upright. They half-run/half-fly when on land, much as they do while taking off from water. At sea, they are extremely agile, well adapted to taking off from the surface, and are swift fliers. Like shearwaters, they will rest on the sea in rafts, and they can perish in severe storms.

Northern Hemisphere Alcids

The predominant pelagic birds of the northern hemisphere are the 25 species of alcids, members of the *Alcidae* family, part of the large, diversified *Charadriiformes* order, which includes shorebirds, buttonquails, pratincoles, jaegers, gulls and terns. The alcids include the puffins, auklets, murrelets, guillemots, and dovekie. These black-and-white birds range in size from the six-ounce dovekie (170 gms) to the tufted puffin, which weighs about 1.6 pounds (725 gms).

The dovekie is very similar in size, appearance, and lifestyle to the diving-petrels, but the two have different origins, a case of convergent evolution. The dovekie is an alcid, while the diving-petrels are in the order of seabirds, which share a common ancestor with loons (Pacheco *et al.* 2011).

The largest member of the alcid family was the great auk, which was driven to extinction in about 1840. This flightless bird, about the size of a gentoo penguin, had a penguin-like lifestyle. Within the alcids, this represented convergent evolution with the penguin order (*Sphenisciformes*).

Puffins are the largest alcids. They have short wings, feed underwater and winter at sea in the northern hemisphere. The three species breed in the colder regions of coastal Asia, North America and Europe, from about the 45th parallel north to the Arctic Circle.

The young of large and medium-sized alcids leave the breeding colony and do not return until they are at least two years old. In fall, alcids move out to sea, to areas of rich food sources, sometimes hundreds of miles from their breeding locations. They molt their feathers at this time. During molt they have a rapid loss of wing feathers, some by as much as 40 percent, and are flightless for some period, but they always have enough wing feathers to propel themselves underwater (Bridge 2004).

Penguins

Penguins are, perhaps, the most unusual birds on earth; they are more like seals than birds in appearance and behavior. They adapted to life on the open sea and returned to land only to breed and molt. Like many seabirds, they are counter-shaded with black and white plumage. Their anatomy for life at sea required a number of adjustments from their ancestors: the eye increased spectral sensitivity for underwater sight, feather density increased for thermal needs, the wings adapted for underwater propulsion, bone density increased to reduce buoyancy, and legs shortened and adapted for incubating (Ksepka and Ando 2011). Penguins are southern hemisphere birds, ranging from the colder waters of the Antarctic sea to the more temperate southern oceans; only the Galápagos penguin is found north of the equator.

The number of species of penguins varies depending on the sources and state of current research. Today, there are 18 species in six genera, with the crested penguins, genus *Eudyptes*, the most diverse with seven species.

In general, penguins lay two eggs, but the crested penguins will raise only one chick. The emperor and king penguins lay only one egg. Penguins breed in the spring and summer, but emperor penguin is unique in that it breeds in fall, and raises its chick during winter. After breeding, juvenile penguins molt in February and adults molt during March and April. Penguins molt all at one time and are confined to land for this period. They lose considerable weight during molt.

After breeding the emperor penguins will find an ice flow to haul out—move from the ocean onto the ice—and molt. They then return to the sea where they remain until the following April when their breeding cycle begins again. During this time they will restore their fat reserves to prepare for the next breeding season.

Penguins live the majority of their life at sea. They capture prey underwater, using their wings as flippers that function like those of sea mammals.

The emperor penguin is the largest living penguin, reaching a height of about 4 feet (1.2 m) tall and weighing 50 to 100 pounds (22–45 kg), depending on sex and degree of fat reserve. When emperor penguins feed, they normally dive to about 130 feet (40 m). However, deeper dives are not unusual, and they have reached a depth of 1850 feet (550m), remaining submerged for 22 minutes (Wienecke *et al.* 2007). Satellite tracking has shown that they can travel as far as 550 miles (880 km) from their breeding site (Ancel *et al.* 1992) when they leave to feed.

The little penguin of Australia, New Zealand, and the Chatham Islands is the smallest penguin, weighing about 3.3 pounds (1.5 kg). It fishes during the day, and comes ashore at night to its burrow to avoid predators. In contrast to the largest penguin, the little penguin travels about 45 miles (73 km) during its daily foraging; most

Figure 20.3 Red-billed Tropicbirds have a more restricted distribution than the other two tropicbirds. All of the tropicbirds have strikingly beautiful plumage; their all-white plumage is unusual among pelagic birds (author's collection).

of its dives do not exceed about seven feet (2 m), but they can and will dive to a depth of 90 feet (27 m) (Bethge *et al.* 1997).

Like other ocean birds, penguins drink seawater and ingest salt water in capturing prey. They also expel salt through their nasal passages by means of salt glands. However, small penguins, through their biology, take additional measures to lower the salt content in the regurgitated food fed to their chicks.

Penguins range over great distances in the southern ocean during the non-breeding season. In 2009, Charles-André Bost and his associates fitted Macaroni penguins with geolocators, and tracked them during their winter foraging, when the birds remained at sea for six months in the central Indian Ocean. The 21 tracked individuals covered a narrow north-south latitude band, but some reached as far as 2100 miles (3400 km) east of their breeding island after a three month period (Bost *et al.* 2009). During this time they fed mainly on crustaceans rather than krill.

Three species of tropicbirds, the *Phaethontiformes* order, inhabit tropical oceans around the globe. Most remarkable for their beauty and grace, they are large white birds with black highlights, red bills, and long, decorative central tail feathers, as shown in Figure 20.3. Their relationship to other sea birds is uncertain, and their plumage is unlike other pelagic birds. They nest on cliffs or under vegetation, and one species has nested in tree holes. They feed on fish and squid by plunge-diving. Like all seabirds, they have webbed feet, but are not very mobile on land because their feet are too far back.

Frigatebirds, Gannets, and Boobies

The large diverse order *Suliformes* includes frigatebirds, gannets, boobies, cormorants, and darters (anhinga). The frigatebirds, gannets, and boobies breed at coastal locations and spend most of their lives on the open ocean. Gannets and boobies are colonial nesting birds. In order to capture prey, they will plunge-dive headlong from a height of 150 feet (45 m), reaching speeds of 55 miles per hour (88 km/hr) to take fish by surprise as deep as 33 feet (11 m), and use their wings and feet to chase prey to depths of 80 feet (24 m) (Ropert-Coudert *et al.* 2009). They attack their prey like a missile, their eyes and body built for the impact of the dive. The wing shape of gannets and boobies are more ideal for flight, and they also serve well for chasing fish underwater. Gannets, boobies, loons, grebes and cormorants rely more on their large webbed feet, which are set far back for underwater propulsion.

Cormorants

Many species of cormorants are coastal birds that feed in ocean waters, and others have adapted to life on inland lakes, marshes, and ponds. A few species, including great and double-crested cormorants, breed at both inland and coastal locations. Darters are birds of tropical inland lakes, rivers and canals. They are similar in many respects to cormorants.

The earliest cormorant fossil dates back to the Eocene-Oligocene boundary, 33.9 MYBP (Johnsgard 1993); there are many fragmentary earlier fossils, but their exact relationship to modern cormorants is uncertain. In addition, other species of ancient birds have a structure similar to modern cormorants.

Taxonomists currently recognize 41 species of cormorants, and their diversity is centered in the southern hemisphere. Some are referred to as shags, because of their tufted crest. The southern hemisphere species have more white feathers around the head, breast, and neck, while the northern hemisphere species have mostly dark plumages. Cormorants have a sharp, well defined nail on the end of their beak to catch and hold fish while swimming or diving. Great cormorants can swim to depths of 65 feet (20 m) and remain submerged for 70 seconds (Johnsgard 1993). Other species have been known to dive even deeper. They can swim very rapidly, up to 5 feet per second (1.5 m/sec), and use their long tail to steer. They consume their prey mostly on the surface of the water, but can swallow small fish while submerged, especially if harassed by gulls.

Cormorants were used by humans for many thousands of years in Japan, Korea, and China to catch fish for human consumption, but they are not used for commercial fishing today. The Guanay Cormorant [Figure 20.4], which breeds on the west coast of South America, is important commercially because it produces guano (manure) rich in nitrogen, phosphate, and potassium; this can be used for fertilizer. Guano has been mined for over 1,500 years from small islands where the cormorants breed. Guanay cormorants' population once totaled millions of birds, and although their population has decreased significantly from historic levels, more recent conservation measures have tried to reduce their losses because of the birds' value to agriculture.

Prior to 1800, the double-crested cormorant was a very abundant species, probably more abundant than today. Their population decreased significantly by 1900 and began to cycle after that time, but after a low in 1970 due to pesticide use, they

Figure 20.4 It had been estimated that in 1955 there were about 25 million Guanay Cormorants, but today the population is about 3 million. One million Guanay Cormorants produce 11,000 tons of guano annually (Diane Henderson).

have increased by sevenfold (Wires and Cuthbert 2006). Because this species' diet of fish and their destruction of vegetation in their colonial nesting areas now interferes with human interests, the current population is beyond public tolerance. Double-cresteds are the most common cormorant species in the interior of the United States, and they are numerous on large lakes, especially in the Great Lakes region. As a result, they have come to be considered a nuisance. Studies of their impact on fish populations and fish farms has prompted legislation to reduce their numbers. In Europe, the great cormorant is also sometimes considered a nuisance species.

Pelicans, Jaegers, and Skuas

The pelicans are in the *Pelecaniformes* order, which includes herons, bitterns, spoonbills, and ibises. Within the pelican family, six species of white or gray pelicans are inland inhabitants, but some winter in coastal locations. The male dalmatian and great white pelicans are very large birds, with a weight range up to 30 pounds (13.5 kg). They soar on wings over 11 feet long (3.4 m), although females are smaller than males.

Although some white pelicans will occasionally feed by plunge diving, they obtain their food by surface feeding while swimming. Most of these six species hunt for food socially in moving semi-circular groups.

There are currently two species of brown pelicans: the Peruvian pelican restricted to western South America, and the brown pelican of North and South America and the Caribbean region. Brown pelicans are coastal birds that are smaller than white pelicans, and they feed close to shore. They are solitary feeders, plunge-diving by heading downwind from as high as 70 feet (20 m), a height very accurately based on the depth the fish. On capturing their prey, they expel any water from their expanded throat (gular pouch) before breaking the surface, and emerge heading upwind, thus making a complete turn under water and taking advantage of the wind to assist in take-off (Bent 1922).

Jaegers and skuas are predatory sea birds, with a hooked bill, claws on their webbed feet, and strong flight muscles. Three species of jaegers and four species of skua make up the *Stercorarius* genus. The three species of jaegers and the great skua breed in the northern hemisphere and winter in the southern latitudes, while the other three skuas inhabit the southern hemisphere—though south polar skua winters above the equator.

All seven species eat various kinds of fish, squid and bird eggs, but also rely on food obtained by stealing from gulls and terns in aerial chases, a process called *kleptoparasitism*. In the breeding season, jaegers will take fish, small mammals and shorebirds. The southern hemisphere skuas prey on penguin eggs and penguin chicks in summer, and small sea birds in their wintering range. Great skua will attack and kill seabirds as large as a herring gull. Brown skua, about the size of a great black-backed gull, is the largest of this genus, and observations have shown that they can recognize individual humans (Lee *et al.* 2016), suggesting that they have high intelligence.

Gulls, Terns, and Noddies

Gulls, of the family *Lauridae*, are birds of the land, inland lakes, waterways and coastal interface. Many that breed inland use coastal habitat for feeding. They breed on every continent, including Antarctica. The term "seagull" is a misnomer, as there are very few gull species that are truly pelagic. Gulls are colonial breeders, and most of the gull species breed in the northern hemisphere—probably because two-thirds of the earth's landmass is in the northern hemisphere. Only one species, the kelp gull, breeds on Antarctica.

There are 51 gull species recognized worldwide, and they are divided into several groups based on genetic analysis (Pons *et al.* 2005). The largest group are the white-headed gulls (*Larus* gulls, 22 species); they include the Holarctic species we are most familiar with in the northern hemisphere. Many of the remaining gulls are hooded, with black heads in breeding season. Most of the ten species of the

Chroicocephalus genus are hooded, and include little, Bonaparte's, and black-headed gulls. The next largest genus, the *Ichthyaetus* gulls, include six medium to large gulls that are found from Russia to the west coast of Africa. They are omnivores, and much of their diet is based on fish except for two of them. Mediterranean and relict gulls forage on invertebrates.

The remaining species of gulls are monotypic except for the two closely related kittiwakes.

The large gulls are omnivorous, opportunistic, scavengers, and predatory. They feed on fish, crustaceans, small birds, eggs of other birds, carrion, and human refuse. Herring gull and other large gulls have often been seen dropping clams from a height onto rocks to break the shells. Research on western gulls has suggested that this is a learned trait, and that the height that gulls drop the clam depends on the weight of the clam; lighter clams are dropped from a greater height (Marion 1982) because it takes more force to break smaller clams.

Since the large gulls are powerful birds with large bills, predation is common among them. Glaucous gulls will feed on small birds such as dovekies; herring gulls will eat the chicks of their own species (Nicolson 2018), and great black-backed gulls plunder seabirds, ducks, and puffins. Herring gulls and ring-billed gulls will attack, capture and eat small migrant passerines caught crossing large bodies of water.

Landfills and rubbish dumps have become a food source for larger gulls as well as other animals. During breeding season and before fledging, adult gulls rely on natural food sources, but after fledging, adults and immature birds visit landfills more frequently (Belant *et al.* 1993). Research on herring gulls found that first-year birds find most of their food by scavenging. Landfills have allowed more immatures to survive bitter winters and times with decreased natural food. This has led to population increases which, in turn, have led to significant losses of more vulnerable coastal species, such as the piping plover. Furthermore, increased populations of large gulls are a threat to airports.

Among the hooded gulls, the tern-like little gull is the smallest, only about 11 inches long (28 cm); this Eurasian species was first found breeding in North America in 1962. Bonaparte's gull is the only gull that breeds in trees in the boreal forest of North America. Franklin's gulls breed on prairie marshes and feed primarily on invertebrates, but also take small shellfish and fish in winter. They can hawk insects in flight, and they scavenge freshly tilled soil for insects and worms. Swallow-tailed gulls of the Galápagos have very large eyes and forage for squid at night. They are the only gull that feeds at night.

Research on black-headed gull done by Graham Phillips (1962) sheds light on how the plumage characteristics of hooded gulls influence their feeding strategies. During breeding, with their black hoods, they feed on insects, mollusks, crustaceans, worms, other invertebrates, and some also catch small fish. During winter, they lose their dark hoods and their heads are mostly white, and they feed on fish by tern-like plunge diving. Phillips' work showed that when gulls approach the water with a dark head, fish flee more rapidly than when they approach with a white head, thus affecting their success rate. He also found that some of the gulls and seabirds in the tropics were exceptions to this rule.

A few small gulls, including Sabine's, Ross's, ivory, and swallow-tailed gulls, and black-legged and red-legged kittiwakes, leave their breeding areas and migrate south to winter on the ocean. Many will migrate to areas where ocean prey is more concentrated—at upwellings near the continental shelf, where they are surface feeders. The majority of gull species migrate short distances, and winter on small lakes or at coastal locations. Franklin's gulls cross over North America and winter off the west coast of South America, closer to the continent than the more pelagic small gulls. Sabine's gull makes the longest migration from the high Arctic to winter off South America and Africa.

Terns and Noddies

Forty-seven species of terns and skimmers make up the remainder of the *Lauridae*. These mostly small, long-winged birds breed near inland lakes and on ocean islands, with some in trees, and several species on sea cliffs. The Caspian tern, the largest, is the size of a medium-sized gull. Most terns are smaller than gulls, with a more streamlined body, buoyant flight, and a forked tail. Terns generally have predominantly white to pale gray bodies, with a gray back. Some ocean terns have dark plumage, which is better for feeding at night. This countershading is beneficial for capturing fish.

Terns can be divided into those that breed inland on marshes and those that inhabit a marine environment. Terns that breed inland near ponds, marshes, estuaries, or bays will also feed on aquatic insects as well as other small animals. Most terns feed by plunge-diving or by snatching fish from the surface. White-winged tern of Eurasia and the black tern of North America are similar species, foraging on the wing by picking up items at or near the surface. Black tern winters off South America, and some spend the winter on the ocean, but its close relative, the white-winged tern, winters on large lakes in Africa and Australia.

Many species of inland terns that migrate to the tropics are found in coastal areas, and feed offshore in a marine habitat. Based on geolocators, common tern migration involves long flights over water. They mostly winter in tropical areas with large numbers off the coast of Peru, and others spread throughout the Gulf of Mexico, Central America, and northwestern South America (Bracey *et al.* 2018).

The marine terns are pelagic and are mostly found in tropical seas, where they breed on islands and feed offshore. These include the noddies and other terns like bridled, sooty, arctic, Aleutian, and perhaps others. Arctic tern breeds in the extreme northern hemisphere, and winters in the Antarctic seas. When monitored by tracking devices from Greenland to the Antarctic, arctic terns carry out the longest migration in the animal kingdom, traveling from 44,100 to 50,700 miles annually (79,000 km–81,000 km) (Fijn *et al.* 2013). The Aleutian tern also breeds in the far north, but winters in the equatorial regions.

The five species of noddies (genus: *Anous*), as well as sooty and bridled tern, do not plunge dive, but feed by capturing small fish driven to the surface by larger fish. They capture fish from flight, never landing on the water (Bent 1921)—picking fish from the surface, or snatching them in the air if the minnows leap above the surface. Five of these species are very dark backed and are more pelagic than other terns.

Sooty terns feed on fish driven to the surface during the day, and on squid at night. They rarely land on water for an extended period, because they have poor waterproofing and will become flightless and waterlogged after about 25 minutes on the surface (Johnson 1979). Immature sooty terns will remain at sea for about four years or more before returning to breed. When covering very long distances, these birds will sleep on the wing.

Phalaropes, Loons, and Others

The only shorebirds that live on the open sea are two species of phalaropes, red and red-necked, very small shorebirds that find food by pecking at the surface of the sea. During winter they are found in flocks on the surface, and congregate in areas with upwellings or tide rips, or where warm and cold currents converge. Their diet at sea is not well known, but they feed while swimming on the surface.

Migration studies of red-necked phalaropes based on geolocators show that there are two populations that occupy different breeding and wintering areas (van Bemmelen *et al.* 2019). The population breeding in the eastern Canadian Arctic migrate 6,000 miles (10,000 km) over sea, and across Central America to the tropical eastern Pacific Ocean from southern Mexico to as far south as Peru. The second population, breeding on the Scandinavian Peninsula and Russia, migrate over land about 3,600 miles (6,000 km) to the Arabian Sea.

Loons, grebes and some species of ducks and geese use the ocean as a feeding location in non-breeding season. Brant are found in coastal locations in winter, and sea ducks are permanent residents of inshore habitats. Some species of ducks, like spectacled eider, spend the winter near pack ice in the Arctic sea. Loons feed in coastal ocean waters up to 330 feet deep (100 m), and range over the continental shelf as far as 60 miles (100 km) from land. Many will remain close to land, since they prefer to feed in depths from the surface up to sixty feet (0–19 m). But as the extent of water clarity decreases near shore due to river discharge, they move more offshore during midwinter (Haney 1990).

21

Shorebirds

It is a joy to discover the world of shorebirds. They are not particularly shy and spend a lot of time feeding, which provides ample time for viewing. In spring, many species have colorful plumages. Throughout their range, we can share the beach with them. Their lifestyle is very different from many other orders of birds.

Shorebirds are members of an ancient order of birds that goes back to the time of dinosaurs. Molecular dating and fossils suggests that the major groups of shorebirds originated in the late Cretaceous period, between 79 and 102 MYBP (Baker *et al.* 2006).

The *Charadriiformes* order is made up of 20 families with about 380 species (Clements *et al.* 2021). The families can be divided naturally into three groups (Mayr 2017):

1. Gulls and allies: gulls, terns, skimmers, skuas and jaegers, pratincoles, alcids, sheathbills, and buttonquail.
2. The medium waders we refer to as shorebirds: sandpipers, phalaropes, snipe, jacanas, and others.
3. Plovers: plovers, stilts and avocets, oystercatchers, thick-knees, and seedsnipes.

Among the 20 families, five are monotypic. The taxonomy of this order of birds has been relatively stable, except for some reordering between families. The families are very diverse in many ways, and are linked together by similarities in anatomical features.

Within the order, there is a significant variation in morphologies, and most of the families are associated with wetland habitats, others with short grass and dry habitats, and a few with mountainous areas. The first group in the *Charadriifores* (gull, etc.) are associated with coastal habitats and were mentioned in Chapter 20.

Shorebird Habitats and Conservation

Almost all shorebirds are long distance migrants that winter in warmer regions, some as far south as Australia and southern South America. As a group, they have the longest annual migration of all birds. They migrate along selective pathways between continents. In the Americas, there are three major migration corridors:

Atlantic coastal, the Great Plains, and Pacific coastal (Myers *et al.* 1987). Similar corridors are found in the East Asian-Australasian flyway (Bamford *et al.* 2008). These corridors lead to large concentrations of shorebirds in selected locations because they rely on specialized habitats where they replenish fat reserves. They are dependent on these stopover sites in order to complete their migration.

During winter and migration shorebird stopover sites include intertidal areas, shallow ponds, pond edges, flooded fields, beaches, and short grass prairies for feeding locations. As they migrate, these birds feed on mollusks by probing in shallow water and mud, and by picking at the water surface. Some prefer wet fields and coastal areas; others, such as the plovers and some shorebirds, prefer grasslands and open ground.

Many factors have led to a decline in these habitats. Agricultural fields have been drained, eliminating the spring surface water. Beaches have been overrun with vehicles, pets, other human disturbances and debris, reducing areas for feeding and breeding. Sea level rise has led to the loss of coastal marshes and shallow areas. Generally, water levels in ponds within refuges are not controlled for shorebirds during spring and fall migration.

The decline in populations of red knots seen in the United States illustrates the point. The number of red knots moving up the east coast of the United States during spring migration has fallen by 80 percent along the Atlantic flyway, due to the decrease in horseshoe crabs from overharvesting, eroding beaches from climate change, and coastal development. The eggs of the crabs are a primary food source for the knots, so some states have put restrictions on the harvesting of crabs.

In the nineteenth century, hunting shorebirds was a common and unregulated activity in the United States. Many species were driven to small populations, a decline from which some have never recovered—and at least one species, the Eskimo curlew, went extinct from overhunting. Shorebirds are not difficult to approach, and they become easy targets when they flock, so they were basically used as target practice. There are many accounts of 50 shorebirds killed with two shots from a double-barreled shotgun. It was not until 1918 that the Migratory Bird Treaty Act legally ended the senseless slaughter of these small birds in the United States, but unregulated hunting continues in other countries.

Shorebirds are one of the groups of birds with very high conservation concerns because many species are in decline. Even with the Migratory Bird Treaty Act protecting native species throughout the United States, hunting of migratory shorebirds along the Atlantic Flyway is legal in many jurisdictions, because the birds migrate through these areas and are not resident breeding species. Shorebirds exist in a gray area of the law that makes them very vulnerable to human predation, even though other regions of the country are working to protect and sustain these fragile bird populations.

Because of this ability to hunt shorebirds in some areas, as well as in countries outside of the United States, there is very little information on shorebird mortality available, even though some countries have bag limits for some species. As shorebirds migrate from one country to another, they face different local regulations. For example, we do know that more than 1,500 whimbrel are killed annually on the

Western Atlantic Flyway at stopover locations in the Lesser Antilles (Watt and Turrin 2016). This region even has locations for shorebird "destination hunts" (Weidensaul 2021).

The biggest problems that shorebirds face in the current century is lack of habitat for feeding along their long migratory route, and continued threats in their wintering and breeding ranges, also from habitat loss. For most migratory shorebirds, the most important conservation measures are the creation and maintenance of wetland and grassland habitats that accommodate their requirements. Conservation of beaches and tidal wetlands is extremely important to shorebird survival. Most costal modifications are carried out to preserve human infrastructure, but they need to encompass the preservation of the natural habitats they encompass as well. The control of invasive plant species, natural predators such as raccoons and foxes, and feral as well as pet cats and dogs are important to safeguarding shorebirds.

Plans and programs to improve regional habitats, food sources, and migration staging sites have been developed by different organizations: the Manomet Center for Conservation Science, Western Hemisphere Shorebird Reserve Network, North American Waterfowl Management Plan, U.S. Shorebird Conservation Partnership, the Prairie Pothole Joint Venture, and the Migratory Bird Habitat Initiative of the Natural Resource Conservation Service (NRCS) (Kaminski and Davis 2014). Coalitions for Shorebird Conservation, a Manomet program, is an important initiative to save shorebirds. Many private organizations and groups in other countries focus on shorebird conservation as well.

These are important efforts, but we face the difficulty of convincing property owners, developers, and private citizens that they play a major role in any effort to conserve these species.

Sandpipers

The sandpipers (*Scolopacidae*) are a large family with 97 species (Clements *et al.* 2021). Compare this with other members of the shorebird clan: the stilts and avocets, oystercatchers, jacanas, and thick-knees each have between eight and twelve species; pratincoles and coursers have 17 species; seedsnipes have four; and painted snipe has three.

Most sandpipers are monotypic and do not have subspecies—with the notable exception of the dunlin, which is widely distributed. This Holarctic wader breeds across the entire northern hemisphere, and its five subspecies winter across the globe in a wide belt at about 30 degrees north latitude.

The sandpipers are a diverse family and include many separate genera with names like curlew, whimbrel, godwit, turnstone, sandpiper, knot, stint, dowitcher, snipe, woodcock and phalarope. With so much diversity, it's no surprise that they have a large range in sizes. The least sandpiper at six inches long (15 cm) and a weight of 0.7 oz. (20 gm) is the smallest, while the largest is the Far Eastern curlew, which is 26 inches long (66 cm) and has a normal weight of 1.5 pounds (800 gm). The bill lengths cover the full range of sizes between the small sandpipers with short bills

Figure 21.1 The Least Sandpiper is the world's smallest sandpiper and has a short, thin, straight, dark bill similar to other small sandpipers (author's collection).

Figure 21.2 The Far Eastern Curlew, one of the largest Holarctic shorebirds, breeds in southwestern Asia and winters as far south as Australia. Note the length of its long, curved bill (author's collection).

[Figure 21.1] and curlews and godwits, which have very long bills [Figure 21.2]; bill shape and length are adapted to the feeding method, prey type, and habitat.

Shorebirds have been referred to as "the Wind Birds" (Matthiessen 1973) because of their amazing ability to fly and migrate long distances. They fly in tight formations to avoid predators and maintain contact in flight. They are extremely fast as they wheel and flip through maneuvers, twisting and turning with precision. They have long wings and are transoceanic and transcontinental migrants. Although all shorebirds can swim and do so occasionally at feeding locations, only the phalaropes land or rest on large bodies of water.

Because they are ground birds and feed in open areas, they are relatively easy to view. During migration they concentrate in certain locations and feed in mixed flocks. They coexist less aggressively than other bird families, due to the high abundance of food in these preferred locations (Choi, *et al.* 2017). This congregating makes it easy to compare species to one another for their appearance, feeding habits, and behavior. Many species are similar and can be difficult to separate, but their size, shape differences, and feeding style provide some guidance to identification (O'Brien *et al.* 2006). They always give bird watchers an enjoyable challenge.

Sandpiper Breeding

Most sandpipers breed in the Northern Hemisphere, and more than half breed on open tundra—a featureless plane of wet grasses, sedges and mosses. Although shorebirds concentrate at stopover sites along their migratory pathways, they breed over a widely dispersed area across the Arctic and subarctic regions. A few, such as whimbrels and godwits, breed in the semi-open areas where the boreal forest meets the tundra. Sandpipers also breed along rivers and in grasslands, wetlands, coastal marshes and a few on mountain meadows. Almost all shorebirds nest on the ground. However, three species of shorebirds nest in trees.

Because of their Arctic habitat and long migration, sandpipers have a short breeding season. On their breeding ranges, the male defends a territory against other males. In most cases, both male and female incubate the eggs. After the young hatch, both parents might care for the precocial chicks, but in some species, only one adult will take care of the hatchlings until they are totally independent.

Nest sites are located near feeding areas and sources of fresh water, because chicks cannot metabolize salt for the first few weeks of life (NRCS 2000). During breeding season in the Arctic, many species feed on aquatic insects, aquatic and terrestrial invertebrates, very small fish, plants, seeds, and berries. A few large shorebirds that inhabit uplands will feed by searching for insects; some will eat berries, seeds, and plant matter.

The bills of shorebirds have adapted to the various ground habitats. Some smaller shorebirds peck at the water's surface. Phalaropes draw minute quantities of food suspended in water into their mouths by capillary pressure from surface tension. Some studies show that by probing at different depths, prey selection varies by bill size and the size of the shorebird. Shorter billed shorebirds probe shallower water for insects and larvae, and long-billed birds probe deeper and capture larger prey, such as marine worms, leaches, and mollusks (Kober and Bairlein 2009). Some with long bills have tactile and taste sense organs in the bill tip and find prey by touch and smell. Others rely more on sight, and some use both methods. The spoon-billed sandpiper of Asia feed much like other shorebirds by probing, but it also appears to use its bill to strain nutrients after probing (Dixon 1981).

Shorebirds and plovers will feed during the day or night on migration and during breeding, but there are variations in night feeding. Some species will feed less at night in spring and autumn. Some species prefer daytime, while others feed both day and night, and still others during twilight and night. Twilight feeding is common among most shorebirds. Possible reasons for nighttime and twilight feeding are to avoid predators, to continue to feed if they do not find enough during the day, or possibly because nighttime provides more prey species. Certainly, tidal cycles provide feeding opportunities at night. Some long-billed species change from visual to tactile foraging between day and night (McNeil and Rodriguez 1996). On their wintering range, many species will defend territories both day and night.

Shorebirds are only one of many bird types that reside on the ground. Ground birds in general have developed a variety of defensive measures. The most common defense is to take flight—they will fly a short distance, land and eventually return. Protective coloration is an important defense mechanism. Some blend into their environments so well that we are unaware of their presence until one flushes from underfoot.

Most sandpipers are socially monogamous. Once they have established a territory, males defend their territory against other males. Picture the small semipalmated sandpiper carrying out mock battles with other males, facing and chasing each other with wings held up while making noisy chatter (Bent 1929, Poole 2008). Male least sandpipers are also physically aggressive towards rivals in their breeding territory. It is unfortunate that these displays only take place on the Arctic tundra, because we are unable to appreciate the complexity of their displays and their songs in our own environment.

Songs and calls play a major role in the lives of sandpipers, both for communicating and for breeding. Songs and calls are not learned—they are inborn. During migration and at feeding locations, the birds interact with brief, high-pitched, chattering notes, whistles, or ringing sounds, and are generally not antagonistic to each other. On their breeding ground on the open tundra, however, they carry out flight displays with more complex songs than the calls given at migratory stopover sites or wintering grounds. These flight displays usually involve an ascending flight, flying in circles, uttering songs, and finishing by fluttering to the ground. Some species use their wings or tail to produce sounds.

The ruff, great snipe, and buff-breasted sandpiper are the only shorebirds that gather at a lek for males to display and compete for females, much like the behavior of prairie chickens and sage grouse. The long, colorful feathers around the neck of the ruff are erected in display posture. Buff-breasted sandpipers display by raising and flashing the satin white underside of their wings. Pectoral sandpipers also have extremely elaborate displays, in which the male flies in circles above the female, lands near her and inflates his esophagus with erected breast feathers, drops his wings, raises his tail and utters booming sounds. The male ruff, buff-breasted, and pectoral sandpipers are polygamous. Spotted sandpiper females will compete for males, and mate with several males through the breeding season.

Two species that breed across the United States provide examples of these rituals: American woodcock and Wilson's snipe. The American woodcock carries out an elaborate flight display beginning after sundown. It takes flight and circles upward to about 300 feet (90 m) while making twittering sounds; then as it descends, it makes a louder, high-pitched, three-syllable sound and repeats it until it lands. These sounds are made by the outer tail feathers and have an eerie, almost other-worldly quality. In the case of the Wilson's snipe, the bird flies high overhead in circles and gives a rolling series of pulsating tremolo sounds, best described as a winnowing.

Birds will give warning calls if a predator is approaching, and these calls can be understood by other species. Anyone who has tried to approach shorebirds knows that killdeers and yellowlegs will act as sentries by giving alarm calls. Distraction display is a common form of nest defense among many species of birds, including some passerines, ducks, and especially shorebirds. They will give a broken wing display by dragging the wing as they lead an intruder away until they are far enough from their nest site, and then fly off. Some species of birds will gather together as a mob and attack the intruder, but this is not common in shorebirds. When hawks or falcons hunt them, shorebirds will form a flock and evade the attacker with their swift and complex flight, which makes it hard for an aerial predator to identify a specific target.

Shorebirds molt twice annually. The only species that is flightless during molt is the bristle-thighed curlew (Marks 1993). The winter or basic plumage of shorebirds is generally gray, and some species do not change their coloration between seasons. Breeding plumage usually includes dark colors of brown or gray with rufous highlights, designed to aid in concealment on the open tundra. In most shorebird families, the male and female have the same or very similar plumage; phalaropes are the exception to this.

Within the 90 or so species of sandpipers, we find there is a diversity of interesting and unique behaviors. Some of the larger sandpipers, like the curlews and godwits, can perch on tree limbs, but only a few of the small sandpipers—notably spotted and least—have been seen doing this. The solitary, green and wood sandpipers are the only species of this family that breed in trees. All three of these species use the abandoned nests of other passerines, and wood sandpiper can also nest on the ground.

Among sandpipers, some unexpected behaviors emerge. The only members of the shorebird order that routinely spend time swimming are the phalaropes, even though virtually all the shorebirds are capable of swimming. Spotted sandpiper can wade in shallow water and completely submerge like a dipper, and has been seen diving into water from flight to escape predators (Sutton 1925). Some shorebirds like buff-breasted sandpipers and upland sandpipers behave more like plovers than sandpipers in terms of their habitat and feeding habits. The unusual breeding practices of the ruff, described in Chapter 10, are possibly the most unusual mating system in the animal kingdom (Lamichhaney *et al.* 2016).

Other Members of the Charadriiformes

The third subgroup of the *Charadriiformes* are the families of the plovers, lapwings, stilts and avocets, oystercatchers, thick-knees, and seedsnipes.

The plovers and lapwings (*Charadriidae*) total 67 species, divided into two main subfamilies. Like sandpipers, they are also ground birds of open areas. They have short bills and large eyes and prefer shortgrass plains, plowed fields or dried mud flats. They have the habit of alternating running and stopping to look, and finally picking at the surface of the ground with their short bills. Unlike shorebirds that find their prey by their tactile sense, plovers hunt for invertebrates by sight. They range in size from the smaller types, such as snowy plover at about 6 in long (16 cm) to a large group of medium-sized birds like the golden plovers, which are Holarctic. The lapwings are the largest plovers. The gray-headed lapwing of Asia is among the largest at 14 inches (360 mm) long and 9 ounces (300 gm).

Each continent has plovers unique to that region, and these do not undertake long distance migration, unlike Holarctic shorebirds that breed in the Arctic and winter in subtropical areas. The lapwings, for example, are mostly tropical or subtropical (Bamford 2003). Most of the plovers, dotterels, and lapwings that confined themselves to a single continent are boldly marked. The Tawny-throated Dotterel [Figure 21.3] is an example of a beautiful bird with a plumage that makes it hard to find in its grassland habitat.

Figure 21.3 The Tawny-throated Dotterel, a South American resident in the plover family, is often seen in an upright stance in its grassland habitat (Kyle Elliott and Mélanie Guigueno).

Plovers make a simple ground nest depression. As a family, they defend their nest by using the broken-wing distraction display.

Stilts and avocets are very long-legged waders that occupy shallow ponds that are deeper than those typically frequented by shorebirds. Oystercatchers are stoutly built coastal birds with strong, medium-length legs and a very strong bill for feeding on bivalves, mussels and worms. They are either all black or have a black upper body and white breast and belly. In South America, some species of oystercatchers have adapted to living in wet, grassy fields.

Thick-knees are large plover-like birds found mostly in the arid and tropical regions. They have large eyes and feed primarily at night. Pratincoles are long-winged birds with short legs, and they feed on insects by aerial capture, but can also feed from the ground. In behavior, they are like nighthawks, with graceful flight, but can twist and turn in pursuit of prey. Closely related to the pratincoles are the longer legged ground feeding coursers, inhabitants of mostly open areas, although some are woodland species. Coursers are restricted to Africa and a small area of the Middle East, but pratincoles are more widespread through Europe, Asia, Africa and Australia.

Seedsnipes (*Thinocoridae*) are unusual shorebirds of South America, and are the only member of the *Charadriiformes* order that are herbivores; they use their short bills they feed on seeds, buds, and bits of plants. They are long-winged, have short legs, and are grouse-like in their natural habitat. In flight, they fly low with rapid wingbeats, use a zigzag pattern when flushed, and fly fast in steady flight. They inhabit grasslands and Andean mountain plateau habitats throughout the year. Three of the four species are found in mountainous habitats, and the rufous-bellied seedsnipe is found above 13,000 feet (4,000 m) elevation. Their habitat has harsh climate, and the high-elevation species will descend with heavy snow. Like most members of this family, they are well camouflaged. Their plumage and some aspects of their behavior are similar to the Australian Plains-Wanderer, described in Chapter 10. Because of their unusual characteristics, researchers have questioned their placement in the *Charadriiformes* order.

Many authors have written excellent summaries of the sandpiper family, so we refer you to Hayman *et al.* (1986) and Warnock and Warnock (2001) for more thorough accounts. In his book, Matthiessen (1973) provides fascinating insights into their lives.

22

Birds of Prey

Hawks, buzzards, falcons, eagles, and owls are usually referred to as birds of prey, but they are not the only birds that feed on mammals, birds, reptiles, amphibians, or fish. Many other species of birds hunt vertebrates: shrikes, skuas and jaegers, gulls, storks, herons, motmots, trogons, hornbills, some ground cuckoos, many species of waterfowl that feed on fish, and others. These species are all "birds of prey," but in this chapter, we will talk specifically about hawks and owls.

The term "raptor" usually refers to all types of hawks, eagles, and falcons, but most authors include owls as raptors as well. All of these large birds of prey share traits that make them skilled hunters: extremely good eyesight for locating prey, a hooked bill for tearing flesh, the ability to carry out aerial attacks, and powerful talons for grabbing and killing. Other species usually lack some or all of these qualities, so their diets focus on insects, seeds, nectar, and fruit.

Conservation of Raptors

Trees play an important role in the habitat requirements of almost all raptors, except for the few that breed on grasslands. Birds of prey depend on trees for nest sites; studies in California show that red-tailed hawks prefer larger and taller trees for nesting (Vreeland 1997). Trees also serve as perch sites, where raptors rest and scan for prey.

Different raptor species use different types of treed habitats. Tropical raptors, some temperate raptors such as American and Eurasian goshawk, and most owls prefer forested habitats; most are found in deciduous forests, but some prefer conifers. Many species will use forest edges as a vantage point for scanning, though raptors in this habitat type are the most threatened and show the highest levels of decline. The loss of tropical and temperate forests and its effect on bird populations has been discussed in Chapter 12.

Open savannas with scattered trees are one of the most common and important habitats for some species of raptors. This is especially true in the savannah grasslands of Africa, which support so many hawks and eagles. North America has a great deal of open country dominated by farmers' fields, but most of these large agricultural fields are devoid of scattered trees, limiting their benefit to raptors.

Species like osprey and the fish eagles rely on a diet of fish, so they need trees near lakes, rivers, and coastal areas. Finally, a few species have specialized habitats—such as snail kite, which is restricted to freshwater marshes that support large snails.

Another important habitat requirement is size, a controlling factor in prey abundance. As an example, red-tailed hawks can have a home range from 0.5 to 10 square miles (1.3–25 sq km) (Johnsgard 1990). The average home range size for red-shouldered hawk is 0.46 square miles (1.2 sq km) in southern California (Bloom *et al.* 1993) and 0.51 sq mi (1.3 sq km) in Georgia for breeding pairs in a forested habitat (Howell and Chapman 1997). Some raptors require a great deal more space to maintain an adequate supply of prey: the home range size for golden eagles in California, for example, varies from 24 to 1,700 square miles (62–4,400 sq km), and averages about 650 square miles (1683 sq km) (Katzner *et al.* 2012). Home range size can depend on whether birds are breeding, and it may be different for males and females during non-breeding season. Human-altered habitats also can influence breeding range.

Raptors face serious challenges within their habitats: habitat loss and alteration, global expansion of agriculture, pesticides, ingestion of toxic lead shot, intentional killing, electrocution on utility poles, and collisions with manmade structures like wind turbines, vehicles, trains, powerlines, and windows. These are not the only challenges, but they are the most prevalent.

In the United States, there are 20 species of raptors (excluding owls). Long-term survey trends based on migration counts indicate increasing population for nine species, decreasing numbers for five species, and mixed trends for six species. Most noteworthy was evidence of widespread declines in populations of American kestrels. Golden eagles, prairie falcon, and rough-legged hawk all show net declines as well (Farmer *et al.* 2008).

A team of ornithologists led by Christopher McClure (2018) studied the status of the distributions, threats, and conservation of raptors worldwide. They found that 18 percent of raptors are threatened with extinction, and 52 percent have declining global populations. On a worldwide basis, 50 percent of owl species are declining. South and Southeast Asia have the highest diversity and the largest number of threatened raptor species, closely followed by East Africa. In South America, raptor populations are declining. Old World vultures are the most threatened group among raptors; 80 percent of Old World vultures are in serious decline, and 50 percent are critically endangered.

Protection of land and water are the most important priorities recommended for global raptor conservation (McClure *et al.* 2018). Identifying and protecting important sites, and supporting and enforcing legislative actions to protect raptors are important activities that should be taken at the country and state level. State, local, and conservation organizations have taken beneficial actions including monitoring population trends, educating the public about raptors, and offering economic incentives to protect land and birds that require it. Curtailing or reducing the use of poisons and insecticides like neonicotinoids, toxic pollutants, and lead shot, are all actions that individuals can practice to reduce the mortalities of raptors. Obviously, intentional killing of raptors needs to cease as well.

Hawks, Eagles, and Falcons

All diurnal raptors except owls were originally placed in the *Falconiformes* order. The interrelationship of the species within this order has been significantly revised in the past decade, and changes in the various subfamilies are sometimes reported based on molecular analysis. More recently, these birds are separated into three orders:

1. The largest group, consisting of 252 species, is the *Accipitriformes* which includes the osprey, hawks, eagles, kites, harriers, and Old World vultures.
2. The New World vultures, *Cathartiformes*, are a separate sister order to the *Accipitriformes*.
3. The third order of raptors are the *Falconiformes,* which includes 65 species and are divided into two or three subfamilies (Wink 2018).

Ornithologists continue to study the subfamilies of these three groups. As recently as 2018, the American Ornithological Society revised the subfamilies of the *Accipitriformes* (Chesser *et al.* 2018).

Hunting techniques among raptors vary depending on the prey and habitat type. All birds rely heavily on their binocular vision, which is adapted to the task of finding their specific kind of prey. Owls and some hawks also rely on sound to locate their quarry. After identifying their target, they use chase and surprise to capture their prey, which they subdue as rapidly as possible with their talons and bill in order to avoid injury to themselves.

Raptors rarely hunt prey larger than themselves, but golden eagles are able to kill fox, deer, and young mammals that might weigh more than themselves. Bald eagle has been recorded killing a small deer by drowning. A few species of raptors hunt cooperatively: for example, wedge-tailed eagles of Australia will work together to hunt kangaroos.

How exactly did these predacious birds get that way? Their abilities most likely developed with the spread of small mammals, but there is no doubt that predatory instincts arouse from their dinosaur ancestors. Examples of predatory maniraptors are well known: the seven-pound (3 kg) buitreraptor, which was discovered in 2004 in Argentina, might have been able to fly (Markovicky *et al.* 2005). It came from a predatory family of dinosaurs and is believed to have been a carnivore.

It is likely that predatory birds evolved simultaneously at numerous times and places. The earliest fossils of raptors closely related to the osprey come from the Eocene epoch in England, and were widely distributed in prehistoric times (Mayr 2017). The teratorns of North and South America, previously mentioned in Chapter 8, date back to the Oligocene epoch. The Secretarybird of Africa is an early relative of the hawks and eagles (*Accipitriformes*); two species of related taxon, the pelargopappus, were reported from the Oligocene and early Miocene epochs of France (Mayr 2017a). Although currently restricted to Africa, the Secretarybird had a much wider distribution in past history. There are records of a member of the *Accipitriformes* throughout Europe and North America, and some dating back to the Eocene epoch (Mayr 2017). Two species of seriemas of South America of the *Cariamiformes* order are living raptors with a deeply rooted history in prehistoric times.

Among the many fossils, an early type of falcon was found on Seymour Island in the Antarctic Peninsula (Cenizo *et al.* 2016). This specimen supported the hypothesis of a Neotropical or Austral origin of falcons (*Falconiformes*). Further analysis and the earliest fossil records of falcons suggest that they likely originated in South America (Mayr 2017).

With their excellent dispersal capabilities, raptors have developed long-distance annual migration from South and Central America to North America, and from Africa to Europe and Asia. Migration has evolved multiple times in birds of prey, with the earliest evidence of it occurring in true hawks (*Accipitridae*) during the middle Miocene epoch (Nagy and Tokoli 2014), well after the advent of raptors.

Families of Hawks and Eagles

Eagles

The large eagles, buzzards, and *buteo* hawks are the most well-known among the birds of prey because they are well diversified and tend to occupy open habitats. They are found everywhere on earth except Antarctica. Many species are migratory and, like all raptors, the female is larger than the male. These raptors kill their prey with their talons and tear up their prey with their bills. They have an expanded pouch known as a crop near the throat, which allows for temporary storage of food.

"Eagle" is a generic term applied to larger hawk-like raptors, but eagles are divided into several families. *Aquila* eagles, referred to as "true" eagles, are large birds that prey on moderately sized mammals. The *Aquila* eagles are generally brown in color with feathered legs and require large territories, usually over 50 square miles (130 sq km) and often much larger. There are currently ten species of *Aquila* eagles, including golden eagle, found in North America, Europe and Asia; Bonelli's eagle of Europe, Africa, India and South Asia; and wedge-tailed eagle of Australia, among others. Five of the *Aquila* eagles are residents of or winter visitors to Africa. Some authors include the three species of spotted eagles (*Clanga*) of Eurasia, and the five species of medium-sized to small sized *Hieraaetus* eagles of Australia, Eurasia and Africa among the *Aquila* family.

Golden eagles usually capture small mammals such as ground squirrels and marmots, and also prey on larger mammals including fox, young pronghorn antelope and deer. Wedge-tailed eagles of Australia prey on small mammals and wallabies, and they are among the raptors that will hunt in groups to kill large kangaroos.

Very closely related to the *Aquila* eagles, the *Haliaeetus* eagles are ten species of fish-eating eagles that include bald eagle of North America, white-tailed eagle of northern Eurasia, white-bellied sea-eagle of southern Asia and Australia, and the African Fish-Eagle [Figure 22.1], among others. The ten species of this genus are referred to as sea eagles and primarily feed on fish as well as other prey.

Three of the *Haliaeetus* eagles—the Philippine Eagle, the Harpy Eagle, and the Steller's Sea-Eagle—are among the three largest eagles in the world based on weight;

Figure 22.1 The African Fish-Eagle, of the genus *Haliaeetus* (sea eagles), has widespread distribution in Africa and is similar to North America's Bald Eagle (author's collection).

these birds have 8-foot (2.5 m) wingspans. Of these, the Steller's Sea-Eagle is the heaviest, often exceeding 20 pounds (9 kg).

The osprey is a monotypic family and one of the very few species with a worldwide range. Some sources divide the osprey into three separate species, but this bird is usually considered as one species. It is completely adapted to feeding on fish, captured over water using shallow aerial dives with talons extended. The feet, talons, head and wing shape are all adapted to catching and holding onto slippery fish.

Snake-eagles have long, thick, scale-covered legs and hunt from a prominent perch in woodlands, or by soaring. They scan for prey that they capture by plummeting from above. After they grab the snake, they crush or remove its head, thus eliminating the threat of being bitten by their prey. The Brown Snake-Eagle captures very deadly snakes as long as 6 feet (2m), including cobra, puffadder, and mamba.

There have, however, been rare instances where snake-eagles succumb to their dangerous prey. Two known cases from Africa include one snake-eagle being blinded by a spitting cobra, while another was crushed in the coils of a python (Siyabona Africa 2017).

The snake-eagles (*Circaetus*) and serpent-eagles (*Spilornis*) make up two groups that include 13 species. Snake-eagles are primarily found in Africa. The Crested Serpent-Eagle is widespread, and most species of serpent-eagle reside in Southeast Asia. Snake-eagles also feed on lizards, rodents, insects, birds and occasionally on fish. The bateleur of Africa is not related to these two groups; it preys on vertebrates, and large snakes make up about 20 percent of its diet.

Hawk-eagles are divided into two groups: the *Nisaetus* with 11 species of the Old World, and the *Spizaetus* with four species from Central and South America. Hawk-eagles, fierce predators of the tropical forests, are medium-sized compared to the large eagles, and usually have a decorative crest. They are woodland species with slender bodies, rounded wings and long tails. Some of these hawk-eagle species are endangered. The hawk-eagle category includes many other medium-sized species, a number of which are monotypic.

Buteos

Buteos, referred to as hawks in North America and buzzards in Europe, are known for their broad wings and aerial soaring, which makes them popular among hawk watchers. They are a large genus of raptors with 29 species. Their size ranges from as large as some hawk-eagles, with a body length of about 27 inches long (70 cm), to as small as 12 inches long (30 cm). They feed mostly on small mammals; many are also opportunistic foragers and will take snakes, lizards, and sometime insects. They are more associated with open areas. Some *Buteos* inhabit tropical or subtropical regions, while a few, including broad-winged hawk, nest in deciduous forests and hunt from perches on forest edges or openings.

Most *Buteo* hawks are migratory, preferring overland routes to take advantage of thermals and updrafts for soaring. They have broad wings rounded at the tip—except in certain species, like the Swainson's Hawk, which has more pointed and proportionally longer wings. The Swainson's Hawk winters in southern South America and migrates annually to western North America through the Isthmus of Panama, where three million migrating raptors are seen annually. This lengthy route is the longest migration known for a raptor; one way, the distance is about 7,000 miles (11,300 km). The Swainson's Hawk also stands out as unusual because it will feed on insects, particularly grasshoppers.

Closely related to *Buteos* are the *Buteogallus* hawks, large tropical hawks that include the various black hawks: Common, Great and Cuban. Nine species within this genus are generally very dark-colored hawks, with most sporting a black tail with one or two white bands. Many *Buteogallus* hawks feed on crustaceans, but the largest, the solitary eagle, prefers small mammals and reptiles, and inhabits mountainous areas. The long-legged Savanna Hawk will feed by walking and flushing prey, as well as capturing small birds in flight.

Sixteen species of hawks in eight different genera are very much like *Buteo* hawks in terms of size and general behavior. These include such well-known species as Harris's Hawk, White-Tailed Hawk, Roadside Hawk, White Hawk, and Crane Hawk.

Finally, the four buzzards of the *Butastur* genus, tropical raptors of Africa and Southeast Asia, are much smaller hawks. They have characteristics between those of a *Buteo* and an *Accipiter*. Many of them prefer open country and some feed on insects.

Accipiters, Harriers, and Kites

Accipiters are the largest genus of the *Accipitridae* family. Generally, Accipiters have long legs and short, rounded wings for rapid acceleration. They are known for feeding on birds, but a few also prey on mammals. Generally ambush predators, they kill with their talons like larger hawks, because they have lower bite force than do falcons (Sustaita and Hertel, 2010). The family includes 47 species, and its members are referred to as sparrowhawks, goshawks or simply hawks. Three goshawks from other families are nonetheless closely related to the Accipiters: chestnut-shouldered, red, and Doria's goshawks.

Accipiters are closely related to harriers, even though they are structurally different from them—and from all other hawks, thanks to a detail of a nerve attachment site on the shoulder (coracoid) joint. The Accipiters range in size from the smallest, the little sparrowhawk at 9 inches long (22 cm), to largest, the American and Eurasian goshawks, in which the female is 25 inches long (64 cm). The long-tailed Doria's goshawk of New Guinea is 31 inches long (79 cm).

The northern goshawks, American and Eurasian (recently split), are found all across North America, Europe, and Asia. These large, powerful hawks, known for their swiftness and strength, hunt birds and small mammals within and along the forest edges. Goshawks are formidable hunters and relentless in their pursuit; among birds, they prey on grouse, pheasants, domestic fowl, and ducks. One account tells of a goshawk that followed a chicken into a kitchen in Connecticut, and seized the hen on the kitchen floor in the presence of a man and his daughter. The father had to beat the hawk off with his cane (Bent 1937). Goshawks kill prey up to five pounds in weight—heavier than themselves, but they do not carry game back to their nest if it weighs more than about two pounds. A twenty-year study carried out in Northern Arizona found that northern goshawks showed extensive annual variation in breeding success. The highest fledging success was in wet years, and lowest in dry years (Reynolds *et al.* 2017).

Harriers of the genus *Cicus* include 14 species that are widely distributed birds of grasslands and marshes. The ones that breed in the colder regions are migratory. They have a rounded face like an owl, and use sound to find prey. These small hawks have buoyant flight close to the ground as they search for food on long dihedral wings, with a long, narrow tail. They prey on birds, small mammals and reptiles. Most nest on the ground, and some roost communally.

About 30 species of kites and honey-buzzards fall within the *Accipitridae* family, the latter of which are close in taxonomy to some kites. The kites are either unrelated or loosely related to each other, except in a few cases where several species share the same genus; they are distributed among 19 genera. Nagy and Tokolyi (2014) studied the historical patterns of range evolution in hawks and showed, for instance, that the *Elanus* kites (such as Black-Winged Kite) have a Neotropical origin, whereas the *Milvus* kites (such as Black Kite) have an Australian origin.

It appears that the kites and honey-buzzards have evolved to exploit specific habitats and food sources. As an example, oriental honey-buzzard feeds on wasp grubs as well as honey and other insects, an unusual diet among raptors, but plentiful in its Southeast Asian range. The Swallow-Tailed Kite feeds on dragonflies, the Letter-winged Kite hunts rodents, and the Double-toothed Kite prefers arthropods, lizards, and large insects.

Other than the fact that kites tend to be small and lightly built, there are no typical characteristics. Many feed on insects as well as small rodents, lizards, frogs, salamanders, and small birds, and some are scavengers. Three species feed on snails: hooked-billed, slender-billed and snail kites. Several species feed on a variety of prey and are able to catch fish. A few kites hunt by hovering in place and dropping on unsuspecting prey.

The kites and honey-buzzards tend to occupy warmer regions, and many are tropical. Most of them have long, pointed wings, but a few have short paddle-shaped wings. Kites are well known for their graceful fight: they are buoyant in the air, frequently gliding, and agile enough to catch insects on the wing, and some have the ability to soar. The *Pernis* honey-buzzards carry out a roller-coaster flight.

A few species of kites breed in loose colonies: Mississippi Kite, Black Kite, Red Kite, Black-shouldered Kite, Snail Kite, and Letter-winged Kite, as well as a few other species of hawks (Kemp and Newton 2003). Tropical and sub-tropical kites like double-toothed kite are forest birds, but a few kites—such as black-breasted kite and letter-winged kite—are found in open country in Australia.

New and Old World Vultures

Based on morphology and other features, we know that the New World and Old World vultures are not directly related to each other, and recent work has corroborated this and provided further understanding. The most recent taxonomic analysis shows that New and Old World vultures represent a case of convergent evolution (Wink 1995). Based on the fossil record, we know that vulture-like birds have evolved a number of times in history.

All vultures feed on carrion, though there have been documented cases of black vultures killing small animals, and there are also claims of them attacking small sheep. These reports are uncommon, and it is unclear if the animal "attacked" was already dead. Vultures are not built to be hunters—in particular, their feet are not adapted for predation. Their hooked bill enables cutting of flesh, which is typical of predators, but they also rely on other predators to open the flesh. Their heads are featherless, which allows them to feed inside a carcass without getting offal on their feathers, which could bring parasites. Their intestines are adapted to digest rotting meat and bones.

Some vultures find their prey by sight and by watching other scavengers. Black vultures and California condors will watch common ravens, while turkey vultures and greater and lesser yellow-headed vultures find their prey by scent. Vultures cover large areas of the sky and watch each other for a signal if one bird finds a carcass.

The California and Andean condors are the largest vultures, and are among the largest flying birds in the world with wingspans ranging from 9 to 10 feet (2.7–3.1 m). The Andean is slightly larger than the California condor. Although they have the same lifestyle, they actually look quite different, and are in different genera.

New and Old World vultures differ genetically as well as behaviorally. New World vultures have weak feet, do not build nests, have no voice box, and are not restricted to open country, all qualities opposite to their Old World counterparts. Of the 16 species of Old World vultures in nine genera, the *Gyps* genus has eight species, four of which are referred to as griffons; the remaining eight species are monotypic.

All these vultures find their prey by sight. Unique among these is the palm-nut vulture, which consumes palm tree fruit. Most interesting, the bearded vulture (also called lammergeier) will crush bones by dropping them from a height, then consume them to extract the bone marrow.

Falcons

The last order of diurnal raptors is the *Falconiformes*. This order is made up of eleven genera; the more well-known are referred to as falcons, forest-falcons, caracaras, falconets, and kestrels. The largest order of *Falconiformes*, the *Falco* genus, includes 38 species, all referred to as falcons.

Falcons do not build nests; they nest on the ground, on cliffs, in cavities, or in abandoned nests left by other birds. Because of their long, thin, pointed wings, falcons are capable of fast flight and are active fliers. They kill with their beak, making them different from hawks and eagles, and many feed by aerial capture of birds. Some, however, also feed on small mammals, snakes, small lizards, and large insects.

Many falcons are migratory, and unlike *Accipiters* and hawks, they will routinely cross large bodies of water. We know quite a bit about the long-distance migration of the Amur falcon, for example, because of the groundbreaking tracking work undertaken by R. Suresh Kumar. Kumar radio-tracked three of these 5.1 oz (180 gm) birds from Nagaland, in northeast India, to their wintering range in southern Africa. Two birds stopped briefly in India, but one flew non-stop, covering a distance of 3,500 mi (5,600 km) in 5 days and 10 hrs. Along their route, all three crossed the Arabian Sea, a distance of about 1,860 mi (3,000 km) in an average of 77 hours (Kumar 2015). The Amur falcon is not the only one to achieve this: the Red-footed Falcon also carries out a long-distance migration.

One of the most well-known members of this order is the peregrine falcon, which has the distinction of the fastest flight speed recorded for any animal; it can achieve 200 mph (320 kmph) in a dive. Peregrines usually attack their prey by gaining altitude, then striking from above at high speed. They will injure their prey in the attack and return if necessary for the kill. They are also capable of subtle tactics through which they can corner a bird and inflict injury with short, slow flight. They have used this technique to kill ring-billed gulls, which are about their own size. The peregrine falcon has 19 subspecies and is one of only a few species that has world-wide distribution.

The smallest falcon is the pygmy falcon of Africa, with a body length of 8 in (20 cm), but the four species of falconets (*Microhierax*) of Southeast Asia, India and Borneo are even smaller—only about 6 in (15 cm) long. These small forest birds feed mostly on insects and hunt like flycatchers. On the other end of the spectrum, the Gyrfalcon [Figure 22.2] is the largest true falcon, averaging about 22 inches long (55 cm). Gyrfalcons are distributed across the High Arctic tundra, making them the world's northernmost falcon. They take larger prey, including ground squirrels, ducks and ptarmigan. Unusual among falcons, the Gyrfalcon has three color morphs: white, gray and blackish-brown.

The seven species of forest-falcons (*Micrastur*) are very unlike true falcons in wing shape, habitat, and prey capture, making them more like *Accipiters* than falcons. Recent analysis suggests that they evolved much later than the true falcons

Figure 22.2 Gyrfalcons are large falcons, and they occur in three color morphs. The species is distributed around the northern hemisphere, and most are found in tundra habitat (author's collection).

(Fuchs *et al.* 2011). They occupy the humid, tropical forests in the lowland and mid-elevation of Mexico, as well as South and Central America. Their disk-shaped face resembles a harrier's, enabling sound collection, which they use to locate prey. These medium-sized raptors feed on birds, insects, mammals, and reptiles, hunting by wait-and-ambush. Their rounded, short wings and long tail give them great maneuverability.

Collared forest falcon [Figure 22.3] has been observed feeding on lizards and insects disturbed by army ants. The falcon remains perched a short distance above the ground and makes sallies, or stays on the ground, chasing the ant column and foraging among the leaf litter (Mays 1984). Arevalo and Araya-Salas (2013) give a detailed account of a collared forest falcon initially attacking a yellow-throated toucan. The two birds are about the same length, although the toucan is heavier. The falcon made repeated attacks for 30 minutes, driving the toucan to the ground, where it finally killed it.

Laughing falcon, related to the forest falcons, is larger in size than its cousins and prefers forest edges with open areas. It preys on terrestrial and arboreal snakes, which it hunts like the snake-eagles. Its diet also includes birds, insects and small mammals.

Figure 22.3 The Collared Forest-Falcon is a common inhabitant of tropical forests of Latin America. Like *Accipiters*, the forest-falcons have short, rounded wings for rapid acceleration (author's collection).

Caracara

Caracaras are large members of the *Falconidae* family; the ten living species are divided into five genera. The only species that is not restricted to South and Central America is crested caracara, whose range extends through Florida, Texas, and Arizona, and down into Mexico and South America. Caracaras are unlike the other falcons; their habitats and lifestyles vary considerably. Their wing shape is more like a *Buteo* hawk, and yet they are slow fliers, do not soar, and do not carry out aerial attacks as do the true falcons. Some species are cooperative breeders. Caracaras are considered to be very intelligent among birds, with good observational skills and high problem-solving ability (Meiburg 2021).

As a group, caracaras are known for their foraging versatility. Generally, they find food items through their sense of smell, but also by sight. Most are scavengers and feed on carrion, but they also feed on small vertebrates, nestlings, insects, fruits, and almost anything edible. Three caracara species are even known to catch and eat fish: the crow-sized Chimango Caracara of the scrubland and grasslands of southern South America can hover above water and catch fish at the water surface, by plunging with talons extended (Sazima and Olmos 2009). Observers report that it sometimes uses shrimp as prey (Lopez-Idiaquez *et al.* 2019).

Closely related to chimango caracara, the more northern Yellow-headed Caracara is also widespread and is omnivorous. The Mountain Caracara, found at higher elevations along the west coast of South America, occupies the same environmental niche as the common raven in North America. Red-throated Caracara, a tropical lowland forest bird that carries out cooperative breeding, is not known as a scavenger—it feeds on the larvae of wasps and bees. It will skillfully attack the wasp nest with repeated strikes, avoiding counterattacks by the wasps, and causing sufficient damage to force the wasps to abandon the nest (McCann *et al.* 2013).

Among the diurnal raptors, we see many examples of convergent evolution—different species adapting to their environment in similar ways. The forest falcons have a lifestyle like the *Accipiters*. Snake-eagles, serpent-eagles, and some other eagles and falcons, such as prairie falcons, will all hunt snakes. The Old and New World vultures have parallels in their structure, appearance and behavior. On the cellular level, molecular analysis has shown that the Old World and New World hawk-eagles also represent convergent evolutionary clades (Haring *et al.* 2007).

Owls

Owls are depicted in drawings in the Chauvet-Pont-d'Arc Cave in France, dated to 30,000 years ago, and identifiable owls have been found in numerous petroglyphs throughout the southwestern United States. This indicates that people have always had a special fascination with owls' eerie calls and nocturnal behavior, associating them with darkness as opposed to creatures of the light.

Many cultures have taken note of owls in their mythology. Early Native American cultures viewed owls as the spirits of the dead, and as guides in the afterlife. In the Bible, the owl is referred to as an "unclean" and abominable animal (Leviticus 11:13). Many myths view owls as harbingers of death, and owls have been routinely killed for no reason other than superstition in some cultures—a sharp contrast with our myth of the "wise old owl." There is more mystique about owls than any other group of birds, in spite of the fact that they are rarely ever seen by the average citizen. In reality, owls are raptors, similar in many ways to hawks.

Owls were formerly considered part of the order of hawks and eagles, but today, all owls are placed in the *Strigiformes* order. Hawks and owls have many things in common: similarly shaped beaks for tearing flesh, sharp talons, exceptional binocular vision, adaption to similar habitats, and females that are larger than males. However, there are many more ways in which hawks and owls are different.

In taxonomy, structural analysis and the earliest genetic work placed the owls near the nighthawks (*Caprimulgiformes*) (Johnsgard 2002). Today the similarities in these two groups are viewed as example of convergent evolution. The earliest owl-like fossils were found in Colorado and are dated to the Paleocene epoch, approximately 60 MYBP, near the time of the end of dinosaurs (Rich and Bohaska 1980). In their work, Dyke and Gardiner (2011) used the ages of 93 owl fossils to estimate that they originated before the extinction of dinosaurs. Other fossils resembling modern owls appeared in the Eocene epoch (Kurochkin and Dyke 2011). Examples of owls in the barn owl family have been identified from 25 MYBP. The most recent molecular analyses found the surprising result that owls are the sister group to the diverse clade *Coraciimorphae*, comprised of mousebirds and their relatives, including cuckoos (Prum *et al.* 2015).

Several species of owl from the more recent past occupied islands and were flightless. The Cuban giant owl, previously mentioned, stood three and a half feet tall (1.1 m), had no natural predators, had talons twice the size of great horned owl, and preyed on a large rodent. This prehistoric owl of more than 10,000 years ago is believed to have been flightless, and is closely related to the modern *Strix* owls (wood owls). The Andros Island barn owl and two other closely related owls of the Bahamas were also flightless owls of the pine forests. These owls were related to modern barn owls and were present on the islands when the first Europeans arrived in the sixteenth century; they disappeared with the clearing of forests (Marcot 1995).

Owls and Their Senses

Owls are particularly known for their vision and exceptional hearing, which they use to find prey. They have the best night vision of any creature, their large eyes well adapted to low light gathering, giving them at least three times more capacity for this than humans (Johnsgard 2002). This, however, leaves them with a poor sense of color—though color may be less important to nocturnal birds. The shape of their eyes is adapted for distant focusing in larger owls, and for closer focusing in smaller owls. Owls' audio sensitivity is the result of two structural features: their facial feathers channel and amplify sounds coming into their forward facing ears, and many owl species have asymmetric placement of the ears. Some even have movable ear flaps. These qualities are used in combination to locate the distance and direction of a sound source with great accuracy.

In general, owls are believed to have a poor sense of smell, though there has been very little scientific work on their sense of smell. Measurements of the olfactory bulb in two species of owls indicate that the bulb is in a proportional range that led authors of the study to classify smell as unimportant to these birds (Bang and Cobb 1968). However, a more recent study found the olfactory bulb may play a more important role in nocturnal birds (Healy and Guilford 1990). Further evidence based on field observations suggests that some small owls might use odors to assist in finding small mammals (Lawrence 1997).

Owls kill with their talons and sometimes with their beaks. They kill their prey as rapidly as possible by grasping with their talons, and using the beak to sever the neck vertebrae. They have strong talons, and one toe on each talon is moveable to the front or back to support grasping their prey. Owls swallow small prey whole and tear up larger prey, swallowing the flesh, bones and feathers. They have no crop to store food, and they use enzymes and their gizzard to digest what they eat. Most bird watchers are familiar with owl pellets, created when owls regurgitate the fur and bones of their most recent meal. Hawks and eagles, on the other hand, have much stronger stomach acids and completely digest fur and bones.

Owl Environments

There are about 243 species of owls worldwide, found everywhere on earth except Antarctica (Clements, *et al.* 2021). Owls have exploited every natural terrestrial habitat on earth including forests, marshes, grasslands, desert, and tundra. The total number of recognized species varies with the authors and the degree to which their source material is up to date. The most recent total number ranges from 205 to 268 (Enriquez *et al.* 2017). This underscores the difficulty of separating species and subspecies. Within the *Strigiformes* order, owls are divided into two families: the barn owl types, Tyto owls of *Tytonidae* family, which includes about 18 species; and the much larger strigid family, *Strigidae*.

Owls are easily recognized by their large head, round face with forward facing eyes, upright stance, and short tail, and most have cryptic brown plumage. Their flight feathers make no sound when flapping, because of the structure of their wing feathers: the serration on the leading edge of the wing and the fringes on the trailing edge of the flight feathers reduce noise by suppressing the generation of vortex sounds—sounds that result from the motion of the wings creating air turbulence (Chen *et al.* 2012; Wagner *et al.* 2017).

Some hawks and eagles require mature, intact forest habitats, but a large percentage of owl species live their lives in old growth, closed canopy, undisturbed temperate and tropical forests throughout the world. A total of 83 owl species are associated with old growth forests; this represents 34 percent of all owl species. A third of these species are found on islands, and the remainder are in continental settings (Marcot 1995). Nine species of woodland owls have gone extinct during recorded time, several more are suspected of having gone extinct, and a few are threatened species. These losses resulted from habitat destruction and deforestation.

Many species of birds prefer mature or old growth habitats. In North America, species that require this type of habitat include spotted owl, northern goshawk, marbled murrelet, and red-cockaded woodpecker. In a study of 56 eastern forest species, one-third showed a strong statistical preference for old growth forests, including fifteen species of passerines like Acadian flycatcher, brown creeper, blackburnian warbler, blue-headed vireo and others (Haney 1999). Old growth forests are particularly vulnerable to logging because of the commercial value of the timber—but harvesting these trees destroys the entire ecosystem. Restoring an old growth forest takes centuries, so forest fragmentation and destruction of old growth forests presents a direct threat to these species.

Most owls are nocturnal raptors, and about a third are crepuscular, hunting during twilight. The northern hawk owl is one of the few owls that hunts small mammals and birds only during the day. Owls generally hunt by foraging from a perch, where they watch and listen and drop onto their prey. Most owls plunge after their prey with their talons extended, as do hawks. A few, such as barn, short-eared, and long-eared owls, hunt from flight; these owls have a lighter wing loading that allows more time for aerial hunting. Larger owls take larger mammalian prey, and also prey on smaller owls. Within open habitats, an owl species will hunt the area at night, while a corresponding species of hawk might occupy the same territories during the daytime. In addition, different species of owls that occupy the same areas will take different prey and hunt at different times of the evening. Like hawks, owls are carnivores, but many of the smaller owls like elf, scops-owls, screech, and pygmy owls feed on insects.

Great gray owls can find hidden prey, like a small mammal that is buried under a foot (30 cm) of snow from a distance of up to 22 yards (20 m). They hunt from a perch, listening for the sound of their prey and gliding to the exact spot, then plunging straight down facing forward, creating a depression in the snow, and grabbing the mammal with their beak. Although snowy owls can hunt in this same manner, they usually scan for prey from a perch or from hovering and use swift flight to capture even small prey.

At least six species of owls are circumpolar in their distribution around the northern hemisphere: great gray, snowy, short-eared, long-eared, boreal owl (known as Tengmalm's owl in Eurasia), and northern hawk owl. Barn owl is one of those rare species that has a worldwide distribution. The barn owl moves south in winter and returns in spring in Australia, the United States, and Eurasia.

Most species of owls remain in the same territory year-round, nesting in tree cavities or using old nests, but a few species nest on the ground. However, some owls that breed in the northern regions of North America and Eurasia migrate south for the winter: short-eared and long-eared migrate, as do those that rely more on insects, including elf, flammulated, northern saw-whet, and Eurasian scops-owl. Some species carry out irruptive movements caused by environmental conditions, primarily insufficient prey to sustain them through the winter; and some owls are nomadic during the non-breeding season.

Owl Families

The Tyto owls (*Tytonidae*) are different from the typical owls in that they have a heart-shaped face, variations in the feet and claws, and their wishbone is fused to the sternum, while the *Strigidae* owls' wishbone is not (Johnsgard 2002). They have a light-colored underside and darker upper side, unlike typical owls. In addition to the widely distributed barn owl, the remaining species are found in Africa, Southeast Asia, Australia (five species), and a few islands. Most Tyto owls nest in cavities and will also use buildings or structures. Occasionally barn owls will dig a burrow in a bank for their nest. There are no counterparts within the hawks that are known to nest in a burrow.

There are 29 subspecies of barn owl throughout the world. Many show significant variations in coloration, and some have variations in size and calls as well. Many of the subspecies are restricted to islands. The taxonomy of this group is complex, and splits into different species may take place in the future, based on continuing research.

Barn owls hunt predominately in open grasslands. They hunt mostly from flight, and capture small mammals and occasionally birds: they have been seen killing white terns, chasing lesser nighthawks in Arizona, taking blackbirds and cowbirds in Texas, and hunting storm-petrels on islands in the Gulf of California (Weidensaul 2015). Tests with barn owls show that they can find prey in total darkness by sound alone.

Also included within the Tyto owls are the three species of the smaller bay-owls, genus *Phodilus*, which have a slightly different facial shape. Bay-owls are forest birds with short wings, and they hunt from perches. One of them, the Congo bay-owl, was first discovered in 1951 in the Rift Mountains of The Democratic Republic of The Congo, and was not found again until 40 years later (Butynski *et al.* 1997); there are few specimens. It has never been studied and its taxonomy is uncertain. Two other species of bay-owls are found in Southeast Asia, and one is also seen in Sri Lanka.

The 206 species of *Strigidae* owls, divided into 26 genera, are viewed as typical owls, all of which have cryptic brown coloration for camouflage, which make them difficult to find when they are roosting. About 40 percent of Stringid owls have feathered "ear" tufts, which have nothing to do with hearing.

Michael Perrone, Jr. (1981) considered the role of ear tufts by examining three possible hypotheses for why owls have them, and applied these to 81 species of both diurnal and nocturnal owls. Among these owls, ear tufts were absent from diurnal owls, but 82 percent of the nocturnal owls had ear tufts. When owls with ear tufts are in alert posture, the ear tufts are raised. Of the hypotheses he tested on the value of ear tufts, Perrone's results favored camouflage and species recognition.

The *Strigidae* owls are mostly woodland birds known for their low-piercing, hooting calls. A few are found in open country: in grassy fields like short-eared owl, in desert scrubland like elf owl, and on tundra like snowy owl. Owls with yellow eyes and ear tufts are usually crepuscular, and those with black eyes tend to be more nocturnal (Johnsgard 2002). There have been a number of studies and proposals on how the *Strigidae* are grouped into subfamilies, and this work continues to evolve as new information comes to light. The interrelationship of these genera is complicated by the fact that the various species have adapted to so many different habitats over such a long period of time, creating many differences and variations. Seven genera are larger groups of seven or more species, while the Band-bellied Owl [Figure 22.4] is placed in a genera with just two other Strigid owls. Of the 26 genera, 11 are considered monotypic.

The three groups of large owls, some of which are more than 30 inches (75 cm) in length, include *Bubo*, *Strix*, and *Ketupa* (fish owls). The genus *Bubo* consists of 17 species of the largest owls. The family includes Great-horned Owls, Snowy Owls,

Figure 22.4 The Band-bellied Owl is a strigid owl of the tropical montane and foothills rain-forest on the eastern slope of the South American Andes. Very little is known about this owl (author's collection).

Eurasian Eagle-Owls (which is among the largest of all owls), and other eagle-owls. Blakiston's fish-owl is also close in size and weight. The *Strix* owls, all wood-owls, are large in size and lack ear tufts, and they all have black eyes, except for Great Gray Owls.

Seven species of fish owls of southern and eastern Asia and Africa are separated into two genera, *Ketupa* and *Scotopelia*. These owls are also large in size, ranging up to 24 in (60 cm) in length, and they lack the disk-shaped face and the feather edges that give silent flight. They are nocturnal predators that snatch fish from the surface of lakes, rivers and streams with their bare legs. Their diet also includes frogs and crabs as well as birds and reptiles.

The *Asio* genus includes only seven species of medium-sized owls: long-eared, short-eared, stygian, and marsh owls, most of which have wide distributions. The Short-eared Owl is one of the most widely distributed owls in the world. *Asio* owls are eared owls and hunt in mostly open or semi-open areas.

The *Ninox* owls include 32 species, ranging in size from the smallest, least boobook, to the largest, powerful owl. Twenty-eight of these species are named "boobook," an onomatopoeia for the call of the southern boobook of Australia, as well as its Australian Aboriginal name. Some are called "hawk owl," such as brown boobook, also known as brown hawk-owl—not to be confused with the Northern Hawk Owl, which is not a part of this genus. The Brown Boobook is the only member that has a wide distribution, from India to Southeast Asia. Most of the *Ninox* owls are found in the Philippines, Australia, Indonesia, and the Malay Peninsula. Many are restricted to islands, including one on Madagascar. The *Ninox* owls lack ear tufts, have a slim posture with the head typically narrower than the body, and have a long tail.

Small owls make up two large groups within the *Strigidae*: the screech-owls (*Megascops*) of the Americas with 23 species, and the scops-owls (*Otus*) of Eurasia and Africa with 49 species. Screech and scops-owls are very similar in appearance, and were placed in the same genus until several different DNA sequence studies confirmed that they should be separated. The two groups have different vocalizations: scops-owls have only one type of whistling call, while screech-owls give sharp trills and songs. The largest screech-owls are about 10 inches long (25 cm). The Western Screech-Owl has a large range along western North America and is the most northern breeding screech-owl, found as far north as Seward, Alaska. Its counterpart in South America, the Tropical Screech-Owl, also has a large range, and is the most southern breeding screech-owl, reaching south of Buenos Aires, Argentina.

The *Otis* and *Megascops* owls are closely related, and sixteen have four or more subspecies. Like many species that have a large number of closely related subspecies, there is always a lot of controversy on how many of these subspecies should be classified as full species status. This problem raises much discussion among those with strong viewpoints on taxonomy.

Like most small owls, these two groups are generalists in their diet, feeding on insects and small vertebrates. The screech-owls take a wide variety of prey including insects of all kinds, small mammals, birds, snakes, snails, fish, crayfish, toads, lizards, and frogs. Eastern screech-owls have killed birds as large as ruffed grouse, feral pigeons, and American woodcock. Like most screech-owls, they can capture insects on the wing, and there are many accounts of finding fish in their nest sites. In one such story, a person found sixteen brown bullhead *(Ameiurus nebulosus)*, four of which were alive, in its roosting hole, even though the local pond was covered with ice and two feet of snow. This owl caught the bullhead using the only opening in the ice made by ice fishermen, and it was one mile from the roost site, implying that the

Figure 22.5 Most pygmy-owls, of the genus *Glaucidium*, look very similar. This Austral Pygmy-owl is found in Chile and Argentina (author's collection).

owl flew 32 miles to bring the 16 bullhead to its roost (Bent 1938). In another account, eastern screech-owl preyed on goldfish in water less than four inches deep (10 cm) deep (Prescott 1985).

The pygmy-owls (*Glaucidium*) [Figure 22.5], with 28 species, are another group of small owls that are widely distributed and similar in appearance. The northern pygmy-owl has the most northern range of all the *Glaucidium* owls, and it carries out altitude migration in winter. The remaining species in the genus are found in tropical latitudes, but they are not in Australia. Northern, Andean and cloud-forest pygmy-owls all breed at high elevations. Twelve of the 28 species are called owlets, and all except one of these species are found in Asia or Africa.

The *Glaucidium* owls range in size from 5.5 to 8 in long (14–20 cm), and look chunky because their body is wider than their head. They are formidable predators with large feet and sharp claws, and they feed on insects, small lizards, mammals, and birds. On the back of the head, some have a pair of prominent black marks that appear as though they are "false eyes." These owls are active during the day, and most are found in woodlands and forest edges. They have very similar plumage and are best identified by their calls, and in the tropics by their habitats.

In Greek and Roman mythology, the little owl—one of the five species of *Athene* owls—accompanies the goddess of wisdom, Athena. Athena was also the goddess of the night, so this owl may have been associated with the goddess because of its nocturnal habits.

Most of these small owls nest in holes, embankments, or ground burrows. The burrowing owl of North and South America hunts during the day in open habitats. Three of the five species have large ranges, but the other two species of *Athene* owls are very restricted: the white-browed owl of Madagascar is strictly nocturnal, while the very rare forest owlet of India was rediscovered in 1997 after an absence of over 100 years (King and Rasmussen, 1998).

The *Aregolius* genus is another group of small owls, including four closely related species mostly restricted to the Americas—except for the boreal owl, which breeds across the boreal forests of the northern hemisphere. These species reside in forested habitats. The Northern Saw-whet Owl is well known in the United States. Their fall migration can be variable in timing and in the habitats they occupy but prefer dense vegetation. The Unspotted Saw-whet Owl is found from Mexico to Panama, and the buff-fronted owl resides in mountainous areas from Venezuela to Peru. The Bermuda Saw-whet Owl, an extinct species in this genus, is known from fossil records and early accounts from the seventeenth century (Hume and Walters 2012). This species was similar to the Northern Saw-whet Owl, but it had some different anatomical features, and it might have been diurnal.

Among the 28 genera of owls, eleven genera are monotypic. One of them, the Laughing Owl of New Zealand, is assumed to be extinct, as the last specimen was collected in 1914. The remaining ten owls are unique species and can be divided into two groups: five of these owls are island residents and include the Jamaican Owl and the Bare-legged Owl of Cuba, while the other five are owls that have colonized some special habitat or have unusual characteristics. Four of these reside in North America. The Elf Owl [Figure 22.6] is the smallest owl, about the size of a sparrow.

Figure 22.6 The Elf Owl, the world's smallest owl, weighs about one and a half ounces (author's collection).

It occupies the south and southwestern desert and riparian areas. The Flammulated Owl is a small owl of western conifer forests. The Northern Hawk Owl is a medium-sized owl of the boreal forests and is unusual because it has adapted a lifestyle similar to that of a hawk; its long tail and pointed wings are like those of falcons. The Crested Owl of southern Mexico and Central America is a resident of dense tropical forests. Finally, the Long-whiskered Owlet is a very small owl of the cloud forests of northern Peru. The taxonomy of these ten monotypic genera has been through many changes, and may be reviewed in the future based on DNA evidence.

Inevitably, there will continue to be changes in the systematics of owls based on future research. The genetic makeup of many owl species has not been studied until recently, and ongoing work in this field provides higher levels of understanding that were not available in the past.

Part V

Birds in Our World
Today and In the Future

Introduction

We live at a time when the planet is undergoing significant change. In the words of Adam Nicolson (2018), "we live in the age of loss." The biodiversity of the planet is in decline. This problem is evident to those who study or take an interest in the status of living organisms.

The average person is caught up with life's everyday problems and will rarely ever notice the disappearance of wildlife, the quality of habitats, or the environmental problems we create unless they are directly impacted by them. We live in a world where economic interests have very few constraints—and if there are constraints, they are sometimes ignored, and other times undermined by litigation. Strong socioeconomic pressures increase development and population, and these always lead to an expansion of the human footprint. But this expansion comes at the expense of other life forms, a problem that is not likely to cease in the foreseeable future.

Studies of the past has shown that humanity has had a very destructive effect on wildlife. Nevertheless, we have the opportunity to save what we have left. This would require the political will to set standards and enforce regulations, come to a level of agreement between nations, and make some level of sacrifice on the part of the public. These are all difficult to achieve because only a small percentage of the public supports wildlife, and politicians have radically different agendas.

23

The Decline of Native Birds

BirdLife International maintains a Red List of birds at risk with three risk categories: threatened, endangered, and critically endangered. "Threatened" in this sense means threatened with extinction. Currently 1,469 (more than 1 in 8) of all bird species are threatened, and 222 are critically endangered.

The definition of "critically endangered" is complicated. It basically means that the population of a species is small enough or is decreasing rapidly enough that slight changes will bring about imminent extinction. Conservation efforts have saved some species from extinction, but 80 percent of threatened species are still declining (BirdLife International 2018).

Other creatures besides birds are threatened as well. The Intergovernmental Science-Policy Platform on Biodiversity and Ecosystem Services (IPBES) estimates that one million species are threatened with extinction. IPBEC further estimates that this represents 12 percent of all species on earth. The International Union for Conservation of Nature (IUCN) estimates that 26 percent of all mammals are threatened; about 30 percent of fish species are threatened, and overfishing has reduced the stocks of commercial fisheries by very large percentages. Among the amphibians, 40 percent are threatened. A recent Red List update by the IUCN stated that 31 percent of all animals are threatened, and 6.2 percent are critically endangered (IUCN 2021). Virtually all of these threatened species' declines have been caused by humankind.

The IUCN also defines three levels of "threatened": vulnerable, endangered, and critically endangered. As of 2017, the IUNC lists 14 percent of the world's bird population as threatened—the 1,469 species of birds noted earlier. Threatened species are a worldwide problem, and the vast majority of these species have very little or no support. Seventy-seven of the IUCN-listed species are threatened in the United States.

Thirty of these threatened species are Hawaiian birds, and 13 of them are critically endangered. Recently, some of these species have been declared extinct. The support to save these birds is woefully inadequate. Many of them have very small populations, and one can only wonder why they are not on the endangered species list of the United States.

A group of scientists from the United States and Canada took a broad look at the decrease in population of birds across North America over the past half century (Rosenberg *et al.* 2019). They used various data sources covering this time period and population studies to analyze the change in the number of each species from different biomes (a biogeographical unit composed of many habitats). The populations of

529 bird species were analyzed based on records compiled over the past 48 years, starting in 1970. They integrated data from the North American Breeding Bird Survey, Christmas Bird Counts, the International Shorebirds Survey, aerial surveys of waterfowl, and migratory estimates taken from weather radar, as well as all biomes including grasslands, various types of forests, Arctic tundra, arid lands, coastal habitats, and wetlands.

A summary of their results, which accounted for both increasing and declining species, reveals a net loss in total abundance of 2.9 billion birds across all habitats, and a reduction of 29 percent of the total population since 1970. This is more than one-quarter of all North American birds. Birds were found to be declining in all habitats except marshlands, where ducks and geese were increasing, but 50 percent of other marshland species were in decline. It is likely that ducks and geese are benefiting from conservation efforts from various organizations and agencies such as Ducks Unlimited, the National Wildlife Refuge System, and the Federal Duck Stamp program of the U.S. Fish and Wildlife Service, among others.

The biggest losses were among grassland birds, where more than 50 percent of the population has already been lost, and 74 percent of the species are in decline. Throughout North America, 57 percent of all species are in decline, and there are no ecosystems that are not seeing declining numbers of species. Species types that are in steepest declines include aerial insectivores, shorebirds, native migratory species, and land birds.

A similar study of 378 species of birds in England and the European Union has shown a decline of 17 to 19 percent of the overall breeding bird abundance since 1980, a loss of about 600 million birds (Burns *et al.* 2021). In a statistical analysis based on long-term population studies within EU countries, they found that the highest declines were with species associated with agricultural lands and grasslands. A worldwide study of bird populations based on habitat loss, climate change, and overexploitation has estimated that 48 percent of existing bird species worldwide are known or suspected to be undergoing population declines. Populations are stable for just 39 percent of species. Only 6 percent are showing increasing population trends, and the status of 7 percent is uncertain (Lees *et al.* 2022). This study examined the many factors that are contributing to the major causes of the decline of birds.

Studies of bird populations have shown declines for many years, but the joint studies mentioned above have quantified how startling this problem has become. We can hope that these results would stir more public support for actions needed to reverse these losses and stabilize the remaining population of our avifauna. Past history has shown that we can enact conservation measures that would protect natural fauna; this is now a global need.

BirdLife International and American Bird Conservancy are two organizations that focus on saving threatened species. The Endangered Species Act in the United States has demonstrated positive accomplishments for some species, but political and economic pressures in the United States are constantly trying to undermine and circumvent this law.

In the United States, conservation has contributed greatly to the reintroduction of bald eagle, peregrine falcon, and trumpeter swan in the lower 48 states. Captive breeding has saved the California condor from extinction; nevertheless, it remains an endangered species. Conservation efforts have improved the habitats for species currently or formally considered endangered species, such as Kirtland's warbler and whooping crane. In the United States, a total of 1,271 species of plants and animals are on the endangered species list, and 77 of these are bird species or subspecies (US Fish and Wildlife Service 2023). Since its inception, 1.3 percent of the species have been delisted. The bald eagle is a major success story and is now reduced to a status of "least concern." The Endangered Species Act was never expected to increase populations of listed species so they could come off the list; its intent was to ensure that those that are listed would not go extinct. With dozens of bird species acknowledged as endangered by global authorities but not listed as endangered on the U.S. list, however, these birds remain unprotected as their need grows more dire.

Conservation measures are important to reduce the number of extinctions of all species of plants, insects, birds, mammals, reptiles, and other animals. One way to look at extinctions is by extinction rate, a rate normalized for the number of species present. The concern among scientists is that we may be entering a period of major mass extinction similar to those seen in the five major extinction events from past geological periods. Extinction rates caused by humans are based on models supported by known extinctions over the past five centuries, and on fossil evidence. The base extinction rate for all vertebrates is two extinctions per million species per year (2 e/msy), but the current estimate of the extinction rate for all vertebrates is 200 e/msy (Ceballos *et al.* 2015). In the case of birds, the base extinction rate is 1 e/msy and the rate since the spread of humanity is estimated to be about 100 e/msy. Conservation efforts in the past century have reduced this to about 50 e/msy (Pimm *et al.* 2006). These estimates are crude, but the important point is that conservation measures have helped.

It is virtually impossible to know exactly how many bird extinctions have occurred since humans began to colonize the entire planet. At least 140 species and 138 subspecies are known by science to have perished at the hands of humankind since 1500 AD (Szabo *et al* 2012). This is only a tiny fraction of all the bird species humanity has driven to extinction. It is well recognized that there were waves of extinctions on islands as they became colonized by humans, and that there are far more extinct species than have been described.

David Steadman (2006) has spent years studying the waves of extinctions that have taken place on the Pacific islands following human habitation. His work is based on analysis of fossils, and he estimated that between 1,000 and 2,000 species were driven to extinction. The species that survive today are only a fraction of the original avifauna. What is most amazing about this is that many of these extinctions took place relatively rapidly: the dodo, for example, was first discovered in 1598 and was last recorded in 1662.

Bird Survival and Island Communities

Just how much land and sea needs to be protected to stem the current rate of species losses? Estimates and strategies to achieve the results vary considerably

(Lambert 2020). As of 2020, about 15 percent of lands and 7 percent of the seas were under some form of protection. The amount of protection needed in the face of fragmented landscapes and ever increasing development depends on many factors, but there is no question that current levels of protection are considered inadequate. In 2010, the United Nations Convention on Biological Diversity (CBD) established targets that were ratified by many nations, but many of these targets have not been met. In response to the CBD targets, a number of biologists estimated that 44 percent of the earth's landmass needs to be set aside to safeguard biodiversity (Allen *et. al* 2019). Others argue that an overall percentage of land as a target is inaccurate because some areas have higher biodiversity than others, such as tropical habitats compared to tundra, and some rare species require a greater percentage of their total range to be protected.

Research conducted over the past fifty years has led to further understanding of the complexities of wildlife extinction, and how we can determine what is needed to protect biodiversity. In the normal course of nature, some animals go extinct, and others recolonize those environments at some later time. Most of these insights come from the study of islands. Researchers have used an understanding of the dynamics of island biology to look at how these ideas would apply to creatures on the continents.

The diversity of all life including birds on islands received extensive study—beginning with Alfred Russel Wallace and Charles Darwin—because islands present unique habitats for the adaptation of birds and other vertebrates. Furthermore, from the standpoint of colonization, islands come in two different, basic types:

1. Land bridge islands were formally connected to a larger land mass at the time of shallower seas, and were initially populated by the original continental species.
2. Remote islands, such as those formed by volcanic eruptions or coral atolls, were established absent of previous life forms. In this case, the establishment of plants and animals is dependent on wind and ocean currents.

The resulting biodiversity differs somewhat in these two cases. Another complication is that smaller islands support fewer species of birds and mammal.

Nevertheless, over long time periods, the biodiversity of island communities reaches an equilibrium of the number of species that can coexist. Unusual species of birds, such as the dodo, moa, flightless rails, ducks, and geese, have evolved on islands that lack mammalian predators (Quammen 1996). The human colonization of islands has unquestionably led to the loss of many of these specialized species that cannot survive due to habitat loss, hunting pressures, and introduced foreign species.

On every continent, the exploitation of land by agriculture, urbanization, and commercial interests has led to significant habitat fragmentation, which in turn has resulted in island communities of birds. The same process that has taken place on oceanic islands is now happening on the continents, and we can see the impact on species that depend on large continuous environments, as well as on rare or uncommon birds. As an example, the antbirds of the tropics are forest-dwelling species that do not cross cleared boundaries, so the clear cutting of forests creates separated zones, limiting the birds' movement and curtailing the exchange of natural populations. (This limitation is also common in many mammals.)

Uncommon grassland species in the eastern United States, such as the Henslow's Sparrow and the Upland Sandpiper, have been lost in many of their former habitats because of fragmentation, causing the population density to fall below some threshold within these land parcels. When the available habitat becomes too small to sustain a population, the species is destined to die out in that fragment of its original habitat, creating what biologists call a population sink. When the habitat enables a sustained population, it is referred to as population source (Vickery and Herkert 2001).

It is not likely that humanity will be able to turn back the clock on the land fragmentation. The difficult question is this: How can we save the species that we have in our heavily altered world? Many conservation biologists have worked on this problem, starting with the concept of a minimum viable population size (Quammen 1996), an estimated number of individual birds of one species required for that species to survive over a specific time period. These scientists proposed a process of modeling these habitats and their specific organisms, with the goal of determining if a species can survive in a more limited environment for an extended period of time. Their analysis also provides recommendations on how to manage the habitat for the support of the species. This process of population viability analysis (Beissinger and McCullough 2002) requires substantial data input, and has been used for policy decisions that impact significant environmental changes. Using simulation, it can also provide direction for reducing threats to species of concern. This important tool provides an understanding of the changed environment's impact on diversity, and how a particular species can survive in the fragmented habitats we have created.

24

Can We Support
Environmental Conservation?

Global population continues to grow, and requires continued increase in the use of land and sea resources. Scientists have tried to model population growth with global resource usage, but there is no simple relationship between population size and environmental change (Hunter 2000).

Ultimately, population growth is limited by available land and fresh water. Various models of the carrying capacity of the earth range from a high of 1,024 billion people to a low of two billion people, but most studies estimate a limit of eight billion people (Dovers and Butler 2015). These models assume different standards of living, and the lowest estimate is based on a higher standard and steady-state use of resources. The models are also based on the assumption that there will be no catastrophic environmental changes that might affect food availability.

Human population growth has been slowing in the past decade, but it is still above one percent annually. Today the earth's population is growing at the rate of about 80 million people per year. Continued population growth will put more pressure on natural habitats. As we have discussed in this book, natural habitats are self-organized by nature and include a wide diversity of life, while unnatural habitats are created by humanity, such as agricultural fields, urban development, industrial complexes, landfills, open-pit mines and many others. An important emphasis in conservation is saving natural habitats. Federal and state governments, local municipalities, and individual citizens can all play an important role in wildlife conservation to help save the extant species.

Conservation of Wildlife on Federal Government Land

The U.S. government owns 27 percent of the land in the United States, and four government agencies manage 90 percent of this land. They control the largest portion of land that supports conservation. The four agencies that manage the majority of federal land are the Bureau of Land Management (BLM), U.S. Forest Service (USFS), National Park Service (NPS), and U.S. Fish and Wildlife Service (FWS).

The vast majority of Federal land—72 percent—falls under two agencies, BLM

(40 percent) and USFS (32 percent). The majority of federal land is located in Alaska (56 percent) and the western United States (36 percent), leaving only 18 percent over the remainder of the country. Only five percent of federal land is east of the Mississippi River, and this includes all land devoted to government buildings, such as the properties in Washington, D.C. Many states have very little federal land, particularly those in the Midwest. The six states forming the north-south corridor including North and South Dakota, Nebraska, Kansas, Oklahoma, and Texas contain only 1.4 percent of all federal land. This corridor is extremely important because it is the heart of the central bird migratory route, as well a critical breeding area.

The four agencies that manage federal land all have conservation goals in their charters. BLM is mostly focused on the use of land for energy development, cattle grazing, timber harvesting, and mining, so its conservation activities are mainly directed at maintaining, improving, or restoring the productivity of land. Its Threatened and Endangered Species Program (T&E Program) is directed at conservation and recovery of federally listed species. Since the T&E Program's inception in 2010 and through 2018, the department has awarded about $1 million per year in grants to support the actions of various organizations toward this goal directed at 71 species, many of which were birds. This agency has an annual budget of over $1 billion annually, so this conservation program represent a tiny portion of its resources.

The U.S. Fish and Wildlife Service is responsible for the National Wildlife Refuge System (NWRS), managing 89 million acres; 87 percent of this land is in Alaska. It also manages the Waterfowl Protection Areas within its own lands, as well as 0.3 million acres of land easements on private land. Submerged lands, some of which are marine wildlife refuges, are part of its responsibility as well. These assets are among the most important for both habitats and wildlife. The primary focus of most of FWS wildlife conservation is towards waterfowl, but the NWRS is one of the most important agencies in the effort to protect wildlife. The main challenge this agency faces is underfunding, though the National Wildlife Refuge Association, a non-profit, non-government group, tries to support the NWRS by lobbying for funding, collecting donations, and supporting the refuge system with legal actions against development on NWRS lands. Frankly, for those of us who have spent a lot of time in national wildlife refuges, it is apparent that they are poorly supported to meet their wildlife management goals, primarily because they lack sufficient labor resources.

The U.S. Forest Service (USFS) manages 193 million acres of land in the United States. Eighty-seven percent of this land is west of the Mississippi River and of that, twelve percent is in Alaska. USFS's primary objectives are timber and water resources management. Mining is sanctioned by government statuses and administered by leases on USFS and BLM lands. Environmentalists and the federal government have been in a constant battle over logging practices for many decades. A variety of laws have challenged the government's leniency in the harvesting of forest products, since forest management and wildlife conservation often have conflicting goals. The agency also has objectives and land set aside for public recreation and for its conservation initiatives, which are focused on grasslands, forest, and water management. The Collaborative Forest Landscape Restoration Program of the USFS was initiated in 2009 to restore priority forest landscapes.

The USFS has goals to restore land affected by forest fires, flooding, invasive species and other threats. Changing climate has caused a two-decade-long dry spell to plague western North America, which has led to increased western wildfires. The USFS has had to spend more than half of its budget on fighting forest fires. This agency does not have sufficient resources to support fire management practices throughout their lands, so their budget for wildlife and fisheries management has been decreased for a number of years as resources are reallocated to firefighting. USFS spending for wildlife and fisheries dropped by 32 percent from 2001 to 2019, while spending for forest fire suppression more than doubled over this period.

In some cases, USFS restoration programs are specifically focused on wildlife, possibly in collaboration with a university or other participating groups. In most of the USFS programs, protection of specific species is a part of their charter, but not the exclusive focus.

The land most directly focused on conservation is that of the National Park Service (NPS). Two-thirds of NPS land is in Alaska, and this land is always under pressure from commercial interests. As in the case of the NWRS, the national parks provide an important role in habitat protection and wildlife conservation. The Wildlife Conservation Branch of the NPS provides technical assistance to manage parks, and it conducts some activities in species restoration. Its biological resources are focused on wildlife health, conservation, and landscape restoration.

The many users of federal government land have conflicting goals. The majority of the government's conservation goals are directed at agriculture, and they provide significant subsidies to that industry. Some goals also benefit wildlife and natural habitats, a few directly and most indirectly through agricultural conservation. The management of these land assets is critical to support the continued existence of all species of animals and plants.

The Endangered Species Act of 1975 brought attention to the need to save threatened and endangered species from extinction. The problem of focusing solely on these two groups of species, however, is that the populations of most bird and animal species are now decreasing due to loss of habitats, even though they may not be listed as endangered or threatened on government lists. It would be of great benefit if the focus of federal efforts could also embrace the need to save habitats for these not-yet-endangered species. Studies need to be done on the best methods to achieve this goal, and practices need to be implemented to accomplish it.

Although the four agencies of the government have conservation and environmental quality goals in their charters, in practice the overall direction of each of these agencies is easily changed by the administration currently in power. Western states in particular have been battling for control of federal public lands because their populations have increased so drastically. We need to resist the political pressures to revise the interpretation of laws to undermine conservation efforts in order to support economic goals.

Federal Programs Directed at Conservation of Natural Habitats on Private Land

In the eastern and central parts of the country, including the Great Plains states of the central flyway west of the Mississippi River, most of the land is privately owned—in fact, just 11.7 percent of these 37 states east of the Great Plains are public land. Much of this region is devoted to agriculture, industry, or urban life. Conservation measures in these parts of the country are mostly voluntary unless they fall under some state or federal regulation, or are on state-owned property. In the United States, some groups have strong sentiment against government control through regulations, especially if they have a direct impact on any individual, group, or business—and every regulation will impact some of these categories, creating even greater resistance to doing what is necessary to protect habitat and environment. Yet if we did not have government regulations, we would not have clean water, clean air, and so many of the protections on which we rely. It is a difficult challenge to protect the natural environment when there is strong resistance at both a local and national level.

In order to encourage conservation of private agricultural lands, the United States Department of Agriculture created the Natural Resource Conservation Service (NRCS) in 1933 to try and reduce land degradation. Formation of the NRCS came as a direct response to the "dust bowl" drought of the 1930s, during which crops, livestock, and people died in dust storms from Texas to Nebraska. This agency is chiefly focused on the agricultural benefits of land conservation. Within NRCS, there are a number of programs that mostly support agriculture, but they can also support conservation of natural habitats.

The NRCS has about 40 different programs. The Voluntary Public Access and Habitat Incentive Program (VPA-HIP), for example, provides state and tribal governments with funding to expand or improve habitat for public access, for recreational purposes including hunting and fishing, and for other wildlife-dependent recreation. The Environmental Quality Incentives Program (EQIP) is a voluntary conservation program that provides incentive contracts to support agricultural producers, to promote agricultural production and environmental quality. This program also targets wildlife habitat enhancement in and around working farms.

The Regional Conservation Partnership Program (RCPP) promotes the coordination of NRCS conservation activities with partners to expand their ability to address farm, watershed, and regional natural resource concerns. One of this plan's major projects is the restoration of the Chesapeake Bay area.

There have been many publications on the restoration of wetlands, an important subject in environmental science. Wetlands that were destroyed for commercial or residential use are not likely to be restored, but land used for agriculture can be returned to its natural state. The Agricultural Conservation Easement Program (ACEP) and the Wetlands Reserve Program (WRP) of the NRCS have supported wetland restorations of former pasture wetlands (Kerzman 2020). These restorations provide benefits to wildlife and flood control.

The Conservation Stewardship Program (CSP) of the NRCS is a cost-sharing program with producers for land conservation, but it also has a program to enable wildlife habitat conservation. One of the agency's funding sources, EQIP, covers improvements or creation of wildlife habitats. The NRCS has defined approximately 150 conservation practice standards. Most of these standards apply to the management of land and benefit wildlife indirectly by controlling soil erosion, improving water quality, optimizing fertilizer usage, and controlling grazing. Some of these practices benefit wildlife directly through habitat creation, restoration, and management with other partners.

The Conservation Reserve Program (CRP) is a contract program of the Farm Service Agency, focused on taking highly erodible or other environmentally sensitive cropland out of production for 10 to 15 years. This is an important program for conserving grassland habitats, 85 percent of which are privately owned in the United States. The CRP is intended to reduce soil erosion, protect the ability to produce food, improve water quality, establish wildlife habitats, provide food and nesting cover for birds, and enhance forest and wetland resources. It encourages farmers to use vegetative cover to keep soil from washing away, using cultivated (tame) or native grasses, wildlife plantings, trees, filter strips, or riparian buffers.

CRP has conducted many studies directed at grassland species in collaboration with universities and other partners (Vickery and Herker 2001). Most of these studies have been in the shortgrass prairies region from Nebraska to Texas, where 6.1 million acres have been enrolled in this program. Studies have also been carried out in some eastern states. These studies have examined the effectiveness of grassland management in preserving habitat for grassland species. Recent studies of CRP grass retention in the Great Plains show that even after contract expiration, 62 percent of producers were likely to keep the expired contract land in grass rather than convert it back to crop production (Barnes *et al.* 2021).

The NRCS also works with landowners to plan, fund, and implement conservation practices that result in diverse and healthy forests, including young forest habitat needed by many kinds of wildlife. Analysis of the young forest habitat conservation programs in the eastern states show a high level of interest in re-enrollment (Lutter *et al.* 2019).

The Conservation Effects Assessment Project (CEAP) quantifies the environmental results achieved by CSP and CRP funding recipients, as well as other USDA programs. The NRCS has also worked with science partners on studies to improve the understanding of habitat needs. In some cases, these studies are directed at natural habitats, like the environment needed for shrubland birds.

Using a similar approach to the Conservation Stewardship Program, The Bobolink Project in Vermont and Massachusetts, started in 2007 in Rhode Island by Drs. Stephen Swallow and Allan Strong, has demonstrated considerable success. This program was initially supported by EQIP with five-year contracts for participants, and has used donated funds to provide financial assistance to participating farmers. The program has been very successful in helping grassland birds by having farmers modify their mowing schedules so that nesting birds can successfully raise their young. An important part of this program is that it monitors results on a species basis, so that the effectiveness and value of this approach can be confirmed annually. The Bobolink Project differs from the NRCS programs in that it solicits the landowner directly, whereas the NRCS programs must be independently initiated by the land owner.

All of these programs are based on the strategy of renting private land for the purpose of supporting wildlife though habitat conservation. One of the difficulties with some of these programs is that producers might not be able or willing to continue when the contracts expire. The delayed harvest of grassland required by the Bobolink Project poses this problem, because it impacts hay production, pushing farmers to purchase some of the hay they need to feed livestock—so farmers need annual financial assistance to continue, and they may abandon the program if this funding dries up. In another case, when the price of corn increased due to the demand for ethanol production, some contracts of NRCS programs were terminated by farmers interested in profiting from this corn boom. This also happened with increases in the value of soybeans and other grains.

State Government Support for Wildlife

Most of the land under direct federal government control is in the western states and Alaska. Protection of land east of the Mississippi River and in the six central flyway states relies on efforts of state government and private citizens.

State governments play an important role for environmental management of their protected lands. There are 6,600 state parks in the United States, and these serve many functions—but they are primarily intended for the recreational interests. Some were established for historical reasons, while others protect natural areas and preserve a place of scenic beauty. Some state parks are large enough to provide protection of important habitats as well.

In addition to parks, states manage other types of lands. Most common among these are wildlife management areas that are open to hunting and fishing. Other protected lands include state forests and state recreational areas. All of these provide some benefit for natural habitat conservation, but in many cases, these assets suffer from a lack of permanent qualified personnel and inadequate budgets.

States annually report on total spending for environmental measures. This spending includes many activities, so only some of it is directed at environmental conservation—but it does support the parks and other managed land areas. Some of this spending likely benefits wildlife conservation. If we normalize the environmental budget for all the states by the area of the state, we find a dramatic difference in the amount of money spent on this goal. The most recent data available is for year 2015, and it tells us that Alaska spends the least of all states on conservation, but this is misleading: much of its land is owned by the U.S. government. Unfortunately, some states underfund their parks, forests, and wildlife management areas. The states that spend the least on conservation management are Oklahoma, Nebraska, South Dakota, Montana, New Mexico, North Dakota, Kansas, and Arkansas. These spent as little as $200 per square mile on conservation in 2015.

Small states like Delaware, Rhode Island, and Maryland spend much more than the average on conservation. The states that spend the most are California, New York, and South Carolina, where spending exceeds $20,000 per square mile. The problem with this measure, however, is that states may include many different activities under the title of conservation.

The Endangered Species Act gave broad responsibility to the U.S. Fish and Wildlife Service to study and recommend protective measures for rare and endangered species. Under the United States Constitution, the states have primary responsibility for the protection of their wildlife, but in most states, wildlife and protected habitats are subordinate to industrial and agricultural goals. States can help to protect the natural environments by being active in setting aside land that is environmentally important or sensitive, protecting endangered species, and managing environmental problems. They can also work with the NRCS to initiate programs to improve habitats. It is unfortunate that in recent years, many states have not supported environmental initiatives, and have removed protections that were already in place. As citizens, we can petition our elected representatives to support the environment and wildlife protection.

In the United States, total agricultural land also decreased since 1950 because of a decrease in forestland grazing. Cropland and pasturelands have also decreased, reaching their lowest point in 2007, but these have increased since then. Sixty-two percent of the total land in the lower 48 states is devoted to agriculture, and this includes forest grazing land, grassland pastures, and cropland.

Agricultural practices have changed over the past seventy years as well. Farms have gotten larger, smaller farms have decreased in numbers, and productivity has improved dramatically. Agricultural landscapes have been simplified, but at the loss of diversified habitats within farms (Spangler *et al* 2020). Farming practices have become more intense with the use of pesticides and multiple-cropping. These practices have eliminated fallow lands and hedgerows, and have had a negative effect on all insectivorous birds.

Private land use for urbanization in the United States in the lower 48 states is about 3.7 percent of this area, or 70 million acres in 2012 (Bigelow and Borchers 2017). It is increasing at a high rate and is expected to reach 166 million acres by 2050 (Nowak and Walton 2005).

The only way we can hope to stem the loss of wildlife is to find better ways to use private land to protect important habitats. The basic problem with private land is that landowners almost universally do not embrace the concept of conservation on private property. These properties are viewed as a means to development and economic expansion. Landowner incentive programs are temporary, short-lived, and most are loosely administered. The only real opportunity for conservation on private land is through acquisition by state agencies or land trusts. Properties that are marginal or degraded might be acquired for less cost. Farmland that includes important habitats such as scrub forests or riparian habitats might be more costly to acquire.

The Role of Municipal Governments and Corporations in Wildlife Conservation

In the continued expansion of urban settlements, residents have lobbied municipal governments for more green space. Most park land under municipal control is developed for recreational purposes. Zoning boards provide some standards in the use of private land for private and corporate development, but they are directed more toward the developed environment than habitat preservation. The goals of habitat protection and public recreation can both be achieved in an urban setting, however. Natural habitats on public land serves many purposes for public recreation, like hiking, photography, birding, and other outdoor activities.

Many researchers who study urban ecosystems have provided a blueprint for managing and preserving habitats in cities. Most municipal sites devoted to nature are fragmented or degraded parcels. Mark McDonnell and Amy Hahs (2014) outline four management actions for cities: link management actions with ecological knowledge, protect existing natural habitats, restore degraded habitats, and integrate remnant patches into the urban landscape. Trees and beneficial plants, for example, can provide a link between fragmented parcels.

For municipalities, the protection of land as a resource devoted to nature is dependent on the outright acquisition of land, the purchase of development rights, or the establishment of conservation easements. In some cases, spending public resources on land acquisition would be objectionable to some portion of the public, but the other approaches would be more acceptable, such as partnerships with land trusts.

It is fortunate that some non-profit groups protect important natural habitats. The Nature Conservancy has protected about 195,000 sq miles (506,000 sq km) in the United States through ownership or conservation easements. Other smaller, localized land trusts take a very active role in land protection. Since so much of the land in the United States is in private hands, it is very important to seek public support for these land trusts.

Some corporations have played a leading role in the conservation effort. A few corporations with land holdings have converted some land parcels into wildlife habitats. In the United States, the Wildlife Habitat Council reviews and carries out certification of these sites.

Another important conservation measure in large cities is to turn off or block the lights of large buildings at night during migration season. Research in many large cities has shown that on a global scale, hundreds of millions of birds are killed annually by collisions with buildings (Loss *et al.* 2014). Countermeasures have proved to be very effective in reducing the mortality rate (Hack 2020).

Role of Private Citizens in Conservation

Other causes of bird mortality have been studied besides habitat loss. Private citizens can be active with these other causes in order to help birds in particular, and wildlife in general.

On its website, the Cornell Laboratory of Ornithology lists seven actions everyone can take to help protect birds. These include:

- Making windows safer by breaking up reflections
- Keeping cats indoors
- Reducing cultivated lawns and planting native vegetation that is attractive to birds
- Avoiding pesticides
- Purchasing shade-grown coffee
- Recycling or depositing unwanted plastic in proper containers
- Being supportive of measures to protect the environment.

One of the simplest measures that the public can support is to keep house cats in the house and eliminate feral cats. Globally, house cats are blamed for the extinction of many species (Morelle 2013). House cats are just one of many introduced predators, but the 2019 "3 Billion Birds" study found domestic cats to be one of the primary reasons for bird fatalities across the continent (Rosenberg *et al*, 2019). The expansion of these and other predator numbers are significant sources of loss.

The most crucial measure that we should all support is the management and conservation of habitats. We have discussed many different kinds of habitat in this book, along with their resident species, but there are also many smaller, more complex habitats within these broad environments. Microhabitats including the canopy, shrub layer, leaf litter, native plant varieties, and more are important to supporting a diversity of animals and plants. The quality of these smaller environments is extremely important for conservation. Destroying these microhabitats reduces the diversity of ecosystems, and they take many years to recover. Volunteers can assist at a public site to manage and preserve microhabitats by removing invasive plants, clearing trails, and planting native plants that birds and other wildlife need to survive.

Conservation of migrating birds requires a network of stopover sites along their routes to provide feeding habitats. Many stopover sites bordering water have already been developed for human enjoyment, and those remaining are always under threat of commercial or residential development. Private owners are rarely willing to protect these locations. Birds use many types of sites in migration, including parks, woodlots, fields, marshes and coastal areas; these are on both public and private land. Private land owners can help by maintaining and improving these to support use by birds.

Pollution of the oceans by plastics and chemicals is a well-recognized problem. People can be active in recycling, but it is much more important to limit the use of disposable plastic containers as much as possible and to dispose of used fishing line in proper receptacles.

Corporations pay attention to consumer demands, so people can have a big impact if they avoid buying products from companies that are known polluters. Instead, buy products from companies that are trying to have a positive impact on the environment. We can advocate for reduced lighting of high-rise buildings at night, and for the placement of wind farms that will pose minimal threats to birds.

Private citizens can provide donations and volunteer support for many organizations that do conservation work, such as land trusts that carry out conservation, and institutions such as zoos and aquariums that lead outreach conservation projects.

We can also make an important contribution to habitat restoration by creating natural habitats in our back yards, simply by planting native species that benefit wildlife. Lawns of mowed grass are among the worst things for bird populations, and they consist entirely of non-native monocultures that provide no food or shelter for birds. A great deal of cultivated suburban lawns mostly serve aesthetic purposes; even the homeowners don't use them. As of 2005, Americans have converted 63,000 square miles (163,000 sq km) into lawns (D'Costa 2017), an area roughly the size of Missouri. In his book *Nature's Best Hope: A New Approach to Conservation That Starts in Your Yard,* Doug Tallamy (2020) provides guidelines for selecting native plants to support declining wildlife.

Conservation of Migratory Birds

The majority of bird species on earth are found in the temperate and tropical regions. Most of the temperate and tropical lands are in third world countries with high poverty rates, and conservation, forest preservation, and preservation of birds are low priority. On a worldwide basis, half of the world's habitable land is used for agriculture, and agricultural land areas are increasing in Asia, Africa, India, Oceania, the Middle East, and Brazil (Ritchie and Roser 2019). The rate of deforestation has decreased in the first decade of this century, but forest land is still being lost at the rate of 12.9 million acres per year (5.2 million ha/yr) (Adams 2012).

In the United States, more than 90 percent of the total cumulative bird losses can be attributed to 12 migratory bird families including sparrows, warblers, blackbirds, finches, larks, and aerial insectivores (swallows, nighthawks, and swifts). Species overwintering in temperate regions experienced the largest net reduction in abundance, but proportional loss was greatest among migratory species overwintering in coastal regions, southwestern arid lands, and South America (Rosenberg *et al.* 2019).

Long-distance migratory birds face additional threats beyond non-migratory species. Challenges faced by bird migration have played a role in the loss of species in the United States, and this has also occurred in the long distance migrants from Europe and Asia to Africa, as well as those species that migrate to and from Australia. Saving habitat in foreign countries as well as our own is important if we are going to stem the loss of migratory birds.

The 1916 Migratory Bird Treaty between Canada and the United States has played an important role in the conservation of birds. In Europe, the European Union's Birds and Habitats Directives enabled member states to work together to protect the EU's most vulnerable species and habitat types, but in recent times, these statutes have fallen prey to political attack. There are no comparable treaties for South America, Asia, Africa, and the Pacific flyways.

Many organizations have active programs directed at saving forests and habitats in the United States and around the world. BirdLife International, the African Wildlife Foundation, Ocean Conservancy, and World Wildlife Fund are just a few of many organizations that rely on private funds for habitat conservation around the world. The Audubon Society works with BirdLife International in North and South America to achieve this goal. Foundations that are working to save tropical forests include: Coalition for Rain Forest Nations, Rainforest Alliance, Rainforest Foundation, Saving Nature, The Nature Conservancy, and others.

Our Role in the Future of Birds

Based on their artwork and legends, we know that Aboriginal tribes had a respect for the animals they hunted. They did not destroy the balance of nature because their population never expanded enough to upset the natural equilibrium, nor did they have the technical means to exploit nature that came with the industrial revolution. Since those days, however, humanity has predominately viewed the natural world as a resource solely for human consumption, and we have expanded our population well beyond the point where we can live in balance with nature. Our concept of morality is homocentric (anthropocentric) rather than ecocentric.

The preservation of nature, including such topics as animal welfare, preservation of biodiversity, integrity of ecosystems, and threats of extinction, are always contentious when they limit human needs and interests. There will always be a conflict between the interests of humans and nature. The only way we can resolve these conflicts is by finding a balance through a process of negotiation between those who control resources and those who want to preserve natural resources. It would be unfortunate if the only way to achieve this balance is through legal means.

This problem of saving habitats is a worldwide issue. We cannot hope to save wildlife and natural habitats without a supportive effort of landowners and conservation organizations. It is only through the good will of people that we can hope to conserve habitats and reduce the rate of loss of species.

Many of the solutions to protect nature mentioned in this chapter have been known for some time. The difficulty we face is that there are too few people that actively support the goals of protection of nature and the environment. The most important issue is to get more people involved and advocating for wildlife protection.

Throughout this book, the emphasis has been on preservation of habitats. Scientists have warned us that the future outlook for the diversity of birds is threatened by the double-edged-sword of habitat loss and global warming (Thomas 2011). Global warming presents other, more complex threats, because there are limits to how rapidly bird species can adjust to climate change.

Birds have been in demise for many years, even centuries, but research and action have shown that we can save the diversity of birds remaining on the planet. It is up to us to make the effort to embrace this goal and try to stem the ever-expanding human dependence on the planet's natural environments. The earth is a limited resource, and we need to be more deliberate about providing for the needs of other life forms beside our own.

There was a time when the world opened to birds, and they diversified over a long history. Today, there is no doubt that they evolved from dinosaurs and survived the great extinction event at the end of the Cretaceous. They left behind a history of fossils and bones that trace their past. When the world opened to them at the end of the Cretaceous, their diversity radiated in many directions eventually leading to the birds we are familiar with today. Over this long history of millions of years, the world was their world. But today, they are guests in our world. They have been in demise and decreasing in numbers for many years because we have denied them the habitats and the environment they need for survival. There is an abundance of research that documents the demise of birds. We cannot predict the future, but the outlook is that they will continue to face losses in their numbers unless we are more deliberate in trying to save them and the habitats they need. It is up to us to make the effort to embrace this goal and try to stem the ever-expanding human expansion into the planet's natural environments. The earth is a limited resource, and we need to be more deliberate about providing for the needs of other life forms beside our own.

Birds are a source of beauty, joy, and fascination for many people. They have inspired generations of people because they are easy to see, easy to appreciate, and inspire a sense of wonder. They greatly add to the richness of life on earth. It would be a tragic loss to drive many of these magnificent creatures to extinction simply because we cannot find a place for them in our world. By developing an appreciation for the richness they add to our lives, we might convince some people to find ways to support measures that would allow us to save what we have left.

Appendix 1
Index of Latin Bird Names

Species	Scientific Name
Akiapola'au	*Hemignathus wilsoni*
Akialoa, Kauai	*Akialoa stejnegeri*
Akepa	*Loxops coccineus*
Albatross, Black-browed	*Thalassarche melanophris*
Albatross, Laysan	*Phoebastria immutabilis*
Albatross, Royal	*Diomedea epomophora*
Albatross, Wandering	*Diomedea exulans*
Amakihi, Hawai'i	*Chlorodrepanis virens*
Antpitta, Streak-chested	*Hylopezus perspicillatus*
Antshrike, Black-crested	*Sakesphorus canadensis*
Archaeopteryx	*Archaeopteryx lithographica*
Bateleur	*Terathopius ecaudatus*
Bay-Owl, Congo	*Phodilus prigoginei*
Bentbill, Southern	*Oncostoma olivaceum*
Bird, Terror	*Kelenken guillermoi*
Bittern, American	*Botaurus lentiginosus*
Bittern, Australasian	*Botaurus poiciloptilus*
Bittern, Black-backed	*Ixobrychus dubius*
Bittern, Dwarf	*Ixobrychus sturmii*
Bittern, Great	*Botaurus stellaris*
Bittern, Least	*Ixobrychus exilis*
Bittern, Little	*Ixobrychus minutus*
Bittern, Pinnated	*Botaurus pinnatus*
Bittern, Stripe-backed	*Ixobrychus involucris*
Black Hawk, Common	*Buteogallus anthracinus*
Black Hawk, Cuban	*Buteogallus gundlachii*
Black Hawk, Great	*Buteogallus urubitinga*
Blackbird, Red-winged	*Agelaius phoeniceus*
Blackcap, Eurasian	*Sylvia atricapilla*
Bluethroat	*Luscinia svecica*
Bobolink	*Dolichonyx oryzivorus*
Boobook, Brown	*Ninox scutulata*

Species	Scientific Name
Boobook, Least	*Ninox sumbaensis*
Boobook, Southern	*Ninox boobook*
Booby, Nazca	*Sula granti*
Budgerigar	*Melopsittacus undulatus*
Bufflehead	*Bucephala albeola*
Bullfinch, Eurasian	*Pyrrhula pyrrhula*
Bunting, Snow	*Plectrophenax nivalis*
Bustard, Great	*Otis tarda*
Calyptura, Kinglet	*Calyptura cristata*
Canvasback	*Aythya valisineria*
Capuchinbird (Calfbird)	*Perissocephalus tricolor*
Caracara, Chimango	*Milvago chimango*
Caracara, Crested	*Caracara cheriway*
Caracara, Mountain	*Phalcoboenus megalopterus*
Caracara, Red-throated	*Ibycter americanus*
Caracara, Yellow-headed	*Milvago chimachima*
Cardinal, Northern	*Cardinalis cardinalis*
Catbird, Gray	*Dumetella carolinensis*
Cassowary, Southern	*Casuarius casuarius*
Chat, Gray-throated	*Granatellus sallaei*
Chat, Yellow-breasted	*Icteria virens*
Chickadee, Black-capped	*Poecile atricapillus*
Chickadee, Carolina	*Poecile carolinensis*
Cock-of-the-Rock, Andean	*Rupicola peruvianus*
Condor, California	*Gymnogyps californianus*
Coot, Horned	*Fulica cornuta*
Cormorant, Double-crested	*Phalacrocorax auritus*
Cormorant, Great	*Phalacrocorax carbo*
Cormorant, Guanay	*Leucocarbo bougainvillii*
Cotinga, Lovely	*Cotinga amabilis*
Crake, Corn	*Crex crex*
Crake, Dot-winged	*Porzana spiloptera*
Crake, Yellow-breasted	*Hapalocrex flaviventer*
Crane, Terror	*Gastornis gigantea*
Crane, Whooping	*Grus americana*
Creeper, Brown	*Certhia americana*
Creeper, Hawai'i	*Loxops mana*
Crossbill, White-winged	*Loxia leucoptera*
Crow, American	*Corvus brachyrhynchos*
Crow, New Caledonian	*Corvus moneduloides*
Cuckoo, Yellow-billed	*Coccyzus americanus*
Curlew, Eskimo	*Numenius borealis*
Curlew, Bristle-thighed	*Numenius tahitiensis*
Curlew, Far Eastern	*Numenius madagascariensis*
Curlew, Long-billed	*Numenius americanus*

Species	*Scientific Name*
Flowerpiercer, White-sided	*Diglossa albilatera*
Flycatcher, Acadian	*Empidonax virescens*
Flycatcher, Boat-billed	*Megarynchus pitangua*
Flycatcher, European Pied	*Ficedula hypoleuca*
Flycatcher, Fork-tailed	*Tyrannus savana*
Flycatcher, Piratic	*Legatus leucophaius*
Flycatcher, Royal	*Onychorhynchus coronatus*
Flycatcher, Scissor-tailed	*Tyrannus forficatus*
Flycatcher, Sulphur-rumped	*Ficedula superciliaris*
Flycatcher, Vermillion	*Pyrocephalus rubinus*
Flycatcher, Yellow-bellied	*Empidonax flaviventris*
Forest-falcon, Collared	*Micrastur semitorquatus*
Fulmar, Northern	*Fulmarus glacialis*
Gadwall	*Mareca strepera*
Gallinule, Purple	*Porphyrio martinica*
Gannet, Cape	*Morus capensis*
Gannet, Northern	*Morus bassanus*
Godwit, Bar-tailed	*Limosa lapponica baueri*
Godwit, Marbled	*Limosa fedoa*
Goldeneye, Barrow's	*Bucephala islandica*
Goose, Canada	*Branta canadensis*
Goose, Egyptian	*Alopochen aegyptiaca*
Goose, Hawaiian (Nene)	*Branta sandvicensis*
Goose, Magpie	*Anseranas semipalmata*
Goose, Snow	*Anser caerulescens*
Goose, Spur-winged	*Plectropterus gambensis*
Goshawk, Chestnut-shouldered	*Erythrotriorchis buergersi*
Goshawk, Doria's	*Megatriorchis doriae*
Goshawk, Northern	*Accipiter gentilis*
Goshawk, Red	*Erythrotriorchis radiatus*
Grackle Common	*Quiscalus quiscula*
Grebe, Eared	*Podiceps nigricollis*
Grebe, Great	*Podiceps major*
Grebe, Horned	*Podiceps auritus*
Grebe, Little	*Tachybaptus ruficollis*
Grebe, Red-necked	*Podiceps grisegena*
Greenshank, Common	*Tringa nebularia*
Griffon, Ruppell's	*Gyps rueppelli*
Grosbeak, Evening	*Hesperiphona vespertina*
Grosbeak, Kona	*Chloridops kona*
Ground-Finch, Española	*Geospiza conirostris*
Ground-Hornbill, Southern	*Bucorvus leadbeateri*
Grouse, Ruffed	*Bonasa umbellus*
Gull, Black-headed	*Chroicocephalus ridibundus*
Gull, Bonaparte's	*Chroicocephalus philadelphia*

Species	*Scientific Name*
Hummingbird, Giant	*Patagona gigas*
Hummingbird, Rivoli's	*Eugenes fulgens*
Hummingbird, Ruby-throated	*Archilochus colubris*
Hummingbird, Rufous	*Selasphorus rufus*
Hummingbird, Sword-billed	*Ensifera ensifera*
Ifrita, Blue-capped	*Ifrita kowaldi*
'I'iwi	*Drepanis coccinea*
Jacobin, Black	*Florisuga fusca*
Jaeger, Parasitic	*Stercorarius parasiticus*
Jay, Blue	*Cyanocitta cristata*
Jay, Green	*Cyanocorax yncas*
Kagu	*Rhynochetos jubatus*
Kakapo	*Strigops habroptila*
Kaka, New Zealand	*Nestor meridionalis*
Kestrel, American	*Falco sparverius*
Kingbird, Eastern	*Tyrannus tyrannus*
Kioea	*Chaetoptila angustipluma*
Kiskadee, Great	*Pitangus sulphuratus*
Kite, Black	*Milvus migrans*
Kite, Black-breasted	*Hamirostra melanosternon*
Kite, Black-shouldered	*Elanus axillaris*
Kite, Double-toothed	*Harpagus bidentatus*
Kite, Hook-billed	*Chondrohierax uncinatus*
Kite, Letter-winged	*Elanus scriptus*
Kite, Mississippi	*Ictinia mississippiensis*
Kite, Red	*Milvus milvus*
Kite, Slender-billed	*Helicolestes hamatus*
Kite, Snail	*Rostrhamus sociabilis*
Kite, Swallow-tailed	*Elanoides forficatus*
Kittiwake, Black-legged	*Rissa tridactyla*
Kittiwake, Red-legged	*Rissa brevirostris*
Kiwi	*Apteryx*
Knot, Red	*Calidris canutus*
Lapwing, Gray-headed	*Vanellus cinereus*
Lapwing, Masked	*Vanellus miles*
Limpkin	*Aramus guarauna*
Longspur, Smith's	*Calcarius pictus*
Loon, Common	*Gavia immer*
Loon, Red-throated	*Gavia stellata*
Lyrebird, Albert's	*Menura alberti*
Lyrebird, Superb	*Menura novaehollandiae*
Magpie, European	*Pica pica*
Magpie-lark	*Grallina cyanoleuca*
Mallard	*Anas platyrhynchos*
Manakin, Club-winged	*Machaeropterus deliciosus*

Species	Scientific Name
Manakin, Golden-headed	*Ceratopipra erythrocephala*
Marsh-Harrier, African	*Circus ranivorus*
Marsh-Harrier, Eurasian	*Circus aeruginosus*
Martin, Purple	*Progne subis*
Meadowlark, Chihuahuan	*Sturnella lilianae*
Meadowlark, Eastern	*Sturnella magna*
Meadowlark, Western	*Sturnella neglecta*
Merganser, Brazilian	*Mergus octosetaceus*
Merganser, Common	*Mergus merganser*
Merganser, Hooded	*Lophodytes cucullatus*
Moa, Giant	*Dinornis robustus*
Moa, Little Bush	*Anomalopteryx didiformis*
Moa, Mantell's	*Pachyornis geranoides*
Moa, Upland	*Megalapteryx didinus*
Mockingbird, Northern	*Mimus polyglottos*
Monjita, White	*Xolmis irupero*
Mountaineer, Bearded	*Oreonympha nobilis*
Murre, Common	*Uria aalge*
Murrelet, Marbled	*Brachyramphus marmoratus*
Nighthawk, Lesser	*Chordeiles acutipennis*
Night-Heron, Black-crowned	*Nycticorax nycticorax*
Nightingale, Common	*Luscinia megarhynchos*
Nightingale Thrush, Orange-billed	*Catharus aurantiirostris*
Niltava, Rufous-bellied	*Niltava sundara*
Nuthatch, Red-breasted	*Sitta canadensis*
Oilbird	*Steatornis caripensis*
Oriole, Baltimore	*Icterus galbula*
Oriole, Bullock's	*Icterus bullockii*
Oriole, Orchard	*Icterus spurius*
Osprey	*Pandion haliaetus*
Ostrich, Common	*Struthio camelus*
Owl, Andros Island Barn	*Tyto pollens*
Owl, Band-bellied	*Pulsatrix melanota*
Owl, Bare-legged	*Margarobyas lawrencii*
Owl, Barn	*Tyto alba*
Owl, Bermuda Saw-whet	*Aegolius gradyi*
Owl, Boreal	*Aegolius funereus*
Owl, Buff-fronted	*Aegolius harrisii*
Owl, Burrowing	*Athene cunicularia*
Owl, Crested	*Lophostrix cristata*
Owl, Cuban Giant	*Ornimegaionyx oteroi*
Owl, Elf	*Micrathene whitneyi*
Owl, Flammulated	*Psiloscops flammeolus*
Owl, Grass	*Tyto longimembris*
Owl, Great Gray	*Strix nebulosa*

Species	Scientific Name
Owl, Great-horned	*Bubo virginianus*
Owl, Laughing	*Ninox albifacies*
Owl, Little	*Athene noctua*
Owl, Long-eared	*Asio otus*
Owl, Marsh	*Asio capensis*
Owl, Northern Hawk	*Surnia ulula*
Owl, Northern Saw-whet	*Aegolius acadicus*
Owl, Powerful	*Ninox strenua*
Owl, Short-eared	*Asio flammeus*
Owl, Snowy	*Bubo scandiacus*
Owl, Stygian	*Asio stygius*
Owl, Unspotted Saw-whet	*Aegolius ridgwayi*
Owl, White-browed	*Athene superciliaris*
Owlet, Forest	*Athene blewitti*
Owlet, Long-whiskered	*Xenoglaux loweryi*
Oxpecker, Red-billed	*Buphagus erythrorhynchus*
Palila	*Loxioides bailleui*
Parakeet, Carolina	*Conuropsis carolinensis*
Parrot, Gray	*Psittacus erithacus*
Parrot, Night	*Pezoporus occidentalis*
Parrotbill, Maui	*Pseudonestor xanthophrys*
Pelican, Brown	*Pelecanus occidentalis*
Pelican, Dalmatian	*Pelecanus crispus*
Pelican, Great White	*Pelecanus onocrotalus*
Pelican, Peruvian	*Pelecanus thagus*
Penguin, Adelie	*Pygoscelis adeliae*
Penguin, Emperor	*Aptenodytes forsteri*
Penguin, Galápagos	*Spheniscus mendiculus*
Penguin, Humboldt	*Spheniscus humboldti*
Penguin, Little	*Eudyptula minor*
Penguin, Macaroni	*Eudyptes chrysolophus*
Penguin, Rockhopper	*Phalacrocorax atriceps*
Petrel, Black-capped	*Pterodroma hasitata*
Petrel, Blue	*Halobaena caerulea*
Petrel, Hawaiian	*Pterodroma sandwichensis*
Petrel, Herald	*Pterodroma heraldica*
Petrel, Snow	*Pagodroma nivea*
Phainopepla	*Phainopepla nitens*
Phalarope, Red-necked	*Phalaropus lobatus*
Phoebe, Black	*Sayornis nigricans*
Phoebe, Eastern	*Sayornis phoebe*
Pigeon, Rock	*Columba livia*
Pinktail, Przevalski's	*Urocynchramus pylzowi*
Pipit, Richard's	*Anthus richardi*
Pitohui, Hooded	*Pitohui dichrous*

Species	*Scientific Name*
Plains-wanderer	*Pedionomus torquatus*
Plover, Banded	*Vanellus tricolor*
Plover, Egyptian	*Pluvianus aegyptius*
Plover, Mountain	*Charadrius montanus*
Plover, Piping	*Charadrius melodus*
Plover, Snowy	*Charadrius nivosus*
Po'ouli (Black-faced Honeycreper)	*Melamprosops phaeosoma*
Prairie-Chicken, Greater	*Tympanuchus cupido*
Prairie-Chicken, Lesser	*Tympanuchus pallidicinctus*
Prion, Slender-billed	*Pachyptila belcheri*
Puffins, Atlantic	*Fratercula arctica*
Pygmy-Owl, Andean	*Glaucidium jardinii*
Pygmy-Owl, Cloud-forest	*Glaucidium nubicola*
Pygmy-Owl, Northern	*Glaucidium californicum*
Pygmy-Tyrant, Short-tailed	*Myiornis ecaudatus*
Quail, Common	*Coturnix coturnix*
Quail, Gambel's	*Callipepla gambelii*
Quelea, Red-billed	*Quelea quelea*
Quetzal, Eared	*Euptilotis neoxenus*
Quetzal, Resplendent	*Pharomachrus mocinno*
Racket-tail, Peruvian	*Ocreatus peruanus*
Rail, Black	*Laterallus jamaicensis*
Rail, Guam	*Gallirallus owstoni*
Rail, Virginia	*Rallus limicola*
Rail, Yellow	*Coturnicops noveboracensis*
Raven, Common	*Corvus corax*
Redpoll, Common	*Acanthis flammea*
Redpoll, Hoary	*Acanthis hornemanni*
Reedling, Bearded	*Panurus biarmicus*
Rhea, Greater	*Rhea americana*
Robin, American	*Turdus migratorius*
Robin, European	*Erithacus rubecula*
Rubythroat, Siberian	*Calliope calliope*
Ruff	*Calidris pugnax*
Rush-Tyrant, Many-colored	*Tachuris rubrigastra*
Sandpiper, Baird's	*Calidris bairdii*
Sandpiper, Buff-breasted	*Calidris subruficollis*
Sandpiper, Green	*Tringa ochropus*
Sandpiper, Least	*Calidris minutilla*
Sandpiper, Pectoral	*Calidris melanotos*
Sandpiper, Semipalmated	*Calidris pusilla*
Sandpiper, Solitary	*Tringa solitaria*
Sandpiper, Spoon-billed	*Calidris pygmaea*
Sandpiper, Spotted	*Actitis macularius*
Sandpiper, Upland	*Bartramia longicauda*

Species	Scientific Name
Sandpiper, White-rumped	*Calidris fuscicollis*
Sandpiper, Wood	*Tringa glareola*
Sapayoa	*Sapayoa aenigma*
Scops-Owl, Eurasian	*Otus scops*
Screech-Owl, Eastern	*Megascops asio*
Screech-Owl, Tropical	*Megascops choliba*
Screech-Owl, Western	*Megascops kennicottii*
Scythebill, Black-billed	*Campylorhamphus falcularius*
Sea-Eagle, Steller's	*Haliaeetus pelagicus*
Sea-Eagle, White-bellied	*Haliaeetus albicilla*
Secretarybird	*Sagittarius serpentarius*
Seedeater, Morelet's	*Sporophila morelleti*
Seedsnipe, Rufous-bellied	*Attagis gayi*
Seriema, Black-legged	*Chunga burmeisteri*
Seriema, Red-legged	*Cariama cristata*
Serpent-Eagle, Crested	*Spilornis cheela*
Shearwater, Audubon's	*Puffinus lherminieri*
Shearwater, Black-vented	*Puffinus opisthomelas*
Shearwater, Cory's	*Calonectris borealis*
Shearwater, Manx	*Puffinus puffinus*
Shearwater, Sooty	*Ardenna grisea*
Shearwater, Streaked	*Calonectris leucomelas*
Shearwater, Tropical	*Puffinus bailloni*
Shoebill	*Balaeniceps rex*
Shoveler, Northern	*Spatula clypeata*
Shrikejay, Crested	*Platylophus galericulatus*
Shrikethrush, Little	*Colluricincla megarhyncha*
Shrike-tit, Crested	*Falcunculus frontatus*
Skua, Brown	*Stercorarius antarcticus*
Skua, Great	*Stercorarius skua*
Skua, South Polar	*Stercorarius maccormicki*
Smew	*Mergellus albellus*
Snake-Eagle, Brown	*Circaetus cinereus*
Snipe, Great	*Gallinago media*
Snipe, Wilson's	*Gallinago delicata*
Sparrow, Baird's	*Ammodramus bairdii*
Sparrow, Clay-colored	*Spizella pallida*
Sparrow, Grasshopper	*Ammodramus savannarum*
Sparrow, Henslow's	*Ammodramus henslowii*
Sparrow, House	*Passer domesticus*
Sparrow, LeConte's	*Ammodramus lesconteii*
Sparrow, Rock	*Petronia petronia*
Sparrow, Saltmarsh	*Ammospiza caudacutus*
Sparrow, Savannah	*Passerculus sandwichensis*
Sparrow, White-crowned	*Zonotrichia leucophrys*

Species	*Scientific Name*
Turkey, Wild	*Meleagris gallopavo*
Turnstone, Ruddy	*Arenaria interpres*
Tyrant, Long-tailed	*Colonia colnus*
Tyrant, Strange-tailed	*Alectrurus risora*
Tyrant, Streamer-tailed	*Gubernetes yetapa*
Umbrellabird, Bare-necked	*Cephalopterus glabricollis*
Upland, Sandpiper	*Bartramia longicauda*
Veery	*Catharus fuscescens*
Vireo, Blue-headed	*Vireo solitarius*
Vireo, Yellow-throated	*Vireo flavifrons*
Vulture, Bearded	*Gypaetus barbatus*
Vulture, Black	*Coragyps atratus*
Vulture, Greater Yellow-headed	*Cathartes melambrotus*
Vulture, Lesser Yellow-headed	*Cathartes burrovianus*
Vulture, Palm-nut	*Gypohierax angolensis*
Vulture, Turkey	*Cathartes aura*
Wallcreeper	*Tichodroma muraria*
Warbler, Blackburnian	*Setophaga fusca*
Warbler, Blackpoll	*Setophaga striata*
Warbler, Hooded	*Setophaga citrina*
Warbler, Kirtland's	*Setophaga kirtlandii*
Warbler, Marsh	*Acrocephalus palustris*
Warbler, Olive	*Peucedramus taeniatus*
Warbler, Oriente	*Teretistris fornsi*
Warbler, Prairie	*Setophaga discolor*
Warbler, Sedge	*Acrocephalus schoenobaenus*
Warbler, Yellow	*Setophaga petechia*
Warbler, Yellow-headed	*Teretistris fernandinae*
Water-Tyrant, Pied	*Fluvicola pica*
Wattlebird, Yellow	*Anthochaera paradoxa*
Waxwing, Bohemian	*Bombycilla garrulus*
Waxwing, Cedar	*Bombycilla cedrorum*
Waxwing, Japanese	*Bombycilla japonica*
Weaver, Village	*Ploceus cucullatus*
Weebill	*Smicornis brevirostris*
Wheatear, Northern	*Oenanthe oenanthe*
Whimbrel	*Numenius phaeopus*
Whip-poor-will, Eastern	*Antrostomus vociferus*
Woodcock, American	*Scolopax minor*
Woodcreeper, Plain-brown	*Dendrocincla fuliginosa*
Woodpecker, Red-cockaded	*Leuconotopicus borealis*
Woodpecker, Red-bellied	*Melanerpes carolinus*
Wood-Rail, Gray-necked	*Aramides cajaneus*
Wren, House	*Troglodytes aedon*
Wren, Marsh	*Cistothorus palustris*

Appendix 2
Glossary of Terms

aftershaft	an additional feather arising from the shaft in certain birds
air sac	a compartment containing air; in birds, an extension of the lung into the annular spaces within certain bones
allopatric	biological populations of the same species that are geographically isolated
allopreening	to preen or groom the skin or feathers of another bird
altricial birds	birds hatched naked, nest raised, are fed, and require an extended nesting period.
alula	a single feather attached to the thumb of the hand
angle of attack	angle between the chord of an airfoil and the direction of the fluid through which it is moving
anisodactyl	in birds, three toes pointing forward and the fourth pointing backward
anterior	situated in the front of the body
anthropocentric	regarding humankind as the central or most important element of existence
antiphonal song	duets in which the male and female sing alternating songs
Archosaur	a large group of reptiles that includes the dinosaurs, pterosaurs, and crocodilians
associative mating	the tendency to mate with genetically similar individuals.
asynchronous	two events not occurring at the same time
Aves	all living and fossil birds
Avialae	all birds and their extinct ancestors, including all theropod dinosaurs related to birds
batrachotoxin	an extremely potent toxic steroidal alkaloid found in certain species of beetles, birds, and frogs
bifurcate	to cause to divide into two, to branch
biome	a large area that can include a variety of habitats
calamus	the hollow lower part of the shaft of a feather, which lacks barbs; a quill; the unbranched basal section of the rachis
cerebellum	the part of the brain at the back of the skull in vertebrates which regulates muscular activity
clade	group of organisms that evolved from a common ancestor
cladistic tree	the arrangement of organisms in groups based on the most recent common ancestor
cochlea	in birds, a narrow tube that is part of the inner ear involved in hearing

coelurosaurs	a subgroup of theropod dinosaurs most closely related to birds
conspecific	belonging to the same species
coprolite	fossilized feces of prehistoric animals
cortex	the outer layer of the cerebrum composed of folded gray matter that has an important role in consciousness
corticosterone	a hormone secreted by the adrenal cortex
cryptochromes	class of flavoproteins (proteins) that are sensitive to blue light
dihedral	an angle formed by two plane surfaces
dimorphic	adult males and females with different plumages
dorsal	the upper side
DNA	material present in nearly all living organisms as the main constituent of chromosomes, carrying genetic information
ecological niche	the role that a species plays in its environment
Enantiornithes	a large group of extinct birds from the Mesozoic with teeth and clawed fingers on each wing
environment	the surroundings or conditions in which an animal or plant lives or operates
extant	species still in existence (as opposed to extinct species)
femur	in birds, bone of the thigh or upper hind limb which connects to the body skeleton
filoplume	hair-like feathers with a few soft barbs that serve sensory or decorative functions
frugivorous	feeding on fruit
gallinaceous birds	grouse, pheasants, turkeys, etc.
gastrolith	rocks swallowed by some animals and birds stored in the stomach and used to grind food; rock size depends on the size of the species.
genera	plural of genus
genus	a taxonomic category above the rank of species
genome	the complete set of genes or genetic material present in a cell or organism
genomic sequence	the chromosomal DNA
gizzard	muscular enlargement of the digestive tract of birds that assists in grinding of food
gorget	the bright patch of color on the throat, especially in hummingbirds
granivorous	feeding on grain
gregarious	tends to be found in company of others of the same species
habitats	a place where birds or animals naturally live; the home range of any bird or animal
herbivorous	birds whose diet is composed primarily of plant material: seeds, grasses, grain, buds, nuts, fruit, nectar, etc.
heterodactyl	in four-toed birds, the third and fourth toe point forward and the other two backward
humerus	in birds, the arm bone from the shoulder to the elbow. In birds the humerus, ulna, and metacarpus make up the "hand."
hypoxia	oxygen starvation
infraorder	a taxonomic rank below suborder
infrasound	low-frequency acoustic
insectivorous	feeding on insects

invertebrates	animals lacking a backbone, such as worms, snails, squid, crabs, insects, spiders, etc.
kleptoparasitism	a form of feeding in which one bird takes food from another by harassment or theft
K–Pg	The boundary between the Cretaceous and Paleogene Periods in the geological time record, 65 MYBP
lek	a location where birds gather to display and select a mate
lipids	oils, fats or waxes
maniraptora	a clade of coelurosaurian dinosaurs which includes the birds and the non-avian dinosaurs
melanosomes	a structure in animal cells that affects the light-absorbing pigments
mesite	a family of small flightless or near flightless birds endemic to Madagascar
metacarpal	the bones of the hand; in birds, the bone from the wrist to the fused fingers
molt	in birds, to shed and replace feathers
monomorphic	having only one type of characteristic; also the same plumage of adult male and female birds of the same species
monotypic family	a family with one species
monotypic genus	having one type of representation, in birds it means that the species is the sole member of that genus
MYBP	millions of years before present
nectarivores	birds and animals that predominately feed on sugar-rich nectar
nectarivorous	feeding on nectar and plants
Neoaves	all modern birds
neognathae	all of the living branches of birds except the paleognathes
Neornithes	a subclass of Aves comprising all living birds with a well-developed sternum
Neotropical	relating to region composed of tropical southern Mexico and Central and South America
olfactory bulb	structure located in the forebrain that detects odors
omnivores	feeding on all types of food, opportunistic feeders
orders of birds	in taxonomy, a taxonomic order, followed by family, genus, species and subspecies
Ornithischian	one of the two major groups of dinosaurs. The second major division are the saurischians.
Ornithurines	a natural group which includes the common ancestor of Ichthyornis, Hesperornis, and all modern birds. The primitive Ornithurines have a tail structure like that of modern birds.
oscine	passerine birds of a large suborder characterized by a vocal apparatus highly specialized for singing
paleobiology	the biology of fossil animals and plants
paleognathae	two living clades of birds: the tinamous and the five branches of flightless birds (Ratites)
paleontologist	a scientist who studies fossils
passerines	largest order of birds that include the perching birds divided into the oscines (songbirds) and suboscines
pectoral muscles	two large breast muscles attached to the sternum and connected to the wings (arms)

pelagic	the open ocean beyond the continental shelf, pelagic zone consists of very deep ocean
pennae	all the contour feathers of a bird, as distinguished from down feathers
phenotype	the set of characteristics of an individual we can observe and that are the results of its environment
phylogenetic	the past evolutionary history of extant species or taxa, based on present data and usually represented as a tree
phylogeny	evolutionary history of a taxonomic group of organisms
phylum	principal taxonomic category that ranks above class and below kingdom; birds are in the Chordata phylum
plumulaceous	the soft portion of the feather lacking barbs and serving as insulation
polyandry	practice of having more than one male mate at a time
polygamy	pattern of mating in which an animal has more than one mate
polygynandrous	both males and females have multiple mating partners
posterior	situated in the back of the body
precocial birds	birds hatched with open eyes and downy feathers, and that leave the nest within a day or two after hatching
primary feathers	the outermost feathers on the rear of the wing, connected to the "hand," and giving thrust to powered flight
quadruped	an animal that has four feet
rachis	the branched central shaft of the feather
Ratites	five branches of flightless birds: ostriches, cassowaries, kiwis, emus, and rheas
rectrices	tail feathers in birds
remiges	wing feathers in birds
Saurischian dinosaurs	the hip-lizard dinosaurs from which birds evolved
secondary feathers	the innermost feathers on the rear of the wing, connected to the arm, that give lift in flight
semiplume	feather intermediate between down and contour feathers; a feather with both down below and contour above
sternum	keel-shaped breastbone of a bird where the flight muscles are attached
suborder	an intermediate taxonomic rank lower than order
suboscine	a clade of passerines birds distinct from oscines and with a different syrinx muscle arrangement
superfamily	an intermediate taxonomic rank below order but above family
sympatric	organisms whose ranges overlap
syrinx	vocal organ of a bird located at the base of the trachea
taxa	a taxonomic group of any rank, such as a species, family, or class, characterized by a scientific name
taxon	singular of taxa; one family, genus, or species. Any species, etc., with a scientific name.
taxonomy	the classification of animals, systematics. Familiar classes include: families, genus, species, and subspecies.
temperate region	latitudes of the Earth between the subtropics and the Arctic circle
theropods	a group of dinosaurs that are typically bipedal and range in size from small to large
tibia	the large bone in birds between the femur and the transmetatarsus
trachea	the windpipe, a tube that allows air to pass to the lungs

transmetatarsus	in birds, the upper part of the foot connected to the digits that are the toes
transverse	for air and blood flow in birds, the air flow crossing at right angles
tridactyl	having three toes or fingers
ulna	the forearm of a bird, bone between the elbow and the digits of the hand
ultrasound	very high frequency sound
umami	flavors characteristic of broths and cooked meats
unihemispheric sleep	one hemisphere of the brain goes into deep sleep while the other hemisphere remains awake
vangas	a group of shrike-like medium-sized birds distributed from Asia to Africa, including Madagascar
vermiculations	a surface pattern of dense, thin, irregular dark and light lines
Wallacea	Indonesian islands separated by deep-water straits from the Asian and Australian continental shelves.
wing aspect ratio	ratio of the wing length divided by width
wing loading	weight divided by wing area
zygodactyl	in birds, the second and third toe point forward and the other two backward

References

Introduction

IPBES 2019. "United Nations Report: Nature's Dangerous Decline 'Unprecedented'; Species Extinction Rates 'Accelerating.'" www.un.org/sustainabledevelopment/blog/2019/05/nature-decline-unprecedented-report/.

Mynott, J. 2020. *Birds in the ancient world: Winged words*. Oxford University Press.

Paulin, J.B., and D. Drake. 2003. Positive benefits and negative impacts of Canada Geese. Rutgers Cooperative Research & Extension. Fact sheet 1027. Rutgers, The State University of New Jersey New Brunswick.

Chapter 1

Atterholt, J., J.H. Hutchison, and J.K. O'Connor. 2018. The most complete enantiornithine from North America and a phylogenetic analysis of the Avisauridae. *PeerJ* 6:e5910. doi: 10.7717/peerj.5910.

Baron, M.G., D.B. Norman, and P.M. Barrett. 2017. A new hypothesis of dinosaur relationships and early dinosaur evolution. *Nature* 543: 501–506.

Benito, J., P-C. Kuo, K. Widrig, J.W.M. Jagt, and D.J. Field. 2022. Latest Cretaceous ornithurine supports a neognathous crown bird ancestor. *Nature* 621(7938): 100-105. doi:10.1038/s41586-022-05445-y.

Brusatte, S. 2018. *The rise and fall of the dinosaurs: A new history of a lost world*. William Morrow, New York.

Brusatte, S.L. 2017. Taking wing. *Scientific American* 316(1): 49–55.

Brusatte, S.L., G. Niedzwiedzki, and R.J. Butler. 2011. Footprints pull origin and diversification of dinosaurs stem lineage deep into early Triassic. *Proceedings of the Royal Society B* 78(April 7): 1107–1113.

Brusatte, S.L., G.T. Lloyd, S.C. Wang, and M.A. Norell. 2014. Gradual assembly of avian body plan culminated in rapid rates of evolution across the dinosaur-bird transition. *Current Biology* 24(20): 2386–2392.

Carrano, M.T., J.R. Hutchinson, and S.D. Sampson. 2005. New information on *Segisaurus halli*, a small theropod dinosaur from the Early Jurassic of Arizona. *Journal of Vertebrate Paleontology*, 25(4): 835–849.

Cau, A., T. Brougham and D. Naish. 2015. The phylogenetic affinities of the bizarre Late Cretaceous Romanian theropod *Balaur bondoc* (Dinosauria, Maniraptora): dromaeosaurid or flightless bird? *PeerJ*, 3: e1032. doi: 10.7717/peerj.1032.

Clarke, J.A. 2004. Morphology, phylogenetic taxonomy, and systematics of Ichthyornis and Apatornis (Avialae: *Ornithurae*). *Bulletin of the American Museum of Natural History* 286: 1–179.

Clarke, J.A., C.P. Tambussi, J.I. Noriega, G.M. Erickson, and R.A. Ketcham. 2005. Definitive fossil evidence for the extant avian radiation in the Cretaceous. *Nature*, 433: 305–308.

Colleary, C., *et al*. 2015. Chemical, experimental, and morphological evidence for diagenetically altered melanin in exceptionally preserved fossils. *PNAS* (41): 12592–12597. doi.org/10.1073/pnas.1509831112.

Dyke, G., R. de Kat, C. Palmer, J. van der Kindere, D. Naish, and B. Ganapathisubramani. 2013. Aerodynamic performance of the feathered dinosaur Microraptor and the evolution of feathered flight. *Nature Communications*. doi: 10.1038/ncomms3489.

Ericson, Per G.P., C.L. Anderson, T. Britton, A. Elzanowski, U.S. Johansson, M. Källersjö, J.I. Ohlson, T.J. Parsons, D. Zuccon, and G. Mayr. 2006. Diversification of Neoaves: integration of molecular sequence data and fossils. *Evol. Biology* (2): 543–547. doi: 10.1098/rsbl.2006.0523.

Feduccia, A. 1999. *The origin and evolution of birds*, 2nd Ed. Yale University Press.

Field, D.J., J. Juan, A. Chen, J.W.M. Jagt, and D.T. Ksepka. 2020. Late Cretaceous neornithine from Europe illuminates the origins of crown birds. *Nature*. 579 (7799): 397-401. doi:10.1038/s41586-020-2096-0.

Habib, M.B. 2019. Monsters of the Mesozoic skies. *Sci. American* 321(4): 26–33.

Hone, D. 2016. *The Tyrannosaur chronicles: The biology of the tyrant dinosaurs*. Bloomsbury Sigma.

Hope, S. 2002. *The Mesozoic radiation of Neornithes*. In *Mesozoic birds: Above the heads of dinosaurs*. Eds. L.M. Chiappe, and L.M. Witmer. University of California Press.

Hull, P.M., *et al.* 2020. On impact and volcanism across the Cretaceous-Paleogene boundary. *Science* 367(6475): 266–272. doi: 10.1126/science.aay5055.

Kaiser, G., and G.J. Dyke. 2011. Introduction: Changing the questions in Avian Paleontology. In *Living dinosaurs: The evolutionary history of modern birds.* Ed. G. Dyke and G. Kaiser. Wiley-Blackwell. p. 3–8.

Kurochkin, E.N., G.J. Dyke, and A.A. Karhu. 2002. A new presbyornithid bird (Aves, Anseriformes) from the late Cretaceous of southern Mongolia. American Museum Novitates, No. 3386. http://hdl.handle.net/2246/2875.

Lamb, J.P. Jr. 1997. Marsh was right: Ichthyornis had a beak. *Journal of Vertebrate Paleontology* 17: 59A.

Longrich, N.R., T. Tokaryk, and D.J. Field. 2011. Mass extinction of birds at the Cretaceous-Paleogene (K–Pg) boundary. Proceedings of the National Academy of Sciences 108(37): 15253–15257. doi: 10.1073/pnas.1110395108.

Makovicky, P.J., and L.E. Zanno. 2011. Theropod diversity and the refinement of Avian characteristics. In *Living dinosaurs: The evolutionary history of modern birds.* Ed. G. Dyke and G. Kaiser, 2011. Wiley-Blackwell. p. 9–29.

Mayr, G. 2017. *Avian evolution: The fossil record of birds and its paleobiological significance.* Wiley-Blackwell.

Nesbitt, S.J., C.A. Sidor, R.B. Irmis, K.D. Angielczyk, R.M.H. Smith, and L.A. Tsuji. 2010. Ecologically distinct dinosaurian sister group shows early diversification of Ornithodira. *Nature* 464 (7285): 95–98. Doi:10.1038/nature08718.

O'Connor, J., L.M. Chiappe, and A. Bell. 2011. Pre-modern birds: Avian divergences in the Mesozoic. In *Living dinosaurs: The evolutionary history of modern birds.* Ed. G. Dyke and G. Kaiser, 2011. Wiley-Blackwell. pp. 39–114.

Preston, D. 2019. The day the earth died. *New Yorker.* April 8, 2019, p. 53–66.

Prum, R.O. 2002. Why ornithologists should care about the theropod origin of birds. *The Auk* 119: 1–17.

Prum, R.O., J.S. Berv, A. Dornburg, D.J. Field, J.P. Townsend, E.M. Lemmon and A.R. Lemmon. 2015. A comprehensive phylogeny of birds (Aves) using targeted next-generation DNA sequencing. *Nature* (526): 569–573. doi:10.1038/nature15697.

Sakamoto, M., M. Benton, and C. Venditti. 2016. Dinosaurs in decline tens of millions of years before their final extinction. *Proceeding of the National Academy of Sciences.* 113(18): 5036–5040.

Summer, T. 2017. Devastation detectives. *Science News Magazine* Feb. 4, 2017, p. 16–21.

Vinther, J. 2017. The true color of dinosaurs. *Scientific American* 316(3): 50–57.

Vinther, J., R. Nicholls, S. Lautenschlager, M. Pittman, T.G. Kaye, E. Rayfield, G. Mayr, and I.C. Cuthill. 2016. 3D camouflage in an ornithischian dinosaur. *Current Biology* 26(18): 2456–62.

Voeten, D., J. Cubo, E de Margerie, M. Roper, V. Beyrand, S. Bures, P. Tafforeau, and S. Sanchez. 2018. Wing bone geometry reveals active flight in Archaeopteryx. *Nature Communication* 9: 923. doi:10.1038/s41467-018-03296-8.

Wang, M., Zheng, X. O'Connor, J.K. Lloyd, G.T. Wang, X. Wang, Y. Zhang, X. Zhou, and Z. Zhou. 2015. The oldest record of ornithuromorpha from the early cretaceous of China. *Nature Communications* 6, article number: 6987. doi: 10.1038/ncomms7987.

Wellnhofer, P. 2009. *Archaeopteryx: The icon of evolution.* Pfeil Verlag.

Xu, X., Z. Zhou, and X. Wang. 2000. The smallest known non-avian theropod dinosaur. *Nature* 408: 705–708. doi.org/10.1038/35047056.

Zixiao, Y., B. Jiang, M.E. McNamara, S.L. Kearns, M. Pittman, T.G. Kaye, P.J. Orr, X. Xu, and M.J. Benton. 2018. Pterosaur integumentary structures with complex feather-like branching. *Nature Ecology & Evolution,* 3(1): 24. doi: 10.1038/s41559-018-0728-7.

Chapter 2

Bardeen, C.G., R.R. Garcia, O.B. Toon, and A.J. Conley. 2017. On transient climate change at the Cretaceous–Paleogene boundary due to atmospheric soot injections. *PNAS* 36: E7415-E7424. doi.org/10.1073/pnas.1708980114.

Claramunt, S., and J. Cracraft. 2015. A new time tree reveals Earth history's imprint on the evolution of modern birds. *Science Advances* 1(11). doi: 10.1126/sciadv.1501005. http://advances.sciencemag.org/content/1/11/e1501005.full. See also: Study Uncovers Influence of Earth's History on the Dawn of Modern Birds. American Museum of Natural History. http://phys.org/news/2015-12-uncovers-earth-history-dawn-modern.html.

Donovan, M.P., A. Iglesias, P. Wilf, C.C. Labandeira, and N.R. Cuneo. 2016. Rapid recovery of Patagonian plant–insect associations after the end-Cretaceous extinction. *Nature Ecology & Evolution* 1(12). www.nature.com/articles/s41559-016-0012.

Dyke, G., and E. Gardiner. 2011. The utility of fossil taxa and the evolution of modern birds: commentary and analysis. In *Living dinosaurs: The evolutionary history of modern birds.* Ed. G. Dyke and G. Kaiser, 2011. Wiley-Blackwell.

Dyke, G.J., R.L. Nudds and M.J. Benton. 2007. Modern avian radiation across the Cretaceous-Paleogene Boundary. *The Auk* 124: 339–341.

Feduccia, A. 1995. Explosive evolution in Tertiary birds and mammals. *Science* 267: 637–638.

Feduccia, A. 1999. *The origin and evolutions of birds*, 2nd Ed. Yale University Press.

Field, D.J., A. Bercovici, J.S. Berv, R. Dunn, D.E. Fastovsky, T.R. Lyson, V. Vajda, and J.A. Gauthier. 2018. Early evolution of modern birds structured by global forest collapse at the end-Cretaceous mass extinction. *Current Biology* 28(11): 1825–1831. doi.org/10.1016/j/cub/2018.04.062.

James, H.F. 2005. Paleogene fossils and the radiation of modern birds. *The Auk* 122(4): 1049–1054.

Kaiho, K., N. Oshima, K. Adachi, Y. Adachi, T. Mizukami, M. Fujibayashi, and R. Saito. 2016. Global climate change driven by soot at the K–Pg boundary as the cause of the mass extinction. *Scientific Reports*, 6: 28427. doi: 10.1038/srep28427.

Ksepka, D.T., L. Grande, and G. Mayr. 2019. Oldest finch-beaked birds reveal parallel ecological radiations in the earliest evolution of passerines. *Current Biology* 29: 657–663.

Ksepka, D.T., T.A. Stidham, and T.E. Williamson. 2017. Early Paleocene landbird supports rapid phylogenetic and morphological diversification of crown birds after the K–Pg mass extinction. *PNAS* 114(30): 8047–8052. doi.org/10.1073/pnas.1700188114.

Lindow, B.E. 2011. Bird evolution across the K–Pg boundary and the basal neornithine diversification. In *Living dinosaurs: The evolutionary history of modern birds*. Ed. G. Dyke and G. Kaiser, 2011. Wiley-Blackwell.

Lindow, B.E., and G.J. Dyke. 2006. Bird evolution in the Eocene: climate change in Europe and a Danish fossil fauna. *Biological Reviews* 81(4): 483–499.

Lovette, I.J. 2016. Avian diversity and classification. In *The Cornell Lab of Ornithology handbook of bird biology*. Eds. I.J. Lovette, and J.W. Fitzpatrick, 3rd ed., Wiley-Blackwell. p. 7–61.

Low, T. 2017. *Where song began: Australia's birds and how they changed the world*. Yale University Press.

Lowery, C.J., *et al.* 2018. Rapid recovery of life at ground zero of the end Cretaceous Mass Extinction. *Nature*. 558. (7709). doi: 10.1038/s41586–018–0163–6.

Mayr, G. 2004. Old world fossil record of modern-type hummingbirds. *Science* 304: 861–864.

Mayr, G. 2005a. A tiny barbet-like bird from the lower Oligocene of Germany: the smallest species and earliest substantial fossil record of the *Pici* (woodpeckers and allies). *The Auk* 122: 1055–1063.

Mayr, G. 2005b. The Paleogene fossil record of birds in Europe. *Biological Reviews* 80: 515–542.

Mayr, G. 2014. The origins of crown group birds: molecules and fossils. *Palaeontology* 57(2): 231–242.

Mayr, G. 2017. *Avian evolution: The fossil record of birds and its paleobiological significance*. Wiley-Blackwell.

Prum, R.O., J.S. Berv, A. Dornburg, D.J. Field, J.P. Townsend, E.M. Lemmon, and A.R. Lemmon. 2015. A comprehensive phylogeny of birds (Aves) using targeted next-generation DNA sequencing. *Nature* (526): 569–573. doi: 10.1038/nature15697.

Rosen, M. 2017. The survivors. *Science News Magazine*. Feb. 4, pp. 22–25.

Sibley, C.G., and J.E. Ahlquist. 1990. *Phylogeny and classification of the birds: A study in molecular evolution*. Yale University Press.

Tabor, C.R., C. Bardeen, B. Otto-Bliesner, R. Garcia, and O.B. Toon. 2019. Causes and climatic consequences of the impact winter at the cretaceous-Palogene boundary. Research letter to American Geophysical Union. https://agupubs.onlinelibrary.wiley.com/doi/abs/10.1029/2019GL085572.

Tabor, C.R., C. Bardeen, B. Otto-Bliesner, R. Garcia, O.B. Toon, and C.J. Poulsen. 2016. Simulating the K–Pg with an earth system model. Paper at GSA Annual Meeting in Denver, CO. https://gsa.confex.com/gsa/2016AM/webprogram/Paper279839.html.

Vajda, C., and A. Bercovici. 2014. The global vegetation pattern across the Cretaceous–Paleogene mass extinction interval: A template for other extinction events. *Global and Planetary Change*. November 2014. 122: 29–49. doi.org/10.1016/j.gloplacha.2014.07.014.

Chapter 3

Ackermann, R. 2016. The hybrid origin of "modern" humans. *Evolutionary Biology* 43: 1–11.

Amadon, D. 1949. The seventy-five per cent rule for subspecies. *Condor* 51(6): 250–258.

Barrowclough, G.F., J. Cracraft, J. Klicka, and R.M. Zink. 2016. How many kinds of birds are there and why does it matter. *PLOS ONE*, Nov. 23, 2016. doi:10.1371/journal.pone.0166307.

Borgmann, K. 2021. Sound sleuthing for new species. *Living Bird* 40(1): 48–54.

Burga, A., W. Wang, E. Ben-David, P.C. Wolf, A.M. Ramey, C. Verdugo, K. Lyons, P.G. Parker, and L. Kruglyak. 2017. A genetic signature of the evolution of loss of flight in the Galapagos cormorant. *Science* 356(6341): eaal3345. doi: 10.1126/science.aal3345.

Campagna, L. 2021. A shortcut to speciation. *Living Bird* 40(3): 24–25.

Clements, J.F., T.S. Schulenberg, M.J. Iliff, S.M. Billerman, T.A. Fredericks, J.A. Gerbracht, D. Lepage, B.L. Sullivan, and C.L. Wood. 2021. *The eBird/Clements checklist of birds of the world: v2021*.

Darwin, C. 1871. *The descent of man, and selection in relation to sex*. John Murray Press.

Freeman, B.G., and G.A. Montgomery. 2017. Using song playback experiments to measure recognition between geographically isolated populations: a comparison with acoustic trait analyses. *The Auk* 134: 857–870.

Grant, P.R., and B.R. Grant. 1992. Hybridization of bird species. *Science* 256(5054): 193–197.

Heisman, R. 2022. Into the oriole genome. *Living Bird* 41(2): 28–37.

Hofmeister, N. R., S.J. Werner, and I.J. Lovette. 2021. Environmental correlates of genetic variation in the invasive European starling in North America. *Molecular Ecology* 30(5): 1251–1263. doi.org/10.1111/mec.15806.

Howell, S.N.G. 2021. What isn't a species? *North American Birds* 72(1): 16–25.

Irestedt, M.J. Fjeldsa, L. Dalen, and Per G.P. Ericson. 2009. Convergent evolution, habitat shifts and variable diversification rates in the ovenbird-woodcreeper family (Furnariidae). *BMC Evolutionary Biology* 9: 268–281. doi: 10.1186/1471-2148-9-268.

Jarvis, E.D., *et al.* 2014. Whole-genome analyses resolve early branches in the tree of life of modern birds. *Science* 346: 1320–1331.

Leonard, P. 2021. The secret to starling success: It's in their genes. *Living Bird* 40(2): 16.

Mares, M.A. 2002. *A desert calling: life in a forbidding landscape.* Harvard University Press.

Mason, N.A., and S.A. Taylor. 2015. Differentially expressed genes match bill morphology and plumage despite largely undifferentiated genomes in a Holarctic songbird. *Molecular Ecology.* doi: 10.1111/mec.1314.

Mayr, E. 1946. The number of species of birds. *The Auk* 63(1): 64–69. doi:10.2307/4079907.

McLaren, I.A., and A.G. Horn. 2006. The Ipswich sparrow, past, present, and future. *Birding* 38(5) :52–59.

Minor, N. 2016. Extravagant plumages and a concert of genes. *Birding* 48(3) :26–29.

Mora, C., D.P. Tittensor, S. Adl, A.G.B. Simpson, and B. Worm. 2011. How many species are there on earth and in the ocean? *PLOS Biology* 9(8): e1001127 doi:10.1371/journal.pbio.1001127.

Ottenburghs, J. 2018. Exploring the hybrid speciation continuum in birds. *Ecology and Evolution* 8(24): 13027–13034. doi.org/10.1002/ece3.4558.

Patten, M.A. 2015. Subspecies and philosophy of science. *The Auk* 132: 481–485.

Prum, R.O. 2017. *The evolutions of beauty: How Darwin's forgotten theory of mate choice shapes the animal world and us.* Doubleday.

Quammen, D. 1996. *The song of the dodo: Island biogeography in an age of extinctions.* Scribner.

Rheindt, F.E., D.M. Prawiradilaga, H. Ashari, Suparno, C.Y. Gwee, G.W.X. Lee, M.Y. Wu, and N.S.R. Ng. 2020. A lost world in Wallacea: description of a montane archipelagic avifauna. *Science.* January 10, 367(6474): 167–170. doi: 10.1126/science.aax2146.

Sackton, T.B., P. Grayson, A. Cloutier, Z. Hu, J.S. Liu, N.E. Wheeler, P. P Gardner, J.A. Clarke, M. Clamp, S.V. Edwards, and A.J. Baker. 2019. Convergent regulatory evolution and loss of flight in paleognathous birds. *Science* 364(6435): 74–78. doi: 10.1126/science.aat7244.

Sexton, C. 2018. Global vegetation database details plant life across the world. *Earth.com.* https://www.earth.com/news/global-vegetation-database-plants/.

Snow, D.W. 1976. *The web of adaptation: Bird studies in the American tropics.* Quadrangle-New York Times Book Co.

Snyder, K.T., and N. Creanza. 2019. Polygyny is linked to accelerated birdsong evolution but not to larger song repertoires. *Nat. Commun.* 10: 884. doi.org/10.1038/s41467-019-08621-3.

Turbek, S.P., M. Browne, A.S. Di Giacomo, L. Campagna *et al.* 2021. Rapid speciation via the evolution of pre-mating isolation in the Iberá Seedeater. *Science* March 26: 361(6536). doi: 10.1126/science.abc0256.

Walsh, J., S.M. Billerman, V.G. Rohwer, B.G. Butcher, and I.J. Lovette. 2020. Genomic and plumage variation across the controversial Baltimore and Bullock's oriole hybrid zone. *The Auk* 137:1–15. doi: 10.1093/auk/ukaa044.

Uyeda, J.C., M.W. Pennell, E.T. Miller, R. Maia, and C.R. McClain. 2017. The evolution of energetic scaling across the vertebrate tree of life. *The American Naturalist* 190(2): 185–198.

Zimmer, C., and D.J. Emlen. 2013. *Evolution: Making sense of life.* Roberts and Co.

Chapter 4

Ackerman, J. 2016. *The genius of birds.* Penguin.

Alù, A. 2022. Tricking light. *Scientific American* 327(5): 42–51.

Arnold, K.E., I.P.F. Owens, and N.J. Marshall. 2002. Fluorescent signaling in parrots. *Science* 295: 92.

Balanoff, A.M., G.B. Bever, T.B. Rowe, and M.A. Norell. 2013. Evolutionary origins of the avian brain. *Nature* 501: 93–96. doi: 10.1038/nature12424.

Beason, R.C. 2004. What can birds hear? USDA National Wildlife Research Center—Staff Publications. Paper 78. U. of Nebraska Press. https://digitalcommons.unl.edu/icwdm_usdanwrc/78.

Benedict, L., and K. Odom. 2017. Listening to nature's divas. *Birding* 49(2): 37–43.

Biedermann, P.H.W., K. Delhey, A. Peters, and B. Kempenaers. 2008. Optical properties of the uropygial gland secretion: No evidence for UV cosmetics in birds. *The Science of Nature* 95(10): 939–46. doi: 10.1007/s00114-008-0406-8.

Birkhead, T. 2012. *Bird sense: What it's like to be a bird.* Walker & Co.

Boles, W.E. 1990. Glowing parrots—need for a study of hidden colours. *Birds Int.* 3: 76–79.

Brower, L.C. 1969. Ecological chemistry. *Scientific American* 220(2): 22–29.

Brusatte, S.L. 2017. Taking wing. *Scientific American* 316(1): 49–55.

Cabanac, M., and S. Aizawa. 2000. Fever and tachycardia in bird (Gallus domesticus) after simple handling. *Physiology & Behavior* 69: 541–545.

Catchpole, C.K., and P.J.B. Slater. 2008. *Birdsong: Biolocical themes and variations.* 2nd ed. Cambridge University Press.

Cheng, H. 2007. Morphopathological changes and pain in beak trimmed laying hens. *World's Poultry Science Journal* 62(1): 41–52. doi.org/10.1079/WPS200583.

Clarke, J.A., S. Chatterjee, Z. Li, T. Riede, F. Agnolin, F. Goller, M.P. Isasi, D.R. Martinioni, F.J. Mussel, and F.E. Novas. 2016. Fossil evidence of the avian vocal organ from the Mesozoic. *Nature* 538: 502–505. doi:10.1038/nature19852.

Cockrem, J.F. 2007. Stress, corticosterone responses in avian personalities. *Journal of Ornithology* (Supplement) 148: S169-S178. doi: 10.1007/s10336-007-0175-8.

Collet, J., T. Sasaki, and D. Biro. 2021. Pigeons retain partial memories of homing paths years after learning them individually, collectively or culturally. *Proc. R. Soc. B* 288: 20212110. https://doi.org/10.1098/rspb.2021.2110.

Colombelli-Négrel, D., M.E. Hauber, C. Evans, A.C. Katsis, L. Brouwer, N.M. Adreani, and S. Kleindorfer. 2020. Prenatal auditory learning in avian vocal learners and non-learners. *Phil. Trans. of the Royal Soc. B* 376: 20200247. https://doi.org/10.1098/rstb.2020.0247.

Crick, H., and G.L. Maclean. 2003. Sandgrouse. In *Firefly encyclopedia of birds.* Ed. C. Perrins. Firefly Books. p. 284–287.

Davis, J.N. 2007. Color abnormalities in birds: A proposed nomenclature for birders. *Birding* 39(5): 36–46.

Debose, J.L., and G.A. Nevitt. 2007. Investigating the association between pelagic fish and DMSP in a natural coral reef system. *Marine and Freshwater Research,* 58(8): 720–724.

Dooling, R. 2002. Avian hearing and the avoidance of wind turbines. National Renewable Energy Laboratory Report, Golden, CO.

Doyle, D. 2018. Do Eastern Whip-poor-wills sing antiphonally? *Birding* 50(1): 36–43.

du Toit, C.J., A. Chinsamy, and S.J. Cunningham. 2020. Cretaceous origins of the vibrotactile bill-tip organ in birds, *Proc. Royal Society B: Biological Sciences.* doi: 10.1098/rspb.2020.2322.

Duncan, C.J. The sense of taste in bids. *Ann. appl. Biology* 48(2): 409–414.

Eaton, M.D., and S.M. Lanyon. 2003. The ubiquity of avian ultraviolet plumage reflectance. Royal Society *Proceeding: Biological Sciences* 270(1525): 1721–1726.

Emery, N.J. 2006. Cognitive ornithology: the evolution of avian intelligence. *Phil. Trans. R. Soc. B* 361: 23–43. doi: 10.1098/rstb.2005.1736.

Emery, N.J., and N.S. Clayton. 2005. Evolution of the avian brain and intelligence. *Current Biology* 15: No 23 R946–R950.

Evans, D.H. 2016. Avian anatomy. In *The Cornell Lab of Ornithology handbook of bird biology.* Ed. Lovette, I.J., and J.W. Fitzpatrick, 3rd ed., Wiley-Blackwell. p. 169–213.

Farnsworth, A. 2005. Flight calls and their value for future ornithological studies and conservation research. *The Auk* 122(3): 733–746.

Galeotti, P., and D. Rubolini. 2004. The niche variation hypothesis and the evolution of colour polymorphism in birds: a comparative study of owls, nightjars and raptors. *Biological Journal of the Linnean Society* 82: 237–248.

Galeotti, P., D. Rubolini, P.O. Dunn, and M. Fasola. 2003. Colour polymorphism in birds: causes and functions. *Journal of Evolutionary Biology* 16(4): 635–646. doi: 10.1046/j.1420–9101.2003.00569.x.

Galvan, I, P.R. Camarero, R. Mateo, and J.J. Negro. 2016. Porphyrins produce uniquely ephemeral animal colouration: a possible signal of virginity. *Scientific Reports* 6, article no. 39210. www.nature.com/articles/srep39210#additional-information.

Gentle, M., and S. Wilson. 2004. Pain and the laying hen. In *Welfare of the laying hen,* ed. G.C. Perry. Center for Agriculture and Bioscience International.

Greenewalt, C.H. 1960. *Hummingbirds.* Doubleday.

Güntürkün, O. 2020. The surprising power of the avian mind. *Scientific American* 322(1): 48–55.

Hawkes, L.A., S. Balachandran, N. Buthayar, P.J. Butler, B. Chua, D.C. Douglas, P.B. Frappell, Y. Hou, W.K. Milsorm, S.H. Newman, *et al.* 2013. The paradox of extreme high-altitude migration in bar-headed geese *Anser indicus. Proc. Biol. Sci.* 280: 122114. doi:10.1098/rspb.2012.2114.

Hill, G.E. 2010. *Bird coloration.* National Geographic Society.

Hilty, S. 1994. *Birds of tropical America: A watcher's introduction to behavior, breeding, and diversity.* Chapters Publishing.

Howell, S.N.G. 2003. All you ever wanted to know about molt: Part II: Finding order amid the chaos. *Birding* 35(6): 640–649.

Howell, S.N.G. 2010. *Peterson reference guide to molt in North American birds.* Houghton Mifflin Harcourt.

Hurlburt, G.R., R.C. Ridgely, and L.M. Witmer. 2013. Relative size of brain and cerebrum in Tyrannosaurid dinosaurs: An analysis using brian-endocast quantitative relationships in extant alligators. In

Tyrannosaurid Paleobiology. Eds. J.M. Parrish, R.E. Molnar, P.J. Currie, and E.B. Koppelhus. Chapter 6. Indiana University Press. p. 134–154.

Hutt, G.D., and L. Ball. 1938. Number of feathers and body size in Passerine birds. *The Auk* 38(4): 651–657.

Kaplan, G. 2015. *Bird minds: Cognition and behaviour of Australian native birds.* CSIRO Publishing.

Kelly, M.L., R.A. Peters, R.K. Tisdale and L.A. Lesku. 2015. Unihemispheric sleep in crocodilians? *Journal of Experimental Biology* 218: 3175–3178.

Kiama, S.G., J.N. Maina, J. Bhattacharjee, K.D. Weyrauch. 2006. Functional morphology of the pecten oculi in the nocturnal spotted eagle owl (Bubo bubo africanus), and the diurnal black kite (Milvus migrans) and domestic fowl (Gallus gallus var. domesticus): A comparative study. *Journal of Zoology.* 254(4): 521–528. doi.org/10.1017/S0952836901001029.

Koch, C. 2016. To sleep with half a brain. *Scientific American Mind.* https://www.scientificamerican.com/article/sleeping-with-half-a-brain/

Kroodsma, D.E. 2005. *The Singing Life of Birds.* Houghton Mifflin Harcourt.

Laland, K. 2018. An evolved uniqueness. *Scientific American* 319(3): 33–39.

Larkin, R.P. 1982. Spatial distribution of migrating birds and small-scale atmospheric motion. In *Avian Navigation.* Eds. R. Papi and H.G. Walraff. Springer-Verlag.

Laybourne, R.C. 1974. Collision between a vulture and an aircraft at an altitude of 37,000 feet. *Wilson Bulletin* 86(4): 461–462.

Lesku, J.A., N.C. Rattenborg, M. Valcu, A.L. Vyssotski, S. Kuhn, F. Kuemmeth, W. Heidrich, and B. Kempenaers. 2012. Adaptive sleep loss in polygynous Pectoral Sandpiper. *Science.* 337(6102): 1654–1658.

Letzner, S., O. Güntürkün, and C. Beste. 2017. How birds outperform humans in multi-component behavior. *Current Biology* 27(18): R996-R998. doi.org/10.1016/j.cub.2017.07.056.

Lotem A., E. Schechtman, and G. Katzir. 1991. Capture of submerged prey by little egrets, Egretta garzetta garzetta: strike depth, strike angle and the problem of light refraction. *Anim. Behav.* 42 (3): 341–346.

Lovette, I.J. 2016. How birds evolve. In *The Cornell Lab of Ornithology handbook of bird biology.* Eds. I.J. Lovette and J.W. Fitzpatrick, 3rd ed., Wiley-Blackwell.

Lovette, I.J., and J.W. Fitzpatrick. 2016. *The Cornell Lab of Ornithology handbook of bird biology.* 3rd ed. Wiley.

Makovicky, P.J., and L.E. Zanno. 2011. Theropod diversity and refinement of avian characteristics. In *Living dinosaurs: The evolutionary history of modern birds.* Ed. G. Dyke and G. Kaiser. Wiley-Blackwell.

Martin, R. 2009. What is binocular vision for? A bird's eye view. *Journal of Vision* 9:14.

Martinez-Gonzalez, D., Lesku J.A., and N.C. Rattenborg. 2008. Increased EEG spectral power density during sleep following short-term sleep deprivation in pigeons (*Columba livia*): evidence for avian sleep homeostasis. *Journal of Sleep Research* 17: 140-153.

McKellar, R.C., B.D.E. Chatterton, A.P. Wolfe, and P.J. Currie. 2011. A diverse assemblage of late Cretaceous dinosaur and bird feathers from Canadian amber. *Science* 333, 1619–1622.

Mullen, P., and G. Pohland. 2008. Studies on UV reflection in feathers of some 1000 bird species: are UV peaks in feathers correlated with violet-sensitive and ultraviolet-sensitive cones? *Ibis* 150: 59–68.

Nassau, K. 2001. *The physics and chemistry of color,* 2nd ed. Wiley.

Nemeth, Z., Y. Luo, J.C. Owen, and F.R. Moore. 2017. Seasonal variation in CREB expression in the hippocampal formation of first-year migratory songbirds: implication for the role of memory during migration. *the Auk* 134: 146–152. doi:10.1642/AUK-16-133.1.

Nevitt, G.A., R.R. Veit, and P. Kareiva. 1995. Dimethyl sulphide as a foraging cue for Antarctic Procellariiform seabirds. *Nature* 376: 680–682.

Nicolson, A. 2018. *The seabird's cry: The lives and loves of the planet's great ocean voyagers.* Henry Holt.

Norell, M.A., J.M. Clark, L.M. Chiappe, and D. Dashzeveg. 1995. A nesting dinosaur. *Nature* 378: 774–76. doi:10.1038/378774a0.

Olson, C.R., M. Fernandez-Peters, C.V. Portfors, and C.V. Mello. 2018. Black Jacobin hummingbirds vocalize above the known hearing range of birds. *Current Biology* 28(5): R193-R194. doi: 10.1016/j.cub.2018.01.041.

Olsson, I.A.S., and L.J. Keeling. 2005. Why in earth? Dustbathing behaviour in jungle and domestic fowl reviewed from a Tinbergian and animal welfare perspective. *Applied Animal Behaviour Science* 93(3–4): 259–282. http://dx.doi.org/10.1016/j.applanim.2004.11.018.

Overington, S.E., A.S. Griffin, D. Sol, and L. Lefebvre. 2011. Are innovative species ecological generalists? A test in North American birds. *Behavioral Ecology.* doi:10.1093/beheco/arr130.

Owens, I.P.F., and I.R. Hartley. 1998. Sexual dimorphism in birds: why are there so many different forms of dimorphism? *Proc. Roy. Soc. B* 265: 397–407. doi: 10.1098/rspb.1998.0308.

Pittaway, R. 2000. Plumage and molt. *Ontario Birds* 18(1): 27–43. https://sora.unm.edu/sites/default/files/27-43%20OB%20Vol%2018%231%20Apr2000.pdf.

Prum, R.O. 2017. *The evolution of beauty.* Doubleday.

Pyle, P. 1997. *Identification guide to North American birds, part I: Columbidae to Ploceidae.* State Creek Press.

Pyle, P. 2008. *Identification guide to North American birds, part II: Anatidae to Alcidae.* State Creek Press.

Pyle, P., and M. McPherson. 2017. Why so many White Eared Grebes? *Birding* 49(5): 54–61.

Radford, P. 1984. Hearing in birds. *J. Wildlife Sound Recording Society* 14(7).

Rajchard, J. 2009. Ultraviolet (UV) light perception by birds: a review. *Veterinarni Medicina* 54(8): 351–359.

Ralph, C.L. 1969. The control of color in birds. *American Zoologist* 9: 521–530.

Rattenborg, N.C. 2017. Sleeping on the wing. *Interface Focus* 7(1): 2016.0082. doi.org/10.1098/rsfs.2016.0082.

Rattenborg, N.C., Lima, S.L., and C.J. Amlaner. 1999. Facultative control of avian unihemispheric sleep under the risk of predation. *Behav. Brain Res.* 105(2): 163–72.

Riebel, K., K.J. Odom, N.E. Langmore, M.L. Hall. 2019. New insights from female bird song: towards an integrated approach to studying male and female communication roles. *Biol. Lett.* 15: 20190059. http:// dx.doi.org/10.1098/rsbl.2019.0059.

Rowland, H.M., M.R. Parker, P. Jiang, D. Reed, and G.K. Beauchamp. 2015. Comparative taste biology with special focus on birds and reptiles. Chapter 43 in *Handbook of Olfaction and Gustation*, 3rd Edition. Ed. Richard L. Doty. Wiley-Blackwell, p. 957–982.

Scanes, C.G., and S. Dridi, eds. 2021. *Sturkie's avian physiology*, 7th ed. Academic Press.

Scheid, P., and J. Piiper. 1972. Cross-current gas exchange in avian lungs: Effects of reversed parabronchial air flow in ducks. *Respiration Physiology* 16(3): 304–321. doi.org/10.1016/0034–5687(72)90060–6.

Scott, G.R., L.A. Hawkes, P.B. Frappell, P.J. Butler, C.M. Bishop, and W.K. Milson. 2015. How Bar-Headed Geese fly over the Himalayas. *Physiology* 30(2): 107–115. doi: 10.1152/physiol.00050.2014.

Skutch, A.F. 1999. *Trogons, laughing falcons, and other neotropical birds.* Texas A&M University.

Sol, D., R.P. Duncan, T.M. Blackburn, P. Cassey, and L. Lefebvre. 2005. Big brains, enhanced cognition, and response of birds to novel environments. *Proc. Nat. Acad. Science.* 102 (15) 5460–5465. doi.org/10.1073/ pnas.0408145102.

Stager, K.E. 1964. *The role of olfaction in food location by the turkey vulture (Cathartes aura).* Contributions in Science 81, Los Angeles County Museum, June 30, 1964 (PDF).

Thorpe, W.H. 1973. Duet-singing birds. *Scientific American* 229(2): 70–79.

Tinbergen, N. 1959. Comparative studies of the behavior of gulls (Laridae): A progress report. *Behaviour* 15(1): 1–70.

Van Tyne, J., and A.J. Berger. 1976. *Fundamentals of Ornithology.* Wiley.

Verbeurgt, C., F. Wilkin, M. Tarabichi, F. Gregoire, J.E. Dumont, and P. Chatelain. 2014. Profiling of Olfactory Receptor Gene Expression in Whole Human Olfactory Mucosa. *PLOS ONE.* doi.org/10.1371/journal. pone.0096333.

Ward, P., and R. Berner. 2011. Why where there dinosaurs? Why are there birds? In *Living dinosaurs: The evolutionary history of modern birds*, ed. G. Dyke and G. Kaiser, 2011. Wiley-Blackwell.

Williams, C.L., Hagelin, J.C., and G.L. Kooyman. 2015. Hidden keys to survival: the type, density, pattern and functional role of emperor penguin body feathers. *Proceedings of the Royal Society B.* 282: 20152033.

Williams, T.D., ed. 2020. *What is a bird?* Princeton University Press.

Wolfe, J.D., T.B. Ryder, and P. Pyle. 2010. Using molt cycles to categorize the age of tropical birds: an integrative new system. *Journal Field Ornithology.* 81(2): 186–194, 2010 doi: 10.1111/j.1557–9263.2010.00276.x.

Xing, L., R. McKellar, M. Wang, *et al.* Mummified precocial bird wings in mid–Cretaceous Burmese amber. 2016. *Nat Commun.* 7:12089. Doi.org/10.1038/ncomms12089.

Zelenitsky, D., F. Therrien, R. Ridgely, A. McGee, and L. Witmer. 2011. Evolution of olfaction in non-avian theropod dinosaurs and birds. *Proceedings of the Royal Society B: Biological Sciences.* doi: 10.1098/ rspb.2011.0238.

Chapter 5

Black, J.M. Ed. 1996. *Partnerships in birds: The study of bird monogamy.* Oxford University Press.

Caitlin A.S., and M.R. Servedio. 2017. Evolution of a mating preference for a dual-utility trait used in intrasexual competition in genetically monogamous populations. *Ecology and Evolution* 7(19): 8008–8016. doi: 10.1002/ece3.3145.

Collias, N.E., and E.C. Collias. 1962. An experimental study of the mechanisms of nest building in a Weaverbird. *The Auk* 79:568–595. doi: 10.2307/4082640.

Davis, L.S. 2019. *A polar affair: Antarctica's forgotten hero and the secret love lives of penguins.* Pegasus Books.

Dyke, G., M. Vremir, G. Kaiser, and D. Naish. 2012. A drowned Mesozoic bird breeding colony from the Late Cretaceous of Transylvania. *Naturwissenschaften* 99: 435–442. doi:10.1007/s00114–012–0917–1.

Grace, J.K., K. Dean, M.A. Ottinger, and D.J. Anderson. 2011. Hormonal effects of maltreatment in Nazca Booby nestlings: implications for the "cycle of violence." *Hormones and Behavior* 60: 78–85.

Griggio, M., F. Valera, A. Casas, and A. Pilastro. 2005. Males prefer ornamented females: a field experiment of male choice in the rock sparrow. *Animal Behaviour*, 69: 1243–1250. doi:10.1016/j.anbehav.2004.10.004.

Handel, C,M., and R.E. Gill, Jr. 2000. Mate fidelity and breeding site tenacity in a monogamous sandpiper, the black turnstone. *Animal Behaviour* 60: 471–481. doi:10.1006/anbe.2000.1505.

Johnson, K., and N.T. Burley. 1997. Mating tactics and mating systems of birds. In *Ornithological Monographs No. 49*, Avian Reproductive Tactics, ed. J.M. Hagen.

McGowan, C.P., J.J. Millspaugh, J.J., M.R. Ryan, C.D. Kruse, and G. Pavelka. 2009. Estimating survival of precocial chicks during the prefledging period using a catch-curve analysis and count-based age-class data. *Journal of Field Ornithology*. 80(1): 79–87. doi: 10.1111/j.1557-9263.2009.00207.x.

Moeliker, C.W. 2001. The first case of homosexual necrophilia in the mallard *Anas platyrhynchos* (Aves: Anatidae). *Deinsea* 8: 243–247.

Naef-Daenzer, B., and M.U. Gruebler. 2016. Post-fledging survival of altricial birds: ecological determinants and adaptation. *Journal of Field Ornithology* 87(3): 227–250. doi: 10.1111/jofo.12157.

Nicolson, A. 2018. *The seabird's cry: the lives and loves of the planet's great ocean voyagers*. Henry Holt.

Nolan Jr., V. 1976. *The Ecology and behavior of the Prairie Warbler Dendroica Discolor*. Ornithological Monographs No. 26. American Ornithologists' Union. p. 189.

Norell, M.A., J.M. Clark, L.M. Chiappe, and D. Dashzeveg. 1995. A nesting dinosaur. *Nature* 378: 774–776.

Russell, D.G.D., W.J.L. Slader, and D.G. Ainley. 2012. Murray Levick (1876–1956): unpublished notes on the sexual habits of the Adélie penguin. *Polar Record* 48(4) 387–393. doi: 10.1017/S0032247412000216.

Sweeney, M. 2008. Infanticide and brood reduction in birds. Unpublished notes. https://www.google.com/search?q=Sweeny+%22Infanticide+and+brood+reduction+in+birds%22&tbm=isch&source=univ&sa=X&ved=2ahUKEwiJk_W-6-TiAhXNxlkKHUZlCwMQsAR6BAgAEAE&biw=1260&bih=686.

Tanaka, K., D.K. Zelenitsky, and F. Therrien. 2015. Eggshell porosity provides insight on evolution of nesting in dinosaurs. *PLOS ONE* 10(11): e0142829. doi: org/10.1371/journal.pone.0142829.

Varricchio, D.J., F. Jackson, J.J. Borkowski, and J.R. Horner. 1997. Nest and egg clutches of the dinosaur *Troodon formosus* and the evolution of avian reproductive traits. *Nature* 385: 247–250.

Verhulst, S., and J. Nilsson. 2008. The timing of birds' breeding seasons: a review of experiments that manipulated timing of breeding. *Philos. Trans. Royal Society B: Biological Science* 363(1490)399–410. doi: 10.1098/rstb.2006.2146.

Verner, J., and M.F. Willson. 1966. The influence of habitats on mating systems of North American passerine birds. *Ecology* 47(1): 143–147.

Walters, J.R. 1984. The evolution of parental behavior and clutch size in shorebirds. In *Shorebirds: breeding behavior and population*. Eds. J. Burger & B.L. Olla. Chapter 7. Plenum Press.

Yetter, A.P., S.P. Havera, and C.S. Hine. 1999. Natural-Cavity Use by Nesting Wood Ducks in Illinois. *Journal of Wildlife Management* 63(2): 630–638. doi.org/10.2307/3802652.

Chapter 6

Able, K.P. 1970. A radar study of the altitude of nocturnal passerine migration. *Bird-Banding* 41: 282–290.

Able, K.P. 1993. Orientation cues used by migratory birds: a review of cue-conflict experiments. *Trends Ecol. Evol.* 8: 367–371. doi: 10.1016/0169-5347(93)90221-A.

Able, K.P. 1999. The scope and evolution of bird migration. In *Gathering of angels: Migration of birds and their ecology*, ed. K.P. Able. Comstock Books.

Alerstam, T. 1990. *Bird migration*. Cambridge University Press.

Alerstam, T. 2009. Flight by night or day? Optimal daily timing of bird migration. *Journal of Theoretical Biology*. 258(4): 530–6.

Alerstam, T., M. Rosén, J. Bäckman, P.G.P. Ericson, and O. Hellgren. 2007. Flight Speeds among bird species: Allometric and phylogenetic effects. *PLOS Biol* 5(8): e197. doi:10.1371/journal.pbio.0050197.

Álvarez, J.C., J. Meseguer, E. Meseguer, and A. Perez. 2001. On the role of the alula in steady flight of birds. *Ardeola* 48(2): 161–173.

Beason, R.C. 2005. Mechanisms of Magnetic Orientation in Birds. *Integr. Comp. Biol.*, 45:565–573.

Bonadonna, F., C. Bajzak, S. Benhamou, K. Igloi, P. Jouventin, H.P. Lipp, and G. Dell'Omo. 2005. Orientation in the wandering albatross: interfering with magnetic perception does not affect orientation performance. *Proceedings of the Royal Society B Biological Sciences*, 272(1562): 489–495. doi:10.1098/rspb.2004.2984.

Boyle, W.A. 2017. Altitudinal bird migration in North America. *The Auk* 134(2): 443–465. doi.org/10.1642/AUK-16-228.1.

Chaves-Campos, J. 2004. Elevational movements of large frugivorous birds and temporal variation in abundance of fruits along an elevational gradient. *Orinitologia Neotropical* 15: 435–445.

Deetjen, M.E., A.A. Biewener, and D. Lentink. 2017. High-speed surface reconstruction of a flying bird using structured light. *Journal of Experimental Biology* 220: 1956–1961 doi:10.1242/jeb.149708.

Dial, K.P. 2003a. Wing-assisted incline running and the evolution of flight. *Science* 299: 402–404.

Dial, K.P. 2003b. Evolution of avian locomotion: correlates of flight style, locomotion modules, nesting biology, body size, development, and the origin of flapping flight. *The Auk* 120(4):941–953.

Diehl, R.H., J.M. Bates, D.E. Willand, and T.P. Gnoske. 2104. Bird mortality during nocturnal migration over Lake Michigan: a case study. *Wilson Journal of Ornithology* 126(1): 19–29. https://doi.org/10.1676/12-191.1.

Drymon, J.M., K. Feldheim, A. M.V. Fournier, E.A. Seubert, A.E. Jefferson, A.M. Kroetz, and S.P.

Powers. 2019. Tiger sharks eat songbirds: reply. *Ecology* 100(10): 27 August 2019. https://doi.org/10.1002/ecy.2870.

Ehrlich, P.R., D.S. Dobkin, and D. Wheye. 1988. *Birder's handbook: a field guide to the natural history of North American birds.* Simon & Schuster.

Erickson, W.P., G.D. Johnson, and D.P. Young Jr. 2005. A summary and comparison of bird mortality from anthropogenic causes with an emphasis on collisions. USDA Forest Service Gen. Tech. Rep. PSW-GTR-191. p. 1029–1042.

Gagliardo, A., J. Bried, P. Lambardi, P. Luschi, M. Wikelski, and F. Bonadonna. 2013. Oceanic navigation in Cory's shearwaters: evidence for a crucial role of olfactory cues for homing after displacement. Journal of Experimental Biology 216: 2798–2805. doi: 10.1242/jeb.085738.

Gagliardo, A., P. Ioalè, M. Savini, and M. Wild. 2008. Navigational abilities of homing pigeons deprived of olfactory or trigeminally mediated magnetic information when young. J. Exp. Biology 211: 2046-2051. doi:10.1242/jeb.017608.

Gauthreaux, S.A. Jr. 1999. Neotropical migrants and the Gulf of Mexico: the view from aloft. In Gathering of angels: Migration of birds and their ecology, ed. K.P. Able, 1999. Comstock Books.

Gill, R.E., T.L. Tibbitts, D.C. Douglas, C.M. Handel, D.M. Mulcahy, J.C. Gottschalck, N. Warnock, B.J. McCaffery, P.F. Battley, and T. Piersma. 2009. Extreme endurance flights by landbirds. *Proceedings of the Royal Society B.* 276(1656): 447–457. doi.org/10.1098/rspb.2008.1142.

Hall-Karlsson, K.S.S., and T. Fransson. 2008. How far do birds fly during one migratory flight stage? Ringing & Migration 24(2): 95–100. doi:10.1080/03078698.2008.9674381.

Harrington, B.A. 1999. The hemispheric globetrotting of the White-rumped Sandpiper. In *Gathering of angels: Migration of birds and their ecology,* ed. K.P. Able. Comstock Books. p. 119–133.

Hedenstrom, A., G. Norevik, K. Warfvinge, A. Andersson, J. Beackman, and S. Akesson. 2016. Annual 10-month aerial life phase in the Common Swift *Apus apus. Current Biology* 26: 3066–3070 http://dx.doi.org/10.1016/j.cub.2016.09.014.

Helbig, A.J. 1991. Inheritance of migratory direction in a bird species: a cross-breeding experiment with SE- and SW-migrating blackcaps (*Sylvia atricapilla*). *Behav Ecol Sociobiol* 28: 9–12. doi.org/10.1007/BF00172133.

Herrera, A.M., S.G. Shuster, C.L. Perriton, M.J. Cohn. 2013. Developmental basis of phallus reduction during bird evolution. *Current Biology* 23(12): 1065–1074. doi.org/10.1016/j.cub.2013.04.062.

Hilty, S. 1994. Birds of Tropical America. Chapters Publishing.

Hiscock, H.G., S. Worster, D.R. Kattnig, C. Steers, Ye Jin, D.E. Manolopoulos, H. Mouritsen, and P.J. Hore. 2016. The quantum needle of the avian magnetic compass. *Proceedings of the National Academy of Science.* Published ahead of print April 4, 2016. doi.org/10.1073/pnas.1600341113.

Jaeger, A., C. J. Feare, R. W. Summers, C. Lebarbenchon, C. S. Larose, and M. LeCorre. 2017. Geolocation reveals year-round at-sea distribution and activity of a superabundant tropical seabird, the Sooty Tern *Onychoprion fuscatu. Frontiers in Marine Science* 4. doi: 10.3389/fmars.2017.00394.

Klaassen, R.H.G., M. Hake, R. Strandberg, B.J. Koks, C. Trierweiler, K-M. Exo, F. Bairlein, and T. Alerstam. 2014. When and where does mortality occur in migratory birds? Direct evidence from long-term satellite tracking of raptors. *Journal of Animal Ecology* 83: 176–184.

Liechti, F., W. Witvliet, R. Weber, and E. Bächler. 2013. First evidence of a 200-day non-stop flight in a bird, *Nature* website, Nature Communications, article no. 2554. doi: 10.1038/ncomms3554.

Loss, S.R., T. Wall, S.S. Loss, P. Marra. 2014. Bird–building collisions in the United States: Estimates of annual mortality and species vulnerability. *The Condor* 116(1) 8–23. https://doi.org/10.1650/CONDOR-13-090.1.

Loss, S.R., T. Will, and P.P. Marra. 2013. The impact of free-ranging domestic cats on wildlife of the United States. *Nature Comm.* doi: 10.1038/ncomms2380.

Martínez, A. 2023. Migrating bird, a bar-tailed godwit, flies from Alaska to Australia without stopping. NPR Morning Edition. https://www.npr.org/2023/01/05/1147052902/migrating-bird-a-bar-tailed-godwit-flies-from-alaska-to-australia-without-stoppi.

Mayor, S.J., R.P. Guralnick, M.W. Tingley, J. Otegui, J.C. Withey, S.C. Elmendorf, M.E. Andrew, S. Leyk, I.S. Pearse, and D.C. Schneider. 2017. Increasing phenological asynchrony between spring green-up and arrival of migratory birds. *Scientific Reports* 7, article no. 1902. doi:10.1038/s41598-017-02045-z.

Mehlman, D.W., S.E. Mabey, D.N. Ewert, C. Duncan, B. Abel, D. Cimprich, R.D. Sutter, and M. Woodrey. 2005. Conserving stopover sites for forest-dwelling migratory landbirds. *The Auk* 122(4): 1281–1290. doi.org/10.1642/0004–8038(2005)122[1281:CSSFFM]2.0.CO;2.

Murray, B.G.J. 1989. A critical review of the transoceanic migration of the blackpoll warbler. *The Auk* 106 (1): 8–17. doi: 10.2307/4087751.

Muheim, R., J.B. Phillips, and M.E. Deutschlander. 2009. White-throated sparrows calibrate their magnetic compass by polarized light cues during both autumn and spring migration. *Journal of Experimental Biology* 212:3466–3472. doi: 10.1242/jeb.032771.

Muheim, R., J.B. Phillips, and S. Akesson. (2006). Polarized light cues underlie compass calibration in migratory songbirds. *Science.* 313(5788): 837–839.

Muheim, R., S. Sjöberg, and A. Pinzon-Rodriguez. 2016. Polarized light modulates light-dependent magnetic compass orientation in birds. *Proceedings of the National Academy of Sciences* 113(6): 1654–1659. doi:10.1073/pnas.1513391113.

Newton, I. 2007a. Speed and duration of journeys. In *The migration ecology of birds*. Academic Press.

Newton, I. 2007b. Weather-related mass-mortality events in migrants. *Ibis* 149(3): 453–467. doi. org/10.1111/j.1474–919X.2007.00704.x.

Ostrom, J.H. 1979. Bird Flight: How did it begin? Did birds begin to fly "from the trees down" or "from the ground up?" Reexamination of *Archaeopteryx* adds plausibility to an "up from the ground" origin of avian flight. *American Scientist* 67(1): 46–56.

Ouwehand J., and C. Both. 2016. Alternate non-stop migration strategies of pied flycatchers to cross the Sahara desert. *Biol. Lett.*12:20151060. dx.doi.org/10.1098/rsbl.2015.1060.

Perdeck, A.C. 1958. Two types of orientation in migrating starlings, *Sturnus vulgaris*, and chaffinches, *Fringilla coelebs*, as revealed by displacement experiments. *Ardea* 46: 1–37.

Perdeck, A.C. 1967. Orientation of starlings after displacement to Spain. *Ardea* 55: 194–202.

Prum, R.O. 2017. *The Evolution of Beauty*. Doubleday.

Pulido, F. 2007. The genetics and evolution of avian migration. *BioScience*, 57(2): 165–174. doi.org/10.1641/B570211.

Reynolds, K.V., A.L.R. Thomas, G.K. Taylor. 2014. Wing tucks are a response to atmospheric turbulence in the soaring flight of the steppe eagle *Aquila nipalensis. Journal of the Royal Society Interface*, 11(101): 20140645. doi: 10.1098/%u200Brsif.2014.0645.

Schwartz, M. D., R. Ahas, and A. Aaso. 2006. Onset of spring starting earlier across the Northern Hemisphere. *Global Current Biology* 12(2): 342-351. doi.org/10.1111/j.1365-2486.2005.01097.x.

Senner, N.R., M. Stager, M.A. Verhoeven, Z.A. Cheviron, T. Piersma and W. Bouten. 2018. High-altitude shorebird migration in the absence of topographical barriers: avoiding high air temperatures and searching for profitable winds. *Proceedings of the Royal Society B* doi.org/10.1098/rspb.2018.0569.

Somveille, M., A. Manica, S.H. Butchart, and A.S. Rodrigues. 2013. Mapping global diversity patterns for migratory birds. *PLOS One* 8(8): e70907.

Sparks, T., C. Humphrey, N. Elkins, R. Moss, S. Moss, and K. Mylne. 2002. Birds, weather and climate. *Weather* 57: 399–410. doi.org/10.1256/wea.142.02.

Stiles, F.G. 1988. Elevational movements of birds on the Caribbean slope of Costa Rica: Implications for conservation. In *Tropical rain forest: diversity and conservation*. Eds. F. Almeda and C.M. Pringle. California Academy of Sciences.

Storer, R.W. 1960. Evolution in the diving birds. *Int Ornith Congr* 12: 694 –707.

Tucker, V.A. 1993. Gliding birds: reduction of induced drag by wing tip slots between the primary feathers. *Journal of Experimental Biology* 180(1): 285–310.

Thaxter, C.B., S. Wanless, F. Daunt, M.P. Harris, S. Benvenuti, Y. Watanuki, D. Grémillet, and K.C. Hamer. 2010. Influence of wing loading on the trade-off between pursuit-diving and flight in common guillemots and razorbills. *Journal of Experimental Biology* 213: 1018–1025. doi:10.1242/jeb.037390.

Thourp, K., R.A. Holland, A.P. Tottrup, and M. Wikelski. 2010. Understanding the migratory orientation program of birds: Extending laboratory studies to study free-flying migrants in a natural setting. *Integr Comp Biol.* 50(3): 315–322. doi.org/10.1093/icb/icq065.

Waldron, P. 2014. Why birds fly in V formation. *Science AAAS*, Jan. 15, 2014.

Warrick, D.R. 1998. The turning- and linear-maneuvering performance of birds: The cost of efficiency for coursing insectivores. *Can. J. Zool.* 76: 1063–1079.

Weimerskirch, H. 2004. Wherever the wind may blow. *Natural History Magazine*, October.

Wissa, A., A. Kyungwon Han, and M.R. Cutkosky. 2015. Wings of a feather stick together: Morphing wings with barbule-inspired latching. In *Living Machines 2015*. Eds. S.P. Wilson *et al.* pp. 123–134. doi: 10.1007/978–3-319–22979–913.

Xu, J., L.E. Jarocha, L.E., T. Zollitsch, *et al.* 2021. Magnetic sensitivity of cryptochrome 4 from a migratory songbird. *Nature* 594: 535–540. doi.org/10.1038/s41586–021–03618–9.

Zimmerman, L. 1998. *US Fish and Wildlife Service Circular 16: Migration of Birds*. Originally published by F.C. Linclon (1935), revised.

Chapter 7

Able, K.P. 1999. The scope and evolution of bird migration. In *Gathering of angels: Migration of birds and their ecology*. Ed. K.P. Able, 1999. Comstock Books. pp. 1–10.

Curk, T., T. McDonald, D. Zazelenchuk, S. Weidensaul, D. Brinker, S. Huy, N. Smith, T. Miller, A. Robillard, G. Gauthier, N. Lecomte, and J.F. Therrien. 2018. Winter irruptive Snowy Owls (*Bubo scandiacus*) in North America are not starving. *Canadian J. of Zoology*. https://doi.org/10.1139/cjz-2017-0278.

Drilling, N.E., and C.F. Thompson. 1988. Natal and breeding dispersal in house wrens (*Troglodytes aedon*). *The Auk* 105(3): 480-491.

Dufour *et al.* 2021. A new westward migration route in an Asian passerine bird, *Current Biology* 31: 1–7, https://doi.org/10.1016/j.cub.2021.09.086.

Howell, S.N., G., I. Lemington, and W. Russell. 2014. *Rare birds of North America*. Princeton University Press.

Paradis, E., S.R. Baillie, W.J. Sutherland, and R.D. Gregory. 1998. Patterns of natal and breeding dispersal in birds. *Journal of Animal Ecology* 67: 518–536.

Skrade, P.D.B., and S.J. Dinsmore. 2010. Sex-related dispersal in the mountain plover (*Charadrius montanus*). *The Auk* 127(3): 671-677.

Stenhouse, I.J., G. Robertson, and W.A. Montevecci. 2000. Herring Gull *Larus arentatus* predation on Leach's Storm-Petrel Oceanodroma leucorhoa breeding on Great Island, Newfoundland. *Atlantic Seabird* 2(1): 35–44.

Tonelli, B.A., C. Youngflesh, and M.W. Tingley, 2023. Geomagnetic disturbance associated with increased vagrancy in migratory landbirds. *Sci Rep* 13, 414. https://doi.org/10.1038/s41598-022-26586-0.

Watanuki, Y. 1986. Moonlight avoidance behavior in Leach's Storm-Petrels as a defense against Slaty-Backed Gulls. *The Auk* 103(1): 14–22.

Chapter 8

Alvarenga, H., L. Chiappe, and S. Bertelli. 2011. Phorusrhacids: the terror birds. In *Living dinosaurs: The evolutionary history of modern birds*. Ed. G. Dyke and G. Kaiser, 2011. Wiley-Blackwell.

Arrendondo, O. 1976. The great predatory birds of the Pelistocene of Cuba. In *Smithsonian Contributions to Paleobiology* 27; Collected Papers in Avian Paleontology Honoring the 90th Birthday of Alexander Wetmore, p. 169-187.

Baker, A.J., O. Haddrath, J.D. McPherson, and A. Cloutier. 2014. Genomic support for a Moa–Tinamou clade and adaptive morphological convergence in flightless ratites. *Molecular Bio., and Evol.* 31(7): 1686–1689. doi.org/10.1093/molbev/msu153.

Bakker, Robert; *et al.* 1998. Brontosaur killers: Late Jurassic allosaurids as sabre-tooth cat analogues. *Gaia.* 15(8): 145–158.

Brathwaite, D.H. 1992. Notes on the weight, flying ability, habitat, and prey of Haast's Eagle (Harpagornis moorei). *Notornis* 39(4): 239–247.

Brooke, M. de L., and J.A. Horsfall. 2003. Hoatzin. In *Firefly encyclopedia of birds*. Ed. C. Perrins. Firefly Books.

Campbell, K.E. Jr., and E.P. Tonni. 1983. Size and locomotion in teratorns. *The Auk* 100(2): 390–403.

Clarke, J.A., D.T. Ksepka, M. Stucchi, M. Urbina, N. Giannini, S. Bertelli, Y. Narváez, and C.A. Boyd. 2007. Paleogene equatorial penguins challenge the proposed relationship between penguin biogeography, body size evolution, and Cenozoic climate change. *Proceedings of the National Academy of Sciences.* 104(28): 11545–50. doi:10.1073/pnas.0611099104.

Davies, S.J.J. F. 2003. Moas. In *Grzimek's animal life encyclopedia. Vol. 8 Birds I Tinamous and Ratites to Hoatzins* (2nd ed.), M. Hutchins (ed.), Farmington Hills, MI: Gale Group. pp. 95–98.

Fordyce, R.E., and D.T. Ksepka. 2012. The strangest bird. *Scientific American.* 307(5): 56–61.

Fuller, E. 1988. *Extinct Birds*. Facts on File Pub., New York, pp. 20–32.

Hawkins, A.F. A., and S.M. Goodman. 2003. *The natural history of Madagascar*, Eds. S.M. Goodman and J.P. Benstead. University of Chicago Press. pp. 1026–1029.

Holdaway, R,M., M.E. Allentoft, C. Jacomb, C.L. Oskam, N.R. Beavan, and M. Bunce. 2014. An extremely low-density human population exterminated New Zealand moa. *Nature Communications* (5)5436. doi: 10.1038/ncomms6436.

Hume, J.P., and M. Walters. 2012. *Extinct birds*. Poyser.

Jarvis, E.D. *et al.* 2014. Whole-genome analyses resolve early branches in the tree of life of modern birds. *Science* 346: 1320–1331.

Kloess, P., A.W. Poust, and T.A. Stidham. 2020. Earliest fossils of giant-sized bony-toothed birds (Aves: *Pelagornithidae*) from the Eocene of Seymour Island, Antarctica. *Scientific Reports* 10: 18286.

Ksepka, D. 2009. March of the Fossil Penguins. https://fossilpenguins.wordpress.com/2009/10/03/8/

Ksepka, D. 2014. Flight performance of the largest volant bird. *PNAS* 111(29): 10624–10629. https://doi.org/10.1073/pnas.1320297111.

Mayr, G. 2014. A hoatzin fossil from the middle Miocene of Kenya documents the past occurrence of modern-type Opisthocomiformes in Africa. *The Auk* 131: 55–60.

Mayr, G. 2017. *Avian evolution: The fossil record of birds and its paleobiological significance*. Wiley-Blackwell.

Mayr, G., H. Alvarenga, and C. Mourer-Chauvire. 2011. Out of Africa: fossils shed light on the origin of the hoatzin, an iconic Neotropic bird. *Naturwissenschaften* 98: 961–966.

Mitchell, K.J., B. Llamas, J. Soubrier, N.J. Rawlence, T.H. Worthy, J. Wood, M.S. Y. Lee, and A. Cooper. 2014. Ancient DNA reveals elephant birds and kiwi are sister taxa and clarifies ratite bird evolution. *Science.* 344(6186): 898–900. doi:10.1126/science.1251981.

Mullner, A., and K.E. Linsenmair. 2007. Nesting behavior and breeding success of Hoatzins. *J. Field Ornithology* 74(4): 352–361. doi.org/10.1111/j.1557–9263.2007.00123.x.

Pavia, M., H.J.M. Meijer, A. Rossi, and U.B. Gohlich. 2017. The extreme insular adaptation of *Garganornis ballmanni* Meijer, 2014: a giant Anseriformes of the Neogene of the Mediterranean Basin. *Royal Soc. Open Sci.* doi.org/10.1098/rsos.160722.

Paxinos, E.E., H.F. James, S.L. Olson, M.D. Sorenson, J. Jackson, and R.C. Fleischer. 2002. mtDNA from fossils reveals a radiation of Hawaiian geese recently derived from the Canada goose (*Branta canadensis*). *PNAS* 99(3): 1399–1404. doi: 10.1073/pnas.032166399.

Perry, G.L.W., A.B. Wheeler, J.R. Wood, and J.M. Wilmshurst. 2014. A high-precision chronology for the rapid extinction of New Zealand moa (Aves, *Dinornithiformes*). *Quaternary Science Reviews.* 105: 126–135. doi:10.1016/j.quascirev.2014.09.025.

Prum, R.O., J.S. Berv, A. Dornburg, D.J. Field, J.P. Townsend, E.M. Lemmon, and A.R. Lemmon. 2015. A comprehensive phylogeny of birds (Aves) using targeted next-generation DNA sequencing. *Nature* (526): 569–573. doi:10.1038/nature15697.

Strahl, S.D. 1988. The social organization and behaviour of the Hoatzin *Opisthocomus hoazin* in central Venezuela. *Ibis* 130: 483–502.

Tennyson, A.J.D., T.H. Worthy, C.M. Jones, R.P. Scofield, and S.J. Hand. 2010. Moa's Ark: Miocene fossils reveal the great antiquity of Moa (Aves: Dinornithiformes) in Zealandia. *Records of the Australian Museum* 62: 105–114. doi: 10.3853/j.0067–1975.62.2010.1546.

Thomas, B.T. 1996. Family Opisthocomidae (Hoatzin). In *Handbook of the birds of the world, vol. 3: Hoatzin to Auks.* Eds. J. del Hoyo, A. Elliott, and J. Sargatal. Lynx Edition, Barcelona, Spain. p. 24–32.

Witton, M.P., and M.B. Habib. 2010. On the size and flight Diversity of Giant Pterosaurs, the use of birds as Pterosaur analogues and comments on Pterosaur flightlessness. *PLOS ONE*, 5(11): e13982.

Wood, J.R., J.M. Wilmshurst, S.J. Wagstaff, T.H. Worthy, Nicolas J. Rawlence, and A. Cooper. 2012. High-resolution coproecology: Using coprolites to reconstruct the habits and habitats of New Zealand's extinct Upland Moa (*Megalapteryx didinus*). *PLOS ONE.* doi.org/10.1371/journal.pone.0040025.Chapter 9.

Chapter 10

Ancel, A., G. Kooyman, P. Ponganis, *et al.* 1992. Foraging behaviour of emperor penguins as a resource detector in winter and summer. *Nature* 360: 336–339. https://doi.org/10.1038/360336a0.

Baker-Gabb, D., M. Antos, and G. Brown. 2016. Recent decline of the critically endangered Plains-wanderer (*Pedionomus torquatus*), and the application of a simple method for assessing its cause: Major changes in grassland structure. *Ecological Management & Restoration* 17(3). doi:10.1111/emr.12221.

Boles, W. 2002. Lyrebird: Overview. *Pulse of the Planet.* December 06.

Bush Heritage Australia. 2018. Night Parrot. https://www.bushheritage.org.au/species/night-parrot.

Coffin, H.R., J.V. Watters, J.M. Mateo. 2011. Odor-based recognition of familiar and related conspecifics: a first test conducted on captive Humboldt Penguins (*Spheniscus humboldti*). *PLOS ONE* 6(9): e25002. https://doi.org/10.1371/journal.pone.0025002.

Cookson, M., C.H. Richart, and K.J. Burns. 2018. *Common Diuca-Finch* (Diuca diuca), version 1.0. In *Neotropical Birds Online.* Ed. T. S. Schulenberg, Editor. Cornell Lab of Ornithology. https://doi.org/10.2173/nb.codfin1.01.

De Pietri, V., R.P. Scofield, A. Tennyson, S.J. Hand, and T.H. Worthy. 2015. Wading a lost southern connection: Miocene fossils from New Zealand reveal a new lineage of shorebirds (*Charadriiformes*) linking Gondwanan avifaunas. *J. Syst. Palaeontology* 14. doi:10.1080/14772019.2015.1087064.

Dumbacher, J.P., B.M. Beehler, T.F. Spande, and H.M. Garraffo. 1992. Homobatrachotoxin in the Genus Pitohui: chemical defense in birds? *Science* 258(5083): 799–801. doi:10.1126/science.1439786.

Hardy, R.D., and S.P. Hardy. 2008. White-winged Diuca Finch (*Diuca speculifera*) nesting on Quelccaya Ice Cap, Perú. *Wilson Journal of Ornithology,* 120(3): 613–617.

Irons, D.B. 1988. Black-legged Kittiwakes nest on advancing glacier. *Wilson Bulletin* 100: 324–325.

Isack, H.A., and H.U. Reyer. 1989. Honeyguides and honey gatherers: interspecific communication in a symbiotic relationship. *Science* 243(4896): 1343–1346. doi: 10.1126/science.243.4896.1343.

Jetz, W, G.H. Thomas, J.B. Joy, D.W. Redding, K. Hartman, and A.O. Mooers. 2104. Global distribution and conservation of evolutionary distinctness in birds. *Current Biology* 24(9): 919–930. doi.org/10.1016/j.cub.2014.03.011.

Korkmaz, I., F.M. Kukul Güven, S.H. Eren, and Z. Dogan. 2008. Quail consumption can be harmful. *J. Emergency Medicine.* 41(5): 499–502. doi:10.1016/j.jemermed.2008.03.045.

Lamichhaney, S., *et al.* 2016. Structural genomic changes underlie alternative reproductive strategies in the ruff (*Philomachus pugnax*). *Nature Genetics* 48(1): 84–88. doi:10.1038/ng.3430.

Maderspacher, F. 2016. Evolution: Flight of the ratites. *Current Biology.* http://dx.doi.org/10.1016/j.cub.2016.12.023.

Mayr, G. 2017. *Avian evolution: The fossil record of birds and its paleobiological significance.* Wiley-Blackwell.

McCrae, L. 2019. *My penguin year.* HarperCollins.

Miller, E. 2018. Out on a limb: Birding the phylogenetic tree. *Living Bird.* 37(4): 24–25.

Mulder, R.A., and M.L. Hall. 2013. Animal behavior: A song and dance about lyrebirds. *Current Biology* 23(12): R518-R519. doi:10.1016/j.cub.2013.05.009.

Nugent, D.T., *et al.* 2022. Multi-scale habitat selection by a cryptic, critically endangered grassland bird—The Plains-wanderer (*Pedionomus torquatus*): Implications for habitat management and conservation. *Austral Ecology* 47: 698–712. doi:10.1111/aec.13157.

Pauligk, Y. 2020. Plains-wanderer (*Pedionomus torquatus*) captive report: Investigating behaviours, vocalisations, reproduction. Werribee Open Range Zoo, Zoos Victoria, Melbourne.

Pettigrew, J. 2017. Hearing and vision in the Plains-wanderer. *The Wanderer*, Autumn 2017. 1(3): 1.

Plotz, R.D. & Linklater, W.L. 2010. Red-billed oxpeckers really do increase predator awareness in black rhinoceros. *Proceedings of the 13th International Behavioral Ecology Congress* (ISBE), p. 133, Perth Convention Exhibition Centre, Western Australia.

Portugal, S.J., C.P. Murn, E.L. Sparkes, and M.A. Daley. 2016. The fast and forceful kicking strike of the secretarybird. *Current Biology*, 26(2): R58-R59. doi: 10.1016/j.cub.2015.12.004.

Powlesland, R.G., D.V. Merton, and J.F. Cockrem. 2006. A parrot apart: The natural history of the Kakapo (*Strigops habroptilus*), and the context of its conservation management. *Notornis* 53 (1): 3–26.

Prum, R.O. *et al.* 2015. A comprehensive phylogeny of birds (*Aves*) using targeted next-generation DNA sequencing. *Nature* 526:569–573.

Santos, Joao dos. 1891. *Ethiopia Oriental* (Lisbon, 1891) Published posthumously.

Spottiswoode, C.N., K.S. Begg, and C.M. Begg. 2016. Reciprocal signaling in honeyguide-human mutualism. *Science* 353(6297): 378–389. doi: 10.1126/science.aaf4885.

Taylor, H. 2014. Lyrebirds mimicking chainsaws: fact or lie? *The Conversation.* 3 February https://theconversation.com/lyrebirds-mimicking-chainsaws-fact-or-lie-22529.

Chapter 11

Clements, J.F., T.S. Schulenberg, M.J. Iliff, S.M. Billerman, T.A. Fredericks, J.A. Gerbracht, D. Lepage, B.L. Sullivan, and C.L. Wood. 2021. *The eBird/Clements checklist of birds of the world: v2021.*

Perrins, C. 2003. *Firefly encyclopedia of birds.* Firefly Books.

Peters, J.L. 1931. *Checklist of birds of the world.* Harvard University Press, Cambridge, MA. www.biodiversitylibrary.org/bibliography/14581.

Tudge, C. 2008. *The bird: A natural history of who birds are, where they came from, and how they live.* Three Rivers Press.

Vaidya, G., D. Lepage, and R. Guralnick. 2018. The tempo and mode of the taxonomic correction process: How taxonomists have corrected and recorrected North American bird species over the last 127 years. *PLOS ONE* 13(4): e0195736. https://doi.org/10.1371/journal.pone.0195736.

Winkler, D.W., S.M. Billerman, and I.J. Lovette. 2015. *Bird families of the world: An invitation to the spectacular diversity of birds.* Lynx Editions and Cornell Lab of Ornithology.

Chapter 12

Adams, E.E. 2012. World forest area still on the decline. Earth Policy Institute, Rutgers University. http://www.earth-policy.org/mobile/releases/forests_2012.

Austin, O.L., Jr., and A. Singer 1961. *Birds of the world.* Golden Press.

Barker, F.K., A. Cibois, P. Schikler, J. Feinstein, and J. Cracraft. 2004. Phylogeny and diversification of the largest avian radiation. *PNAS* 10 (30): 11040–11045. doi.org/10.1073/pnas.0401892101.

Boles, W.E. 1997. Fossil songbirds (Passeriformes) from the Early Eocene of Australia. *Emu* 97: 43–50. doi: 10.1071/MU97004.

Clements, J.F., T.S. Schulenberg, M.J. Iliff, S.M. Billerman, T.A. Fredericks, J.A. Gerbracht, D. Lepage, B.L. Sullivan, and C.L. Wood. 2021. *The eBird/Clements checklist of birds of the world: v2021.*

Duca, C., T.J. Guerra, and M.A. Marini. 2006. Territory size of three Antbirds (Aves, Passeriformes) in an Atlantic Forest fragment in southeastern Brazil. *Revista Brasileira de Zoologia* 23(3): 692-698.

Dzielski, S.A., B.M. Van Doren, J.P. Hruska, and J.M. Hite. 2016. Reproductive biology of the Sapayoa (*Sapayoa aenigma*), the "Old World suboscine" of the New World. *The Auk* 133(3): 347–363.

Ericson, G.P., L. Christidis, A. Coope, M. Irestedt, J. Jackson, U.S. Johansson, and J.A. Norman. 2002. A Gondwanan origin of passerine birds supported by DNA sequences of the endemic New Zealand wrens. *Proc. R. Soc. Lond. B.* 269(1488): 235–241. doi:10.1098/rspb.2001.1877.

Ericson, G.P., D. Zuccon, J.I. Ohlson, U. Johansson, *et al.* 2006. Higher-level phylogeny and morphological evolution of tyrant flycatchers, cotingas, manakins, and their allies (Aves: Tyrannida). *Molecular Phylogenetics and Evolution* 40(2): 471–83. doi:10.1016/j.ympev.2006.03.031.

Goodale, E., and G. Beauchamp. 2010. The relationship between leadership and gregariousness in mixed-species bird flocks. *Journal of Avian Biology* 41: 99–103.

Goodale, E., P. Ding, X. Liu, A. Martinez, X. Si, M. Walters, and S.K. Robinson. 2015. The structure of mixed-species bird flocks, and their response to anthropogenic disturbance, with special reference to East Asia. *Avian Research* 6: 14 (Article accesses: 4585). doi.org/10.1186/s40657-015-0023-0.

Hackett, S.J., R.T. Kimball, S. Reddy, R.C.K. Bowie, E.L. Braun, M.J. Braun, J.L. Chojnowski, W.A. Cox, K.-L. Han, J. Harshman, C.J. Huddleston, B.D. Marks, K.J. Miglia, W.S. Moore, F.H. Sheldon, D.W. Steadman, C.C. Witt, and T. Yuri. 2008. A phylogenomic study of birds reveals their evolutionary history. *Science* 320: 1763–1767.

Hansen, M.C.P. V. Potapov, R. Moore, M. Hancher, S.A. Turubanova, A. Tyukavina, D. Thau, S.V. Stehman, S.J. Goetz, T.R. Loveland, A. Kommareddy, A. Egorov, L. Chini, C.O. Justice, and J.R.G. Townshend. 2013. High-Resolution Global Maps of 21st-Century Forest Cover Change. *Science* 342(6160): 850–853. doi: 10.1126/science.1244693.

Hilty, S.L. 1994. *Birds of tropical America: A watcher's introduction to behavior, breeding and diversity.* Chapters Publishing.

Hutto, R.L. 1994. The composition and social organization of mixed-species flocks in a tropical deciduous forest in western Mexico. *The Condor* 96: 105–118.

Kroodsma, D., D. Hamilton, J.E. Sanchez, and B.E. Byer. 2013. Behavioral evidence for song learning in the suboscine bellbirds. *Wilson Journal of Ornithology* 125(1): 1–14. doi: 10.2307/41932830.

Ksepka, D.T., L. Grande and G. Mayr. 2019. Oldest finch-beaked birds reveal parallel ecological radiations in the earliest evolution of passerines. *Current Biology* 29: 657–663.

Lambert, J. 2021. The forested farms of the future. *Science News* 200(1): 30–35.

Lindell, C.A., S.K. Riffell, S.A. Kaiser, A.L. Battin, M.L. Smith, and T.D. Sisk. 2007. Edge response of tropical and temperate birds. *Wilson J of Ornithology* 119(2): 205–220.

Low, T. 2014. *Where song began: Australia's birds and how they changed the world.* Yale University Press.

Manegold, A. 2009. The early fossil record of perching birds (Passeriformes). *Palaeontologica Africana.* 44: 103–107.

Mayr, G. 2017. *Avian evolution: The fossil record of birds and its paleobiological significance.* Wiley-Blackwell.

Moyle, R.G., C.H. Oliveros, M.J. Andersen, P.A. Hosner, B.W. Benz, J.D. Manthey, S.L. Travers, R.M. Brown, and B.C. Faircloth. 2016. Tectonic collision and uplift of Wallacea triggered the global songbird radiation. *Nature Communications* doi: 10.1038/ncomms 12708.

Narango, D.L., D.W. Tallamy, and P.P. Marra. 2018. Nonnative plants reduce population growth of an insectivorous bird. *PNAS* 115(45): 11549–11554. doi.org/10.1073/pnas.1809259115.

Natural Resources Conservation Service. 2012. Conservation practices benefit shrubland birds in New England. https://www.nrcs.usda.gov/Internet/FSE_DOCUMENTS/stelprdb1046969.pdf.

Ohlson, J.I., M. Irestedt, Per G.P. Ericson, and J. Fjeldsa. 2013. Phylogeny and classification of the New World suboscines (Aves, Passeriformes). *Zootaxa* 3613: 1-35. doi: 10.11646/zootaxa.3613.1.1.

Oliveros, C.H., *et al.* 2019. Earth history and the passerine superradiation. *PNAS* 116(16): 7916–7925. doi.org/10.1073/pnas.1813201116.

Pillay, R., M. Venter, J. Aragon-Osejo, P. González-del-Pliego, A.J. Hansen, J. EM Watson, and O. Venter. 2022. Tropical forests are home to over half of the world's vertebrate species. *Front Ecol Environ* 20(1): 10–15. doi:10.1002/fee.2420.

Rice, N.H., and A.M. Hutson. 2003. Antbirds. In *Firefly encyclopedia of birds*. Ed. C. Perrins. Firefly Books.

Richard, M., D.W. Tallamy, and A. Mitchell. 2018. Introduced plants reduce species interactions. *Biological Invasions* 21: 983–992.

Ridgely, R.S., and G. Tudor. 1994. *The birds of South America: Volume I: The oscine passerines.* University of Texas Press.

Schlossberg, S., and D.I. King. 2007. Ecology and management of scrub-shrub birds in New England: A comprehensive review. Report submitted to Natural Resources Conservation Service, Resource Inventory and Assessment Division, Beltsville, Maryland, https://www.nrcs.usda.gov/Internet/FSE_DOCUMENTS/nrcs143_013252.pdf.

Skutch, A. F. 1996. *Antbirds and ovenbirds: Their lives and homes.* University of Texas Press.

Terborgh, J., S.K. Robinson, T.A. Parker III, C.A. Munn, and N. Pierpont. 1990. Structure and organization of an Amazonian forest bird community. *Ecological Monographs* 60(2): 213–238.

Terborgh, J. 1992. *Diversity and the tropical rain forest.* W.H. Freeman and Co.

Thiollay, J-M. 1999. Frequency of mixed species flocking in tropical forest birds and correlates of predation risk: an intertropical comparison. *Journal of Avian Biology* 30: 282–294.

Torgasheva, A., *et al.* 2019. Germline-restricted chromosome is widespread among songbirds. *PNAS* 116(24): 11,845–11,850. doi: 10.1073/pnas.1817373116.

United Nations. 2012. State of the world's forest 2012. Food and Agriculture Organization of the United Nations, Rome. UN Report, http://www.fao.org/3/i3010e/i3010e00.htm.

Wong, K. 2019. Winged victory. *Scientific American* 321(5): 58–61.

Chapter 13

Conover, M.R., and D.E. Miller. 1980. Rictal bristle function in Willow Flycatcher. *The Condor*, 82(4): 469–471. https://doi.org/10.2307/1367580.

Delaunay, M.G., C. Larsen, H. Lloyd, M. Sullivan, and R.A. Grant. 2020. Anatomy of avian rictal bristles in

Caprimulgiformes reveals reduced tactile function in open-habitat, partially diurnal foraging species. *J. Anat.* 237:355–366. https://doi.org/10.1111/joa.13188.

Mead, C.J., and A. Lundberg. 2003. Old world flycatchers. In *Firefly encyclopedia of birds*. Ed. C. Perrins. Firefly Books.

Mumme, R.L. 2014. White tail spots and tail-flicking behavior enhance foraging performance in Hooded Warbler. *The Auk* 131(2): 141–149.

Nyffeler, M., C.H. Şekercioğlu, and C.J. Whelan. 2018. Insectivorous birds consume an estimated 400–500 million tons of prey annually, *The Science of Nature* 105: 47. doi: 10.1007/s00114–018–1571–z.

Robinson, S.K., and R.T. Holmes. 1982. Foraging behavior of forest birds: the relationship among search tactics, diet, and habitat structure. *Ecology* 63(6): 1918–1931. doi: jstor.org/stable/1940130.

Sangster, G.; P. Alström, E. Forsmark, U. Olsson. 2010. Multi-locus phylogenetic analysis of Old World chats and flycatchers reveals extensive paraphyly at family, subfamily and genus level (Aves: *Muscicapidae*). *Molecular Phylogenetics and Evolution.* 57(1): 380–392.

Skutch, A.F. 1997. *Life of the flycatcher.* University of Oklahoma Press.

Terborgh, J. 1992. *Diversity and the tropical rain forest.* W.H. Freeman and Co.

Traylor, M.A. Jr., and J.W. Fitzpatrick. 1980. A survey of the Tyrant Flycatchers. *The living bird.* Nineteenth annual of the Cornell Laboratory of Ornithology 1980–81, Cornell University.

Chapter 14

Austin, O.L. Jr., and A. Singer. 1961. *Birds of the world.* Golden Press.

Grant, P.R., and B.R. Grant. 2002. Adaptive radiation of Darwin's finches: Recent data help explain how this famous group of Galapagos birds evolved, although gaps in our understanding remain. *American Scientist* 90: 130–139.

Lerner, H.R., M. Myer, H.F. James, M. Hofreiter and R.C. Fleischer. 2011. Multilocus resolution of phylogeny and timescale in the extant adaptive radiation of Hawaiian honeycreepers. *Current Biology* 21: 1838–44. doi: 10.1016/j.cub.2011.09.039.

Munro, G.C. 1960. *Birds of Hawaii.* Charles E. Tuttle Co.

Pratt, D.H. 2005. *The Hawaiian honeycreepers Drepanidinae.* Oxford University Press.

Pratt, H.D. 2014. A consensus taxonomy for the Hawaiian honeycreepers. Occasional Papers Museum of Natural Science, Louisiana State University, No. 85. doi: 10.31390/opmns.085.

Uppsala University. 2015. Evolution of Darwin's finches and their beaks. *ScienceDaily,* 11 February 2015. www.sciencedaily.com/releases/2015/02/150211141238.htm.

Chapter 15

Ahrens, M. 2018. Brush, grass, and forest fires. NFPA Research, National Fire Protection Agency.

Ebersole, R. 2020. This brutal pesticide creates a "circle of death." So why is it making a comeback? Audubon Magazine, Spring 2020. www.audubon.org/magazine/spring-2020/this-brutal-pesticide-creates-circle-death-so-why.

EPA 2020. EPA takes important step to reduce unnecessary animal testing. www.epa.gov/newsreleases/epa-takes-important-step-reduce-unnecessary-animal-testing#:~:text=WASHINGTON percent20%E2%80%94%20Today%2C%20the%20U.S.%20Environmental%20Protection%20Agency,enough%20other%20information%20to%20safely%20register%20outdoor%20pesticides.

Gibbons, D., C. Morrissey, and P. Mineau. 2015. A review of the direct and indirect effects of neonicotinoids and fipronil on vertebrate wildlife. *Environ. Sci. Pollut. Res.* 22: 103–118. doi: 1007/s11356–014–3180–5.

Goldstein, M.I., B. Woodbridge, M.E. Zaccagnini, and S.B. Canavelli. 1996. As assessment of mortality of Swainson's Hawks on wintering grounds in Argentina. *Journal of Raptor Research* 30: 106–107.

Johnson, D.H., and L.D. Igl. 2001. Area requirements of grassland birds: a perspective. *The Auk* 118(1): 24–34.

Low, T. 2014. *Where song began: Australia's birds and how they changed the world.* Yale University Press.

Macias-Duarte, A., A.O. Panjabi, D. Pool, E. Youngberg and G. Levandoski. 2011. *Wintering grassland bird density in Chihuahuan desert grassland priority conservation areas, 2007–2011.* Rocky Mountain Bird Observatory Technical Report I-NEOTROP-MXPLAT-10-2.

McCraken, J.D. 2005. Where the bobolinks roam: The plight of North America's Grassland Birds. *Biodiversity* 6(3): 20–29.

Minetor, R. 2022. The grassland conundrum: When pastures support livestock, birds have few options during breeding season. *Birding* 54(4): 34–43.

Norment, C.J. 2002. On grassland bird conservation in the northeast. *The Auk* 119(1): 271–279.

Olsen, P., A. Silcocks, M. Weston, and C. Tzaros. 2006. *Birds of woodlands and grasslands.* Australian State of the Environment Committee, Department of Environment and Heritage, Canberra.

Peterjohn, B.G. 2003. Agricultural landscapes: can they support healthy bird populations as well as farm products? *The Auk* 120: 14–19.

Pianka, E.R. Land. www.zo.utexas.edu/courses/thoc/land.html.

Ramankutty, N., A.T. Evans, C. Monfreda, and J.A. Foley. 2008. Farming the planet: 1. Geographic distribution of global agricultural lands in the year 2000. *Global Biogeochemical Cycles*. doi: 10.1029/2007GB002952.

SANBI (South African National Biodiversity Institute). 2010. Government spheres join forces to protect SA grasslands. 12 July 2010.

Vickery, P.D., M.L. Hunter, Jr., and S.M. Melvin. 1994. Effects of habitat area on the distribution of grassland birds in Maine. *Conservation Biology* 8: 1087–1097. doi.org/10.1046/j.1523–1739.1994.080 41087.x.

Waiken, W., S.E. Hawks, J.A. Shaffer, and D.H. Johnson. 2003. Guidelines for finding nests of passerine birds in tallgrass prairie. *Prairie Naturalist* 35(3): 197–211.

Wyoming Game and Fish Department. 2017. *Wyoming state wildlife action plan: Prairie grasslands.*

Zimmerman, D.A., D.A. Turner, and D.J. Pearson. 1996. *Birds of Kenya and Northern Tanzania.* Princeton University Press.

Chapter 16

Armstrong, D.P. 1992. Correlation between nectar supply and aggression in territorial honeyeaters: causation or coincidence? *Behavioral Ecology and Sociobiology* 30(2): 95–102.

Clark, M., and R.D. Wooller. 2003. Honeyeaters and Australian chats. In *Firefly encyclopedia of birds. Ed.* C. Perrins. Firefly Books.

Connor, J. 2010. Not all sweetness and light: the real diet of hummingbirds. *Living Bird* October 15 (4): 34–37. www.allaboutbirds.org/news/not-all-sweetness-and-light-the-real-diet-of-hummingbirds/.

Da Silva, L.P., J.A. Ramos, A.P. Coutinho, P.Q. Tenreiro, and R.H. Heleno. 2016. Flower visitation by European birds offers the first evidence of interaction release in continents. *Journal of Biogeography.* Doi:10.1111/jbi.12915.

Greenewalt, C. 1960. *Hummingbirds.* Doubleday & Co., Inc., Garden City, NY. 250 pp.

Low, T. 2014. *Where song began: Australia's birds and how they changed the world.* Yale University Press, New Haven. pp. 406.

Nicolson, S.W., and P.A. Fleming. 2003. Nectar as food for birds: the physiological consequences of drinking dilute sugar solutions. *Plant Syst. Evol.* 238: 139–153.

Nicolson, S.W., and P.A. Fleming. 2014. Drinking problems on a 'simple' diet: physiological convergence in nectar-feeding birds. *Journal of Experimental Biology* 217: 1015–1023. doi:10.1242/jeb.054387.

Pauw, A., and K. Louw. 2012. Urbanization drives a reduction in functional diversity in a guild of nectar-feeding birds. *Ecology and Society* 17(2): 27. http://dx.doi.org/10.5751/ES-04758-170227.

Sapir, N., and R. Dudley. 2012. Backward flight in hummingbirds employs unique kinematic adjustments and entails low metabolic cost. *Journal of Experimental Biology* 215: 3603–3611. doi: 10.1242/200Bjeb.073114.

Tilford, T. 2014. *The complete book of hummingbirds.* Thunder Bay Press.

USFS Bulletin, 2017. *Maintaining and improving habitat for hummingbirds in Arizona and New Mexico—A land manager's guide.* FS-1039c. https://www.fs.usda.gov/Internet/FSE_DOCUMENTS/fseprd571183.pdf.

Wagner, H.O. 1946. Food and feeding habitats of Mexican hummingbirds. *Wilson Bulletin* 58(2): 69–132.

Yanega, G.M., and M.A. Rubega. 2004. Feeding mechanisms: hummingbird jaw bends to aid insect capture. *Nature* 424(6983): 615. doi: 10.1038/428615a.

Chapter 17

Austin, O.L. Jr., and A. Singer. 1961. *Birds of the world.* Golden Press, NY.

Bent, A.C. 1919. *Life history of North American diving birds.* Smithsonian Institution, US Government Printing Office.

Bent, A.C. 1926. *Life history of North American marsh birds.* Smithsonian Institution, US Government Printing Office.

Clifton, G.T., and A.A. Biewener. 2018. Foot-propelled swimming kinematics and turning strategies in common loons. *J. Exp. Biol.* https://jeb.biologists.org/content/221/19/jeb168831.

Dahl, T.E. 2011. *Status and trends of wetlands in the conterminous United States 2004 to 2009.* US Fish and Wildlife Service Report to Congress.

Dahl, T.E., and C.E. Johnson. 1991. *Status and trends of wetlands in the conterminous United States, mid-1970's to mid-1980's.* U.S. Department of the Interior, Fish and Wildlife Service.

Gjerdrum, C., C.S. Elphick, and M. Rubega. 2005. Nest site selection and nesting success in saltmarsh breeding sparrows: The importance of nest habitat, timing and study site differences. *The Condor* 107: 849–862.

Gjerdrum, C., K. Sullivan-Wiley, E. King, M.A. Rubega, and C.S. Elphick. 2008. Egg and chick fates during tidal flooding of saltmarsh Sharp-tailed Sparrow nests. *The Condor* 110(3): 579–584.

Johansson, L.C., and U.M.L. Norberg. 2001. Lift-based paddling in diving grebe. *Journal of Experimental Biology* 204: 1687–1696.

Johnsgard, P.A. 1975. *Waterfowl of North America*. Indiana University Press.

Leston, L., and T.A. Bookhout. (2015). Yellow Rail (Coturnicops noveboracensis), version 2.0. In *The birds of North America*. Ed. A.F. Poole. Cornell Lab of Ornithology. doi.org/10.2173/bna.139.

Mayr, G. 2004. Morphological evidence for sister group relationship between flamingos (Aves: Phoenicopteridae) and grebes (Podicipedidae). *Zoological Journal of the Linnean Society* 140: 157–169.

Mowbray, T.B., F. Cooke, and B. Ganter. 2020. Snow Goose *(Anser caerulescens)*, version 1.0. In *Birds of the world*. Ed. P. G. Rodewald. Cornell Lab of Ornithology.

Mowbray, Thomas B., C.R. Ely, J.S. Sedinger, and R.E. Trost. 2002. Canada Goose *(Branta canadensis)*. In *The Birds of North America*. Ed. A.F. Poole. Cornell Lab of Ornithology.

Chapter 18

Anderson, A., *et al*. 2013. Bird damage to select fruit crops: The cost of damage and the benefits of control in five states. *Crop Protection* 52: 103–109.

Barnea, A., J.B. Harborne, and C. Pannell. 1993. What parts of fleshy fruits contain secondary compounds toxic to birds and why? *Biochemical Systematics and Ecology* 21(4): 421–429.

Basili, G.D., and S.A. Temple. 1999. Dickcissels and crop damage in Venezuela: defining the problem with ecological models. *Ecological Applications*. 9(2): 732–739. doi:10.2307/2641158.

Bird Life International. Illegal wild bird trade. https://flightforsurvival.org/threat/illegal-trade/.

Blakesley, D., and S. Elliott. 2000. Restoring conservation forests in northern Thailand and the monitoring of frugivorous birds. *OBC Bulletin* 31. https://orientalbirdclub.org/frugivorous-birds/.

Brittingham, M.C., and S.T. Falker. 2010. Controlling birds on fruit crops. Penn State Extension Code UH121. Pennsylvania State University.

Clabaut, C., A. Herrel, T.J. Sanger, T.B. Smith, and A. Abzhanov. 2009. Development of beak polymorphism in the African seedcracker, Pyrenestes ostrinus. *Evolution and Development* 111(6): 636–46. doi: 10.1111/j.1525–142X.2009.00371.x.

Grant, B.R., and P.R. Grant. 2003. What Darwin's Finches can teach us about the evolutionary origin and regulation of biodiversity. *BioScience* 53(10): 965–975. doi:10.1641/0006–3568(2003)053[0965:WDFCTU]2.0.CO;2.

Hilty, S. 1994. *Birds of tropical America: A watcher's introduction to behavior, breeding, and diversity*. Chapters Publishing.

Howe, H.F. 2017. Fruit-eating birds in experimental plantings in southern Mexico. *Journal of Tropical Ecology* 33(1): 83–88. doi.org/10.1017/S0266467416000596.

Izhaki, I., and U. Safriel. 1989. Why are there so few exclusively frugivorous birds? Experiments on fruit digestibility. *Oikos* 54(1): 23–32.

Ji, Q., P.J. Currie, M.A. Norell, and S. Ji. 1998. Two feathered dinosaurs from northeastern China. *Nature* 393(6687): 753–761. doi:10.1038/31635.

Knapton, S. 2017. British obsession with feeding birds is making their beaks grow longer, scientists believe. *Telegraph*. October 2017.

Levey, D.J., and C. Martínez del Rio. 2001. It takes guts (and more) to eat fruit: lessons from avian nutritional ecology. *The Auk* 118(4): 819–831. doi.org/10.1093/auk/118.4.819.

Mueller, T., J. Lenz, T. Caprano, W. Fiedler, and K. Böhning-Gaese. 2014. Large frugivorous birds facilitate functional connectivity of fragmented landscapes. *Journal of Applied Ecology* 51(3): 684–692. doi.org/10.1111/1365–2664.12247.

Muñoz, M.C., H.M. Schaefer, K. Böhning-Gaese, and M. Schleuning. 2016. Importance of animal and plant traits for fruit removal and seedling recruitment in a tropical forest. *Oikos* 126(6). doi: 10.1111/oik.03547.

Perrin, C.M. 2003. Ture tits. In *Firefly encyclopedia of birds*. Ed. C. Perrins. Firefly Books.

Rey, P.J. 2011. Preserving frugivorous birds in agro-ecosystems: lessons from Spanish olive orchards. *Journal of Applied Ecology* 48: 228–237. doi.org/10.1111/j.1365–2664.2010.01902.x.

Reynolds, S.J., J.A. Galbraith, J.A. Smith, and D.N. Jones. 2017. Garden bird feeding: insights and prospects from a north-south comparison of this global urban phenomenon. *Frontiers in Ecology and Evolution* 5: 24. doi: 10.3389/fevo.2017.00024.

Tudge, C. 2008. *The bird: A natural history of who birds are, where they came from, and how they live*. Three Rivers Press.

Van Tyne, J., and A.J. Berger. 1976. *Fundamentals of ornithology*. Wiley.

Zimmer, C., and D. Emlen. 2012. *Evolution: Making sense of life*. Roberts & Co.

Chapter 19

Bent, A.C. 1940. *Life histories of North America cuckoos, goatsuckers, hummingbirds, and their allies.* Bulletin 176. Smithsonian Institution, US Government Printing Office, Washington, D.C. p. 109.

Clements, J.F., T.S. Schulenberg, M.J. Iliff, S.M. Billerman, T.A. Fredericks, J.A. Gerbracht, D. Lepage, B.L. Sullivan, and C.L. Wood. 2021. *The eBird/Clements checklist of birds of the world: v2021.*

Forshaw, J.M., and A.E. Gilbert. 2009. *Trogons: A natural history of the Trogonidae.* Princeton University Press.

Jarvis, E.D., *et al.* 2014. Whole-genome analyses resolve early branches in the tree of life of modern birds. *Science* 346: 1320–1331. Doi: 10.1126/science.1253451.

Kristoffersen, A.V. 2002. An early Paleogene trogon (*Aves: Trogoniformes*) from the Fur Formation, Denmark. *J. Vertebrate Paleontology* 22(3): 661–666. Doi: 10.2307/4524256.

Lindow, B.E., and G.J. Dyke. 2006. Bird evolution in the Eocene: Climate changes in Europe and a Danish fossil fauna. *Biol. Rev. Camb. Philos. Soc.* 81 (4): 483–499.

Mayr, G. 1999. A new trogon from the middle Oligocene of Cereste, France. *The Auk* 116:427–434.

Skutch, A.F. 1999. *Trogons, laughing falcons, and other Neotropical birds.* Texas A&M University Press.

Sodhi, N.S., C.H. Sekercioglu, J. Barlow, and S. Robinson. 2011. *Conservation of tropical birds.* John Wiley & Sons. doi:10.1002/9781444342611.

Taylor, R.C. 1994. *Trogons of the Arizona borderlands.* Treasure Chest Publications.

Chapter 20

American Bird Conservancy. 2015. Sudden death on the high seas, longline fishing: A global catastrophe for seabirds. https://abcbirds.org/wp-content/uploads/2015/05/seabird_report.pdf.

Ancel, A., G. Kooyman, P. Ponganis, *et al.* 1992. Foraging behaviour of emperor penguins as a resource detector in winter and summer. *Nature* 360:336–339. https://doi.org/10.1038/360336a0.

Bearhop, S., *et al.* 2001. Annual variation in Great Skua diets: the importance of commercial fisheries and predation on seabirds revealed by combining dietary analyses. *Condor* 103:802–809. doi:10.1650/0010-54 22(2001)103[0802:AVIGSD]2.0.CO;2.

Belant, J.L., T.W. Seamans, S.W. Gabrey, and S.K. Ickes. 1993. Importance of landfills to nesting herring gulls. *The Condor* 95:817–830.

Bent, A.C. 1921. *Life histories of North American Gulls and Terns.* United States National Museum Bulletin 113. Washington, D.C., pp. 284, 307.

Bent, A.C. 1922. *Life histories of North American Petrels and Pelicans and their Allies.* United States National Museum Bulletin 121. Washington, D.C. 343 pp.

Bethge, P., S. Nicol, B.M. Culik, and R.P. Wilson. 1997. Diving behaviour and energetics in breeding little penguins (Eudyptula minor). *Journal of Zoology* 242(3): 483–502.

Bost, C.A.J. B. Thiebot, D. Pinaud, Y. Cherel, and P.N. Trathan. 2009. Where do penguins go during the inter-breeding period? Using geolocation to track the winter dispersion of the macaroni penguin. *Biology Letters.* 5(4): 473–476.

Bracey, A.S. Lisovski, D. Moore, A. McKellar, E. Craig, S. Matteson, F. Strand, J Costa, C. Pekarik, C. Curtis, G. Niemi, and F. Cuthbert. 2018. Migratory routes and wintering locations of declining inland North American common terns. *The Auk* 135(3): 385–399. doi.org/10.1642/AUK-17-210.1.

Bridge, E.S. 2004. The effects of intense wing molt on diving in Alcids and potential influences on the evolution of molt patterns. *Journal of Experimental Biology* 207(17): 3003–3014.

Brooke, M. 2010. *The Manx Shearwater.* Poyser Monographs, London 246 pp.

Churchill, R., V. Lowe, and A. Sander. 2022. *The Law of the Sea.* 4th ed. Manchester U. Press.

Cimino, M., H. Lynch, V. Saba, and M.J. Oliver. 2016. Projected asymmetric response of Adélie penguins to Antarctic climate change. *Sci. Rep.* 6, 28785. doi.org/10.1038/srep28785.

de León, A., D. Minguez, P. Harvey, E. Meek, J.E. Crane, and R.W. Furness. 2006. Factors affecting breeding distribution of Storm-petrels (*Hydrobates pelagicus*) in Orkney and Shetland, *Bird Study,* 53(1) 64–72. doi:10.1080/00063650609461417.

Edwards, E.W.J., L.R. Quinn, E.D. Wakefield, P.I. Miller, and P.M. Thompson. 2013. Tracking a northern fulmar from a Scottish nesting site to the Charlie-Gibbs Fracture Zone: Evidence of linkage between coastal breeding seabirds and Mid-Atlantic Ridge feeding sites. *Deep-Sea Research II,* 98:438–444. doi. org/10.1016/j.dsr2.2013.04.011.

Fijn, R.C., D. Hiemstra, R.A. Phillips, and J. van der Winden. 2013. Arctic Terns *Sterna paradisaea* from the Netherlands migrate record distances across three oceans to Wilkes Land, East Antarctica. *Ardea* 101: 3–12. doi:10.5253/078.101.0102.

Gramling, C. 2019. The case of the Arctic's missing ice. *Science News* 195(5): 20–24.

Grémillet, D., A. Ponchon, M. Paleczny, M-L. D. Palomares, V. Karpouzi, and D. Pauly. 2018. Persisting worldwide seabird-fishery competition despite seabird community decline. *Current Biology* 28(24): P4009–4013.e2. doi.org/10.1016/j.cub.2018.10.051.

Grémillet, D., C. Peron, *et al.* 2016. Starving seabirds: Unprofitable foraging and its fitness consequences in Cape gannets competing with fisheries in the Benguela upwelling ecosystem. *Environmental Science Marine Biology.* doi:10.1007/S00227-015-2798-2.

Haney, J.C. 1990. Winter habitat of Common Loons on the continental shelf of the Southeastern United States. *The Wilson Bulletin.* 102(2): 253–63.

Haney, J.C., H. Geiger, and J.W. Short. 2014. Bird mortality from the *Deepwater Horizon* oil spill: II. Carcass sampling and exposure probability in the coastal Gulf of Mexico. *Marine Ecology Progress Series* 513: 239–252. doi:10.3354/meps10839.

Harvey, C. 2019. Oceans are warming faster than predicted. *Scientific American E&E News* January 11.

Howell, S.N.G. 2012. *Petrels, Albatross & Storm-Petrels of North America.* Princeton University Press.

ITOPF 2020. Oil Tanker Spill Statistics. 2020. https://www.itopf.org/knowledge-resources/data-statistics/statistics/.

Johnsgard, P.A. 1993. *Cormorants, Darters, and Pelicans of the World.* Smithsonian Institution Press, Washington, D.C.

Johnson, R.L., A. Venter, M.N. Bester, and W.H. Oosthuizen. 2006. Seabird predation by white shark, *Carcharodon carcharias*, and Cape fur seal, *Arctocephalus pusillus pusillus*, at Dyer Island. *African Journal of Wildlife Research.* 36(1): 23–32.

Johnston, D.W. 1979. The uropygial gland of the Sooty Tern. *Condor* 81:430–432.

Jones, T., L.M. Divine, H. Renner, S. Knowles, K.A. Lefebvre, H.K. Burgess, C. Wright, and J.K. Parish. 2019. Unusual mortality of Tufted puffins (*Fratercula cirrhata*) in the eastern Bering Sea. *PLOS ONE.* doi.org/10.1371/journal.pone.0216532.

Klages, N.T.W., J. Cooper. 1992. Bill morphology and diet of a filter-feeding seabird: The broad-billed prion *Pachyptila vittata* at South Atlantic Gough Island. *Journal of Zoology* 227(3): 385–396.

Ksepka, D., and T. Ando. 2011. Penguins past, present, and future: trends in the evolution of the *Sphenisciformes*. In *Living Dinosaurs: The Evolutionary History of Modern Birds*, ed. G. Dyke and G. Kaiser, 2011. Wiley-Blackwell.

Lavers, J.L., I. Hutton, and A.L. Bond. 2019. Clinical pathology of plastic ingestion in marine birds and relationships with blood chemistry. *Environmental Science & Technology* 53 (15), 9224–9231.

Lempidakis, E., E. Shepard, A.N. Ross, and K. Yoda 2022. Pelagic seabirds reduce risk by flying into the eye of the storm. *PNAS* 119(41): e2212925119. Doi.org/10.1073/pnas.2212925119.

Lopes, C.S., M.I. Laranjeiro, J.L. Lavers, A. Finger, and J. Provencher. 2022. Seabirds as indicators of metal and plastic pollution. In *Seabird Biodiversity and Human Activities.* 1st Edition. CRC Press. 20 pp.

Maggini, I., L. Kennedy, A. Macmillan, K. Elliot, K. Dean, and C.G. Guglielmo. 2017. Light oiling of feathers increases flight energy expenditure in a migratory shorebird. *Journal of Experimental Biology.* 220(13)2372–2379. doi: 10.1242/jeb.158220.

Marion, J. 1982. Shell-dropping behavior of Western Gulls (*Larus occidentalis*). *The Auk* 99: 565–569.

Micol, T., and P. Jouventin. 1995. Restoration of Amsterdam Island, south Indian Ocean, following removal of feral cattle. *Biological Conservation* 73(3): 199–206.

Mougeot, F., and V. Bretagnolle. 2000. Predation risk and moonlight avoidance in nocturnal seabirds. *Journal of Avian Biology* 31(3): 376–386.

Nevitt, G.A., and F. Bonadonna. 2005. Sensitivity to dimethyl sulphide suggests a mechanism for olfactory navigation by seabirds, *Biol. Lett.* 1(3): 303–305. doi:10.1098/rsbl.2005.0350.

Nevitt, G.A., K. Reid, and P. Trathan. 2004. Testing olfactory foraging strategies in an Antarctic seabird assemblage. *Journal of Experimental Biology* 207:3537–3544. doi:10.1242/jeb.01198.

Nicoll, M.A., M. Nevoux, C.G. Jones, *et al.* 2017. Contrasting effects of tropical cyclones on the annual survival of a pelagic seabird in the Indian Ocean. *Global Change Biology.* 23(2): 550–565. doi: 10.1111/gcb.13324. PMID: 27178393.

Nicolson, A. 2018. *The seabird's cry: The lives and loves of the planet's great ocean voyagers.* Henry Holt and Co.

NOAA 2021. Seabird interactions and mitigation efforts in Hawaii longline fisheries. 2020 Annual Report. National Marine Fishery Service, Pacific Islands Regional Office, Honolulu, HI.

Pacheco, M.A., F.U. Battistuzzi, M. Lentino, R.F. Aguilar, S. Kumar, and A.A. Escalante. 2011. Evolution of modern birds revealed by mitogenomics: Timing the radiation and origin of major orders. *Molecular Biology and Evolution* 28(6): 1927–1942. doi.org/10.1093/molbev/msr014.

Paleczny, M., E. Hammill, V. Karpouzi, and D. Pauly. 2015. Population trend of the world's monitored seabirds, 1950–2010. *PLOS*, http://dx.doi.org/10.1371/journal.pone.0129342.

Phillips, G. C. 1962. Survival value of the white coloration of gulls and other sea birds. D. Phil. thesis, Oxford. See also: Tinbergen, N. 1963. On adaptive radiation in gulls (Tribe *Larini*). *Zoologische Mededelingen* 39: 209–223.

Pons, J.M., A. Hassanin, and P.A. Crochet. 2005. Phylogenetic relationships within the *Laridae* (*Charadriiformes: Aves*) inferred from mitochondrial marker. *Molecular Phylogenetics and Evolution* 37(3): 686–699.

Raymond, B., S.A. Shaffer, S. Sokolov, E.J. Woehler, D.P. Costa, and L. Einoder, *et al.* 2010. Shearwater foraging in the Southern Ocean: the roles of prey availability and winds. *PLOS ONE* 5(6): e10960. https://doi.org/10.1371/journal.pone.0010960.

Ropert-Coudert, Y., F. Daunt, A. Kato, P.G. Ryan, S. Lewis, K. Kobayashi, Y. Mori, D. Grémillet, and S. Wanless. 2009. Underwater wingbeats extend depth and duration of plunge dives in northern gannets (*Morus Bassanus*). *Journal of Avian Biology* 40 (4): 380–387.

Shaffer, S.A.; Y. Tremblay, H. Weimerskirch, D. Scott, D.R. Thompson, P.M. Sagar, H. Moller, G.A. Taylor, D.G. Foley, B.A. Block, and D.P. Costa. 2006. Migratory shearwaters integrate oceanic resources across the Pacific Ocean in an endless summer. *Proc. Nat. Academy of Sci.* 103(34): 12799–12802. doi:10.1073/pnas.0603715103.

Sullivan, B. 2006. Sons of the Sea. *Living Bird* 25(3): 26–32.

Tavares, D.C., J.F. Moural, A. Merico, and S. Siciliano. 2020. Mortality of seabirds migrating across the tropical Atlantic in relation to oceanographic processes. *Animal Conservation* 23: 307–319. doi/epdf/10.1111/acv.12539.

Towns, D.R., *et al.* 2011. Impacts of introduced predators on seabirds. In *Seabird islands: Ecology, invasion, and restoration*. Eds. Mulder, C.P.H. *et al.* Oxford Academic, online edition. https://doi.org/10.1093/acprof:osobl/9780199735693.003.0003.

Vadrot, A.B.M., A. Langlet, and I.T. Wysocki. 2021. Who owns marine biodiversity? Contesting the world order through the "common heritage of humankind." principle. *Environmental Politics*. doi: 10.1080/09644016.2021.1911442.

Van Bemmelen, R.S.A., *et al.* 2019. A migratory divide among red-necked phalaropes in the Western Palearctic reveals contrasting migration and wintering movement strategies. *Frontiers in Ecology and Evolution* 4 April. doi: org/10.3389/fevo.2019.00086.

Wahl, T.R. 1984. Observations of the diving behavior of the Northern Fulmar. *Western Birds* 15: 131–133.

Weimerskirch, H., and A. Prudor. 2019. Cyclone avoidance behaviour by foraging seabirds. *Sci. Rep.* 9: 5400. https://doi.org/10.1038/s41598-019-41481-x.

Weimerskirch, H., and Y. Cherel. 1998. Feeding ecology of short-tailed shearwaters: breeding in Tasmania and foraging in the Antarctic? *Marine Ecology Progress Series* 167: 261–274. www.int-res.com/articles/meps/167/m167p261.pdf.

Wienecke, B., G. Robertson, R. Kirkwood, and K. Lawton. 2007. Extreme dives by free-ranging emperor penguins. *Polar Biology* 30(2): 133–142. doi:10.2307/4083837.

Wilcox, C., E. VanSebille, and B.D. Hardesty. 2015. Threat of plastic pollution to seabirds is global, pervasive, and increasing. *PNAS* 112(38): 11899–11904. doi:10.1073/pnas.1502108112.

Wires, L.R., and F.J. Cuthbert. 2006. Historic populations of the Double-crested Cormorant (*Phalacrocorax auritus*): Implications for conservation and management in the 21st century. *Waterbirds* 29(1): 9–37. doi. org/10.1675/1524–4695(2006)29[9:HPOTDC]2.0.CO;2.

Wolfaardt, A.C., S. Crofts, and A.M.M. Baylis. 2012. Effects of a storm on colonies of seabirds breeding at the Falkland Islands. *Marine Ornithology* 40:129–133.

Chapter 21

Baker, A.J., S.L. Pereira, and T.A. Paton. 2006. Phylogenetic relationships and divergence times of *Charadriiformes* genera: multigene evidence for the Cretaceous origin of at least 14 clades of shorebirds. *Biol. Lett.* 3(2): 205–209. doi: 10.1098/rsbl.2006.0606.

Bamford, M., D. Watkins, W. Bancroft, G. Tischler, and J. Wahl. 2008. Migratory shorebirds of the East Asian—Australasian Flyway; Population estimates and internationally important sites. *Wetlands International*—Oceania. Canberra, Australia.

Bamford, M.J. 2003. Plovers and lapwings. In *Firefly encyclopedia of birds*. Ed. C. Perrins. Firefly Books.

Bent, A.C. 1927. *Life History of North American Shore Birds, Vol.1*. United States National Museum, Bulletin 142. Washington, D.C.

Choi, C-Y, P F. Battley, M.A. Potter, Z. Ma, D.S. Melville, and P. Sukkaewmanee. 2017. How migratory shorebirds selectively exploit prey at a staging site dominate by a single prey species. *The Auk* 134:76–91. doi:10.1642/AUK-16–58.1.

Clements, J.F., T.S. Schulenberg, M.J. Iliff, S.M. Billerman, T.A. Fredericks, J.A. Gerbracht, D. Lepage, B.L. Sullivan, and C.L. Wood. 2021. *The eBird/Clements checklist of birds of the world: v2021*.

Dixon, J. 1918. The nesting grounds and nesting habits of the Spoon-billed Sandpiper. *The Auk* 35(4): 387–404.

Hayman, P., J. Marchant, and T. Prater. 1986. *Shorebirds: An identification guide to the waders of the world*. Houghton Mifflin.

Kaminski, R.M., and J.B. Davis. 2014. Evaluation of the migratory bird habitat initiative: Report of findings. Research Bulletin WF391. Forest and Wildlife Research Center, Mississippi State University.

Kober K., and F. Bairlein. 2009. Habitat choice and niche characteristics under poor food conditions. A study on migratory nearctic shorebirds in the intertidal flats of Brazil. *Ardea* 97(1): 31–42. doi. org/10.5253/078.097.0105.

Lamichhaney, S., *et al.* 2016. Structural genomic changes underlie alternative reproductive strategies in the ruff (*Philomachus pugnax*). *Nature Genetics* 48(1): 84–88. doi:10.1038/ng.3430.

Marks, J.S. 1993. Molt of Bristle-thighed Curlews in the northwestern Hawaiian Islands. *The Auk* 110(3): 573–587.

Matthiessen, P. 1973. *The wind birds: Shorebirds of North America.* Chapters Publishing.

Mayr, G. 2017. *Avian evolution: The fossil record of birds and its paleobiological significance.* Wiley-Blackwell.

McNeil, R., and J.R. Rodriguez S. 1996. Nocturnal foraging in shorebirds. *International Wader Studies* 8: 114–121.

Myers, J.P., R.I.G. Morrison, P.Z. Antas, B.A. Harrington, T.E. Lovejoy, M. Sallaberry, S.E. Senner, and A. Tarak, 1987. Conservation strategy for migratory species. *American Scientist* 75: 19–26.

NRCS 2000. *Shorebirds.* Fish and Wildlife Habitat Management Leaflet No. 17, July 2000. Wildlife Habitat Management Institute, Madison, MS.

O'Brien, M., R. Crossley, and K. Karlson. 2006. *The shorebird guide.* Houghton Mifflin Co., Boston. 477 pp.

Poole, A.F. Ed. 2008. *The birds of North America.* Cornell Lab of Ornithology. doi.org/10.2173/bna.115.

Sutton, G.M. 1925. Swimming and diving activity of the Spotted Sandpiper (Actitis macularia). *The Auk* 43: 580–581.

Warnock, N., and S. Warnock. 2001. Sandpipers, Phalaropes, and Allies. In *The Sibley guide to bird life and behavior.* Ed. C. Elphick, J.B. Dunning, Jr., and D.A. Sibley. Alfred A. Knopf.

Watts, B.D., and C. Turrin. 2016. Assessing hunting policies for migratory shorebirds throughout the Western Hemisphere. *Wader Study* 123(1): 6–15. doi:10.18194/ws.00028.

Weidensaul, S. 2021. A miracle of abundance. *Living Bird* 40(4): 42–52.

Chapter 22

Arevalo, J.E., and M. Araya-Salas. 2013. Collared Forest-Falcon (*Micrastur semitorquatus*) preying on Chestnut-mandibled Toucan (*Ramphastos swainsonii*) in Costa Rica. *The Wilson Journal of Ornithology* 125(1): 212–216.

Bang, B.G., and S. Cobb. 1968. The size of the olfactory bulb in 108 species of birds. *The Auk* 85(1): 55–61.

Bent, A.C. 1937. *Life history of North American birds of prey: Order Falconiforms (Part1).* Bulletin 167. Smithsonian Institution, Washington, D.C.

Bent, A.C. 1938. Life history of North American birds of prey: Orders *Falconiforems and Strigiformes (Part2).* Bulletin 170. Smithsonian Institution, Washington, D.C.

Bloom, P.H., M.D. McCrary, and M.J. Gibson. 1993. Red-Shouldered Hawk home-range and habitat use in Southern California. *Journal of Wildlife Management* 57(2): 258–65. https://doi.org/10.2307/3809422.

Butynski T.M., U. Agenonga, B. Ndera, and J.F. Hart. 1997. Rediscovery of the Congo Bay Owl. *African Bird Club Bulletin* 4: 32–35.

Cenizo, M., J.I. Noriega, and M.A. Reguero. 2016. A stem falconid bird from the Lower Eocene of Antarctica and the early southern radiation of the falcons. *Journal of Ornithology.* doi: 10.1007/s10336–015–1316–0.

Chen, K., Q. Liu, G. Liao Y. Yang, L. Ren, H. Yang, and X. Chen. 2012. The sound suppression characteristics of wing feather of owl (*Bubo bubo*). *J. Bionic Eng.* 9(2): 192–199.

Chesser, R.T., K.J. Burns, C. Cicero, J.L. Dunn, A.W. Kratter, I.J. Lovette, P.C. Rasmussen, J.V. Remsen, Jr., D.F. Stotz, B.M. Winger, and K. Winker. 2018. Fifty-ninth supplement to the American Ornithological Society's Check-list of North American birds. *The Auk* 135: 798–813. doi: 10.1642/AUK-18–62.1.

Clements, J.F., T.S. Schulenberg, M.J. Iliff, S.M. Billerman, T.A. Fredericks, J.A. Gerbracht, D. Lepage, B.L. Sullivan, and C.L. Wood. 2021. *The eBird/Clements checklist of birds of the world: v2021.*

Dyke, G., and E. Gardiner. 2011. The utility of fossil taxa and the evolution of modern birds: commentary and analysis. In *Living Dinosaurs: The Evolutionary History of Modern Birds.* Ed. G. Dyke and G. Kaiser, 2011. Wiley-Blackwell.

Enriquez, P.L., K. Eisermann, H. Mikkola, and J.C. Motta-Junior. 2017. A review of the systematics of Neotropical Owls (*Strigiformes*). In *Neotropical Owls: Diversity and Conservation.* Ed. P.L. Enriquez. Springer.

Farmer, C.J., L.J. Goodrich, E. Ruelas Inzunza, and J.P. Smith. 2008. Conservation status of North America's birds of prey. In *State of North America's birds of prey*, Chapter 9. Ed. K.L. Bildstein, Nuttall Ornithological Club, Washington, D.C., American Ornithologists' Union. https://www.rpi-project.org/publications/TP-07.pdf.

Fuchs, J., S. Chen, J.A. Johnson, and D.P. Mindell. 2011. Pliocene diversification within the South American forest falcons (*Falconidae: Micrastur*). *Molecular Phylogenetic Evolution* 60(3): 398–407. doi: 10.1016/j.ympev.2011.05.008.

Haney, J.C. 1999. Hierarchical comparison of the breeding birds in old growth conifer-hardwood forest on the Appalachian plateau. *Wilson Bulletin* 111(1) 89–99.

Haring, E., K. Kvaløy, J.-O. Gjershaug, N. Røv, and A. Gamauf. 2007. Convergent evolution and paraphyly of the hawk-eagles of the genus *Spizaetus* (Aves, Accipitridae)—phylogenetic analyses based on mitochondrial markers. *Journal of Systematics and Evolutionary Research*, 45: 353–365. doi.org/10.1111/j.1439–0469.2007.00410.x.

Healy, S., and T. Guilford. 1990. Olfactory-bulb size and nocturnality in birds. *Evolution* 4(2): 339–346.

Howell, D.L., and B.R. Chapman. 1997. Home range and habitat use of Red-shouldered Hawks in Georgia. *Wilson Bulletin* 109(1): 131–144.

Hume, J.P., and M. Walters. 2012. *Extinct birds*. T & AD Poyser.

IUCN red list of threatened species. 2021. https://www.iucnredlist.org/

Johnsgard, P.A. 1990. *Hawks, eagles, and falcons of North America: Biology and natural history*. Smithsonian Inst. Press.

Johnsgard, P.A. 2002. *North American owls: Biology and natural history, 2nd ed.* Smithsonian Institution Press.

Katzner, T., P. Turk, A. Duerr, D. Brandes, T. Miller, and M. Lanozne. 2012. *Golden eagle home range, habitat use, demographic and renewable energy development in the California desert*. Interim report to Bureau of Land Management, California State office. West Virginia University. https://nrm.dfg.ca.gov/FileHandler.ashx?DocumentID=83964&inline.

Kemp, A., and I. Newton. 2003. Hawks, eagles, and old world vultures. In *Firefly encyclopedia of birds*. Ed. C. Perrins. Firefly Books.

King, B.F., and P.C. Rasmussen. 1998. The rediscovery of the Forest Owlet *Athene (Heteroglaux) blewitti*. *Forktail* 14: 53–55.

Kumar, S. 2015. Tracking the incredible journey of the Amur Falcon. *Conservation India* November 9, 2015.

Kurochkin, E.N., and G.J. Dyke. 2011. The first fossil owls (Aves: *Strigiformes*) from the Paleogene of Asia and a review of the fossil record of *Strigiformes*. *Paleontological Journal* 45: 445.

Lawrence, R.D. 1997. *Owls: The silent fliers*. Firefly Books.

Lopez-Idiaquez, D., D. Canal, I. Calleja Gomez, and J. Sarasola. 2019. First record of the Chimango Caracara (*Milvago chimango*) using shrimp as prey. *Journal of Raptor Research*. doi: 10.3356/0892–1016–53.4.436.

Marcot, B.G. 1995. Owls of old forests of the world (PDF). General Technical Reports. U.S. Department of Agriculture, Forest Service, Pacific Northwest Research Station.

Markovicky, P.J., S. Apesteguia, and F.L. Agnolin. 2005. The earliest dromaeosaurid theropod from South America. *Nature* 437: 1007–1011.

Mayr, G. 2017. *Avian evolution: The fossil record of birds and its paleobiological significance*. Wiley-Blackwell.

Mays, N.M. 1985. Ants and foraging behavior of Collared-Forest Falcon. *Wilson bull.* 97(2): 231–232.

McCann, S., O. Moeri, *et al.* 2013. Strike fast, strike hard: the Red-Throated Caracara exploits absconding behavior of social wasps during nest predation. *PLOS ONE* 8(12): e84114. doi.org/10.1371/journal.pone.0084114.

McClure, C.J.W., *et al.* 2018. State of the world's raptors: Distributions, threats, and conservation recommendations. *Biological Conservation* 227: 390–402. doi.org/10.1016/j.biocon.2018.08.012.

Meiburg, J. 2021. *A most remarkable creature: The hidden life and epic journey of the world's smartest birds of prey*. Alfred A. Knopf.

Nagy, J., and J. Tokolyi. 2014. Phylogeny, historical biogeography and the evolution of migration in accipitrid birds of prey (Aves: *Accipitriformes*). *Ornis Hungarica* 22(1): 15–35. doi:10.2478/orhu-2014–0008.

Perrone, M., Jr. 1981. Adaptive significance of ear tufts in owls. *Condor* 83: 383–384.

Prescott, K.W. 1985. Eastern Screech-Owl captures goldfish in patio pond. *Wilson Bulletin* 97: 572–573.

Prum, R.O., I.S. Bev, A. Dornburg D.J. Field, J.P. Townsend, E.M. Lemmon, and A.R. Lemmon. 2015. A comprehensive phylogeny of *birds* (Aves) using targeted next-generation DNA sequencing. Supplementary Information, *Nature* 526: 569–573. doi:10.1038/nature15697.

Reynolds, R.T., *et al.* 2017. Long-term demography of the Northern Goshawk in a variable environment. *Wildlife Monographs* 197: 1–40. doi: 10.1002/wmon.1023.

Rich, P.V., and D.J. Bohaska. 1980. The Ogygoptyngidae, a new family of owls from the Paleocene of North America. *Alcheringa: Australian journal of Palaeontology* 5(2): 95–102 doi:10.1080/03115518108565424.

Sazima, I., and F. Olmos. 2009. The chimango caracara (Milvago chimango), an additional fisher among Caracarini falcons. *Biota Neotropica*, 9(3): 403–405. http://www.scielo.br/pdf/bn/v9n3/v9n3a36.pdf.

Siyabona Africa. 2017. Kruger Park raptor guide—Snake eagles. http://birding.krugerpark.co.za/birding-in-kruger-raptor-guide.html.

Sustaita, D., and F. Hertel. 2010. *In vivo* bite and grip forces, morphology and prey-killing behavior of North American accipiters (Accipitridae) and falcons (Falconidae*). Journal of Experimental Biology* 213:2617–2628. doi: 10.1242/jeb.041731.

Vreeland, J. 1997. Nesting habitat of Red-Tailed Hawks in oak woodlands. *Oaks 'n' Folks*, U. California Agricultural and Natural Resources 12(2). https://oaks.cnr.berkeley.edu/nesting-habitat-of-red-tailed-hawks/.

Wagner, H., M. Weger, M. Klaas, and W. Schroder. 2017. Features of owl wings that promote silent flight. *Interface Focus* 7(1). doi.org/10.1098/rsfs.2016.0078.

Weidensaul, S. 2015. *Peterson Reference Guide to Owls of North America and the Caribbean*. Houghton Mifflin Harcourt.

Wink, M. 2018. Phylogeny of Falconidae and phylogeography of Peregrine Falcons. *Ornis Hungarica* 26(2): 27–37. doi: 10.1515/orhu-2018–013.

Wink, M. 1995. Phylogeny of old and new world vultures (Aves: *Accipitridae* and *Cathartidae*) inferred from nucleotide sequences of the mitochondrial Cytochrome B gene. *Z. Naturforsch* 50c: 68–882.

Chapter 23

Allen, J.R., *et al.* 2019. Conservation attention necessary across at least 44% of Earth's terrestrial area to safeguard biodiversity. bioRxiv.org. doi.org/10.1101/839977.

Beissinger, S.R., and D.R. McCullough, ed. 2002. *Population viability analysis.* University of Chicago Press.

BirdLife International. 2018. *State of the world's birds: Taking the pulse of the planet.* Cambridge, UK: BirdLife International. http://datazone.birdlife.org/userfiles/docs/SOWB2018_en.pdf.

Burns, F., M.A. Eaton, I.J. Burfield, A. Klvaňová, E. Šilarová, A. Staneva, and R. Gregory. 2021. Abundance decline in the avifauna of the European Union conceals complex patterns of biodiversity change. *Authorea.* July 6, 2021. doi: 10.22541/au.162557488.83915072/v1.

Ceballos, G, P.R. Ehrlich, A.D. Barnosky, A. García, R.M. Pringle, and T.M. Palmer. 2015. Accelerated modern human–induced species losses: Entering the sixth mass extinction. *Sci.Adv.*1, e140025 https://advances.sciencemag.org/content/advances/1/5/e1400253.full.pdf.

IUNC (International Union for Conservation of Nature) 2021. The IUCN Red List of Threatened Species. https://www.iucnredlist.org/en.

Lambert, J. 2020. How much nature need protecting? *Science News* 198(2): 18–21.

Lees, A.C., L. Haskell, T. Allinson, S.B. Bezeng, I.J. Burfield, L.M. Renjifo, K.V. Rosenberg, A. Viswanathan, and S.H.M. Butchart. 2022. State of the World's Birds. *Annual Reviews of Environment and Resources* 47(1): 6.1–6.30. doi: 10.1146/annurev-environ-112420–014642.

Nicolson, A. 2018. *The seabird's cry: The lives and loves of the planet's great ocean voyagers.* Henry Holt.

Pimm, S., P. Raven, A. Peterson, C.H. Şekercioğlu, and P.R. Ehrlich. 2006. Human impacts on the rates of recent, present, and future bird extinctions. *PNAS* 130(29): 10941–10946. doi.org/10.1073/pnas.0604181103.

Quammen, D. 1996. *The song of the dodo: Island biogeography in an age of extinctions.* Scribner's.

Rosenberg, K.V., A.M. Dokter, P.J. Blancher, J.R. Sauer, A.C. Smith, P.A. Smith, J.C. Stanton, A. Panjabi, L.M. Helft, M. Parr, and P.P. Marra. 2019. Decline of the North American Avifauna. *Science* 366(6461): 120–124. http://doi.org/10.1126/science.aaw1313 (see also: //www.birds.cornell.edu/home/wp-content/uploads/2019/09/DECLINE-OF-NORTH-AMERICAN-AVIFAUNA-SCIENCE-2019.pdf).

Steadman, D.E. 2006. *Extinction and biogeography of tropical Pacific birds.* University of Chicago Press.

Szabo J.K., N. Khwaja, S.T. Garnett, and S.H.M. Butchart. 2012. Global patterns and drivers of avian extinctions at the species and subspecies level. *PLOS ONE* 7(10): e47080. https://doi.org/10.1371/journal.pone.0047080.

US Fish and Wildlife Service 2023. Environmental Conservation Online System. https://ecos.fws.gov/ecp/report/boxscore.

Vickery. P.J., and J, R. Herkert. 2001. Recent advances in grassland bird research: Where do we go from here? *The Auk* 118(1): 11–15. doi.org/10.1642/0004–8038(2001)118[0011:RAIGBR]2.0.CO;2.

Chapter 24

Adams, E.E. 2012. World forest area still on the decline. Earth Policy Institute, Rutgers University. http://www.earth-policy.org/mobile/releases/forests_2012.

Barnes, J.C., A.A. Dayer, R. Iovanna, and S. Cline. 2021. Supporting durable grassland habitat conservation through farm bill programs using social-ecological systems science. AOS-SCO Joint Virtual Meeting, 9–13 August.

Bigelow, D.P., and A. Borchers. 2017. Major uses of land in the United States, 2012. EIB-178, U.S. Department of Agriculture, Economic Research Service.

D'Costa, K. 2017. An American obsession with lawns. *Scientific American.* May 3, 2017. https://blogs.scientificamerican.com/anthropology-in-practice/the-american-obsession-with-lawns/.

Dovers, S., and C. Butler. 2015. Population and environment: A global challenge. Australian Academy of Science, July 24, 2015. https://www.science.org.au/curious/earth-environment/population-environment.

Hack, B. 2020. The museum ornithologist who made a difference in reducing birds kill on Chicago's buildings. *Living Bird* 39(2): 10–11.

Hunter, L. 2000. *The environmental implications of population dynamics.* RAND Distribution Services.

Kerzman, G. 2020. Fridays on the farm: restoring wetlands and crating habitat. USDA. www.farmers.gov/blog/fridays-on-farm-restoring-wetlands-and-creating-habitat.

Loss, S.R., T. Wall, S.S. Loss, and P. Marra. 2014. Bird–building collisions in the United States: Estimates of annual mortality and species vulnerability. *The Condor* 116(1): 8–23. https://doi.org/10.1650/CONDOR-13-090.1.

Lutter, S.H., A.A. Dayer, and J.L. Larkin. 2019. Young forest conservation incentive programs: Explaining re-enrollment and post-program persistence. *Environmental Management* 2/2019. https://www.springerprofessional.de/en/young-forest-conservation-incentive-programs-explaining-re-enrol/16328054.

McDonnell, M., and A. Hahs. 2014. Four ways to reduce the loss of native plants and animals from our cities and towns. *The Nature of Cities* 14 April 2014. https://www.thenatureofcities.com/2014/04/14/four-ways-to-reduce-the-loss-of-native-plants-and-animals-from-our-cities-and-towns/.

Morelle, R. 2013. Cats killing billions of animals in the US. BBC News Service, 29 January. bbc.com/news/science-environment-21236690.

Nowak, D.J., and J.T. Walton. 2005. Projected urban growth (2000–2050) and its estimated impact on the US forest resource. *Journal of Forestry* December: 383–389.

Rosenberg, K.V., A.M. Dokter, P.J. Blancher, J.R. Sauer, A.C. Smith, P.A. Smith, J.C. Stanton, A. Panjabi, L.M. Helft, M. Parr, and P.P. Marra. 2019. Decline of the North American Avifauna. *Science* 366(6461): 120–124. http//doi.org10.1126/science.aaw1313 (see also: //www.birds.cornell.edu/home/wp-content/uploads/2019/09/DECLINE-OF-NORTH-AMERICAN-AVIFAUNA-SCIENCE-2019.pdf).

Ritchie, H., and M. Roser. 2013. Land Use. *OurWorldInData.org.* https://ourworldindata.org/land-use.

Spangler, K., E.K. Burchfield, and B. Schumacher. 2020. Past and current dynamics of U.S. agricultural land use and policy. *Frontiers in Sustainable Food Systems.* 21 July. doi.org/10.3389/fsufs.2020.00098.

Tallamy, D.W. 2020. *Nature's best hope: A new approach to conservation that starts in your yard.* Timber Press.

Thomas, G.H. 2011. The state of the world's birds and the future of avian diversity. In *Living Dinosaurs: The Evolutionary History of Modern Birds.* Ed. G. Dyke and G. Kaiser, 2011. Wiley-Blackwell.

Vickery, P.J., and J.R. Herkert. 2001. Recent advances in grassland bird research: Where do we go from here? *The Auk* 118(1): 11–15. doi.org/10.1642/0004–8038(2001)118[0011:RAIGBR]2.0.CO;2.

Index